The Most Successful Way of JEE Preparation

DPP

Daily Practice Problems

JEE Main & Advanced

Periodicity & Elements

with Hydrogen, *s* & *p*-Block Elements (Group 13 & 14)

CHEMISTRY Vol-3

The Most Successful Way of JEE Preparation

DPP

Daily Practice Problems

JEE Main & Advanced

Periodicity & Elements

with Hydrogen, *s* & *p*-Block Elements (Group 13 & 14)

CHEMISTRY Vol-3

Dr. RK GUPTA
Retd. Associate Professor
KGPG College, Simbhawali, GZB

arihant
ARIHANT PRAKASHAN (SERIES), MEERUT

arihant

ARIHANT PRAKASHAN (SERIES), MEERUT

Administrative & Production Offices

Corporate Office 'Ramchhaya' 4577/15, Agarwal Road, Darya Ganj, New Delhi -110002
Tele: 011- 47630600, 43518550; Fax: 011- 23280316

Head Office Kalindi, TP Nagar, Meerut (UP) - 250002
Tel: 0121-2401479, 2512970, 4004199; Fax: 0121-2401648

Sales & Support Offices

Agra, Ahmedabad, Bengaluru, Bareilly, Chennai, Delhi, Guwahati, Hyderabad, Jaipur, Jhansi, Kolkata, Lucknow, Meerut, Nagpur & Pune

ISBN : 978-93-13193-42-5

Published by Arihant Publications (I) Ltd.

Production Team

Publishing Managers	Mahendra Singh Rawat & Keshav Mohan	*Inner Designer*	Mazher Chaudhary
Project Head	Sorabh Chaudhary	*Page Layouting*	Sunil Sharma, Hemant
Cover Designer	Shanu Mahad	*Proof Reading*	Salman

For further information about the products from Arihant, log on to www.arihantbooks.com or email to info@arihantbooks.com

/arihantpub /@arihantpub Arihant Publications /arihantpub

ABOUT DPP

Daily Practice Problems

Indian Institutes of Technology better known as IITs have always been the dream destination of each science group students who want to pursue their career as Engineers. These are the institutes which have got very high value not even in India but in the whole world. To get the admission in such institutes one need to face entrance Examinations; JEE Main & JEE Advanced, out of which JEE Main is considered as the preliminary exam & JEE Advanced is the final exam to get into the IITs. Approximately per year 12 lakh students enroll themselves for JEE Main and only the top 2.25 lakh in JEE Main merit can attempt for JEE Advanced, the fact clearly express the degree of toughness of JEE Main & JEE Advanced.

To have success in such an exam one needs to be very planned & systematic in its preparation. A very popular wise saying 'Rome was not built in a day' suggests that to achieve a big target like IITs we need to work continuously. For continuous work, one need to have proper planning that how much work he/she has to do on regular basis, i.e., on daily basis. Each student preparing for JEE has many books but unfortunately none of them gives the students a plan for daily studies, which is the most effective success key for JEE.

Keeping the daily practice needs of the students in mind we have come up with DPP Series (Daily Practice Problems) having approximately 26 books. The primary focus of this series is on achieving success through practice & proper planning. For that we have divided the whole syllabus of JEE Main & JEE Advanced; Physics, Chemistry & Mathematics, into different volumes (around 8 volumes of Mathematics & 9 volumes each of Physics & Chemistry), each volume has Daily Practice Problems Sheets having specific questions on various Topics of the individual chapter of the volume ensuring the complete Practice & Assessment of the topic.

DPP
Daily Practice Problems

In this way we have tried to give you an effective daily planning for JEE keeping in mind the daily hours that a student can devote to his practice, you are just required to practice one sheet of each subject on daily basis. This habit of daily practice will keep you away from exam stress and thoughts like; how much I have prepared/am I preparing properly or not/ do I need to study more on daily basis or what I am doing is ok, etc., which regularly tease your mind and can give you panic at any stage of your preparation.

Some special features that we have tried to incorporate in our Daily Practice Problems (DPP) series are discussed below

1. Micro Level coverage of all the topics of a Chapter/ Unit given in JEE Main & Advanced Syllabus.
2. Each DPP has various types of questions; Subjective Questions, Single/ Multiple Correct Options, Statements type, Comprehension based, Integer Value Answer type, Matching, etc., making you practice & ready for all the question formats of JEE.
3. New & innovative questions in each DPP challenging & improving your problem solving skills.
4. Along with Topical Coverage, Revisal DPPs for JEE Main & JEE Advanced with each Chapter, covering multidisciplinary questions proving comprehensive practice.
5. JEE Main & JEE Advanced Archive (collection of last 13 years' exams questions) with each chapter.
6. Complete Solutions for each DPP.

We hope that DPP series will prove to be very beneficial for the students preparing for JEE, by giving them daily practice in a systematic & planned manner. The suggestions/criticism from the students, teachers & parents about DPP series will be welcomed whole heartedly and we promise to update this series accordingly from time to time.

DPP
Daily Practice Problems

DAILY TASKS...

1 PERIODIC TABLE

2 HYDROGEN

3 ALKALI METALS (Group 1)

DPP

Daily Practice Problems

1. Periodic Table

DPP
DAILY PRACTICE PROBLEMS

DPP-1 Classification of Elements

Only One Option Correct Type

Working Space

Direction *(Q. Nos. 1-20) This section contains 20 multiple choice questions. Each question has four choices (a), (b), (c) and (d), out of which ONLY ONE is correct.*

1 Following are some of the defects of Mendeleev's Periodic Table:

I. Position of hydrogen was not defined.
II. Positions of lanthanides and actinides were not defined.
III. There was no discussion about noble gases.
IV. Isotopes were placed in the same position.

Select correct defects in it.

a. I, III and IV
b. Both II and III
c. I, II and III
d. All of these

2 Periodic properties in the periodic table are repeated after certain number of elements called magic numbers. Variation is shown by

a. 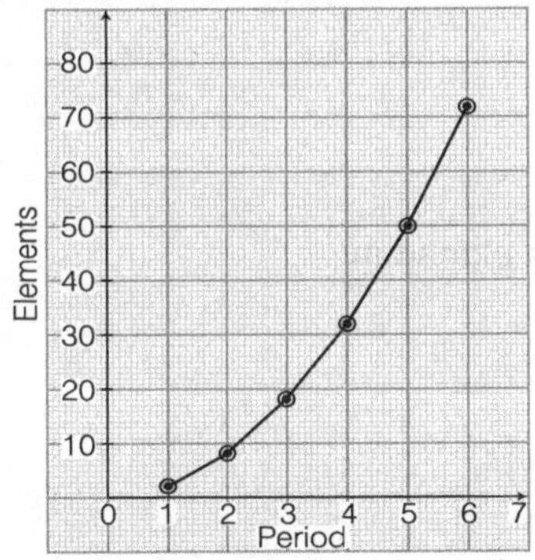

b. 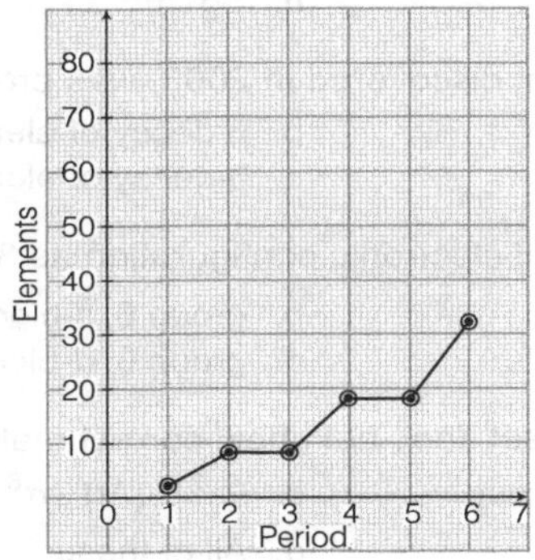

c. 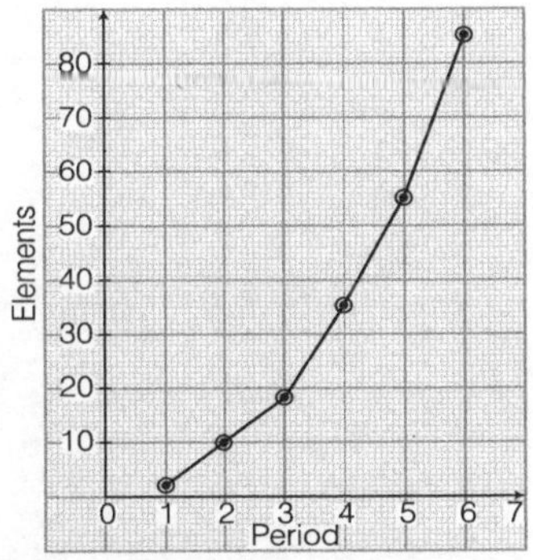

d. None of these

3 Lanthanides are

a. 14 elements in the seventh period (90-103) that are being filled into 5*f*-subshell
b. 14 elements in the sixth period (58-71) that are being filled into 4*f*-subshell
c. 14 elements in the seventh period (58-71) that are being filled into 4*f*-subshell
d. 14 elements in the sixth period (90-103) that are being filled into 4*f*-subshell

Working Space

4 Elements with atomic numbers 113 and 115 discovered in August 2003, have been placed in

a. *s*-block
b. *p*-block
c. *d*-block
d. *f*-block

5 In the long form (Modern form) of Periodic Table, following are some of the facts:

I. Periods are horizontal rows and groups are vertical columns.
II. Transition elements have been placed in 4th to 7th periods.
III. Inner-transition elements have been placed in 6th and 7th periods and grouped at the bottom of the periodic table in 3rd group.
IV. Elements (58-71) are called actinides and elements (90-103) are called lanthanides.

Select the incorrect fact(s).

a. Both III and IV
b. Both I and II
c. Only III
d. Only IV

6 If Aufbau rule is not followed, block of the elements will change in

I. Ca (20)
II. Zn (30)
III. Fe (26)
IV. K (19)

Select the correct answer.

a. All cases
b. Both I and III
c. Both II and IV
d. I, II and IV

7 In 2003, element with atomic number 110 has been named by IUPAC as

a. Hs
b. Mt
c. Ds
d. Sg

8 Element 113, a new element discovered in 2003, was produced *via*

a. α-decay of element 115
b. β-decay of element 114
c. α-decay of element 111
d. β-decay of element 112

9 Element *X* has valence shell electronic configuration as $3d^5 4s^1$. It is placed in

a. group 1, *s*-block
b. group 6, *s*-block
c. group 16, *d*-block
d. group 6, *d*-block

10 Out of the following, which set does not show correct matching?

a. Sc^{3+} : [Ne] $3s^2 3p^6$, zero group
b. Fe^{2+} : [Ar] $3d^6$ 8th group
c. Cr : [Ar] $3d^5 4s^1$, 6th group
d. All of these

11 Consider some facts about group 16 elements

I. They are *d*-block elements.
II. They are called chalcogens.
III. They have $ns^2 np^4$ configuration of valence shell.

Select the correct facts.

a. Both I and II
b. Both II and III
c. Both I and III
d. I, II and III

12 The electronic configuration of transition elements is represented by

a. $(n-1)d^{1-10} ns^1$ or ns^2
b. $ns^2 (n-1)d^{10}$
c. ns^1
d. $ns^2 np^5$

13 Element with electronic configuration $1s^2\ 2s^2\ 2p^6\ 3s^2\ 3p^6\ 3d^{10}\ 4s^2\ 4p^6\ 4d^{10}\ 5s^2\ 5p^3$ belongs to which of the following group of the periodic table?

a. II A (2)
b. V A (15)
c. III A (13)
d. VII A (17)

14 In the periodic table, an element is placed below the element of atomic number 25 in the same group. Its electronic configuration is

a. $1s^2 2s^2 2p^6 3s^2 3p^6 4s^2 3d^5$

b. $1s^2 2s^2 2p^6 3s^2 3p^6 3d^{10} 4s^2 4p^6 5s^2 4d^5$

c. $1s^2 2s^2 2p^6 3s^2 3p^6 3d^{10} 4s^2 4p^6 5s^1 4d^6$

d. $1s^2 2s^2 2p^6 3s^2 3p^6 3d^{10} 4s^2 4p^6 5s^0 4d^7$

15 Element belongs to 5th period and group 16. Its atomic number will be

a. 34
b. 50
c. 52
d. 53

16 A new element discovered has been named Eka-mercury. Its atomic number and symbol respectively are

a. 112, Uub
b. 110, Uun
c. 101, Unu
d. 102, Unb

17 Electronic configurations of elements (ground state or excited states) are given

I. $[Ar]\ 4s^1 3d^1$
II. $1s^2 2s^2 2p^5 3s^1$
III. $[Ar]\ 3d^{10} 4s^1$
IV. $[Xe]\ 4f^{14} 5d^1 6s^2$

These elements respectively belong to .... blocks.

a. *d, s, s, s*
b. *d, p, s, f*
c. *s, p, d, f*
d. *s, p, f, d*

18 Modern periodic law (1913) was based on the X-ray spectra of Henry Moseley and equation.

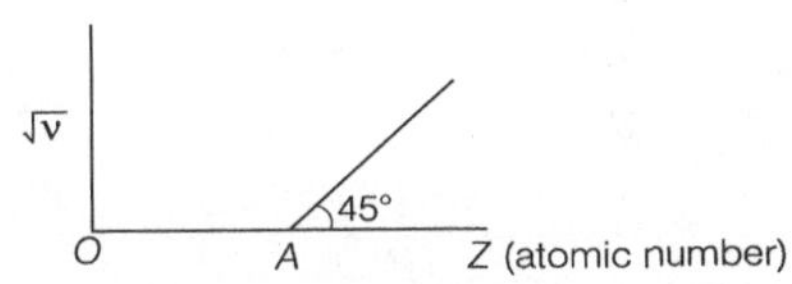

$\sqrt{v} = a(Z - b)$ where, *a* and *b* are constants. If $OA = 1$, then element showing frequency of 225 Hz, then incorrect answer is

a. *p*-block element
b. *d*-block element
c. non-metal
d. of group 16 and 3rd period

19 Which of the following statements related to the modern periodic table is incorroct?

a. The *p*-block has six columns because a maximum of 6 electrons can occupy all the orbitals in *p*-subshell
b. The *d*-block has 8 columns because a maximum of 8 electrons can occupy all the orbitals in *d*-subshell
c. Each block contains a number of columns equal to the number of electrons that can occupy that subshell
d. The block indicates value of azimuthal quantum number (*l*) for the last subshell that received electrons in building up the configuration

20 Select the incorrect statement about the long form of the periodic table.

a. It reflects the sequence of filling the electrons in the order of the energy of sub-shells *s, p, d* and *f*
b. If helps to predict the stable valency states of the elements
c. It reflects the trends in physical and chemical properties of the elements
d. It helps to predict the relative ionicity of the bond between any two elements

Matching List Type

Working Space

Direction *(Q. Nos. 21-25) Choices for the correct combination of elements from Column I and Column II are given as options (a), (b), (c), and (d), out of which one is correct.*

21 Match the set of elements in Column I with the respective series in Column II and select the answer from the codes given below.

	Column I		Column II
i.	Sc - Zn	p.	$6d$
ii.	Y - Cd	q.	$5d$
iii.	La - Hg	r.	$4d$
iv.	Ac - Ds	s.	$3d$
v.	Ce - Lu	t.	$5f$
vi.	Th - Lr	u.	$4f$

Codes

	i	ii	iii	iv	v	vi
a.	s	r	q	p	t	u
b.	p	q	r	s	t	u
c.	s	r	q	p	u	t
d.	p	r	q	s	t	u

22 Match the atomic numbers in Column I with their blocks (s, p, d, f) in Column II and mark the correct option from the codes given below.

	Column I		Column II
i.	43	p.	s
ii.	52	q.	p
iii.	64	r.	d
iv.	88	s.	f

Codes

	i	ii	iii	iv
a.	p	q	r	s
b.	r	q	s	p
c.	r	q	p	s
d.	s	p	q	r

23 Match the atomic numbers in Column I with their groups in Column II and mark the correct option from the codes given below.

	Column I		Column II
i.	11	p.	1
ii.	15	q.	10
iii.	18	r.	15
iv.	28	s.	18

Codes

	i	ii	iii	iv
a.	p	r	s	q
b.	p	s	q	r
c.	s	p	q	r
d.	p	q	r	s

Working Space

24 Match the elements in Column I with their types given in Column II and mark the correct option from the codes given below.

	Column I		Column II
i.	*s*-block	p.	Representatives
ii.	*p*-block	q.	Transition
iii.	*d*-block	r.	Lanthanides
iv.	*f*-block	s.	Actinides

Codes

	i	ii	iii	iv
a.	p	q	r	s
b.	p	q	r	s
c.	p	p	q	r,s
d.	p	p	s	r

25 Match the atomic numbers (in Column I) with their IUPAC nomenclature (in Column II) and mark the correct option from the codes given below.

	Column I		Column II
i.	105	p.	Uub
ii.	107	q.	Uup
iii.	110	r.	Ubp
iv.	112	s.	Unp
v.	115	t.	Uns
vi.	125	u.	Uun

Codes

	i	ii	iii	iv	v	vi
a.	s	t	u	p	q	r
b.	s	t	p	u	r	q
c.	t	p	s	u	q	r
d.	r	q	s	p	t	u

DPP-2 Periodic Trends-I (Radii, Screening Effect, Hydration Energy, etc)

Only One Option Correct Type

Working Space

Direction *(Q. Nos. 1-25) This section contains 25 multiple choice questions. Each question has four choices (a), (b), (c) and (d), out of which ONLY ONE is correct.*

1 Two atoms of the element X are placed in different manners as shown:

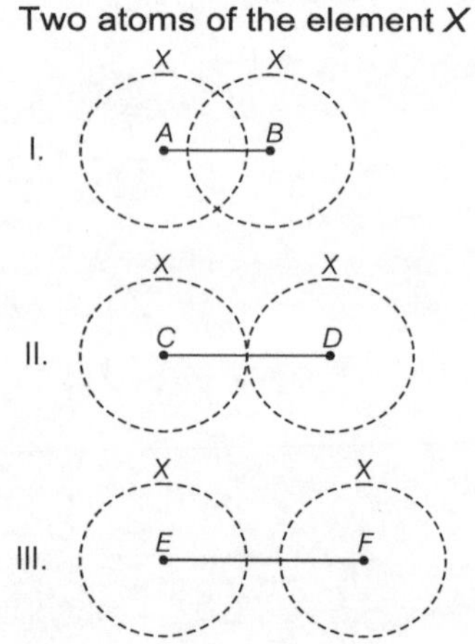

van der Waals' radius is given by

a. $\frac{AB}{2}$ b. $\frac{CD}{2}$

c. $\frac{EF}{2}$ d. None of these

2 In which case, effective nuclear charge is minimum?

a. Be b. Be^+

c. Be^{3+} d. Equal

3 The set representing the correct order of ionic radius is **[AIEEE 2009]**

a. $Li^+ > Be^{2+} > Na^+ > Mg^{2+}$ b. $Na^+ > Li^+ > Mg^{2+} > Be^{2+}$

c. $Li^+ > Na^+ > Mg^{2+} > Be^{2+}$ d. $Mg^{2+} > Be^{2+} > Li^+ > Na^+$

4 The increasing order of the ionic radii of the given isoelectronic species is **[AIEEE 2012]**

a. $Cl^-, Ca^{2+}, K^+, S^{2-}$ b. $S^{2-}, Cl^-, Ca^{2+}, K^+$

c. $Ca^{2+}, K^+, Cl^-, S^{2-}$ d. $K^+, S^{2-}, Ca^{2+}, Cl^-$

5 Atomic radius (r_n) of the nth orbit is given by $r_n = \frac{n^2 a_0}{Z}$, where a_0 is called Bohr's radius. This is applicable for

a. He^+ b. Li^{2+}

c. Be^{3+} d. All of these

Working Space

6 Effective nuclear charges experienced by Li in the following processes are

$$Li\,(g) \longrightarrow Li^+(g) + e^-$$
$$Li\,(g) + e^- \longrightarrow Li^-(g)$$

a. 3, 3
b. 1.3, 1.3
c. 1.3, 0.95
d. 0.95, 0.95

7 In which of the following sets of elements, atomic radii are dependent on the number of orbits but independent of effective nuclear charge?

a. Li, Na, K
b. Li, Na
c. Li, K
d. Na, K

8 The negative ion (anion) is always larger than that of the corresponding atom. Thus is due to

a. increase in number of electrons and force of attraction which pulls valence shell outwards
b. decrease in nuclear charge per electron
c. Both (a) and (b)
d. None of the above

9 Which is the correct order of slze of O^-, O^{2-}, F^- and F?

a. $O^{2-} > O^- > F^- > F$
b. $O^- > O^{2-} > F^- > F$
c. $O^{2-} > F^- > F > O^-$
d. $O^{2-} > F^- > O^- > F$

10 Which of the following factors may be regarded as the main cause of lanthanide contraction?

a. Greater shielding of 5*d*-electrons by 4*f*-electrons
b. Poorer shielding of 5*d*-electrons by 4*f*-electrons
c. Effective shielding of one of 4*f*-electrons by another in the subshell
d. Poor shielding of one of 4*f*-electrons by another in the subshell

11 The relative extent to which the various orbitals penetrate the electron clouds of other orbitals is

a. $s > p > d > f$
b. $s < p < d < f$
c. $s < d < p < f$
d. $d < s < p < f$

12 Screening effect is not observed in

a. He^+
b. Li^{2+}
c. Be^{3+}
d. All of these

13 Variation of valency while moving along a period left to right is shown below

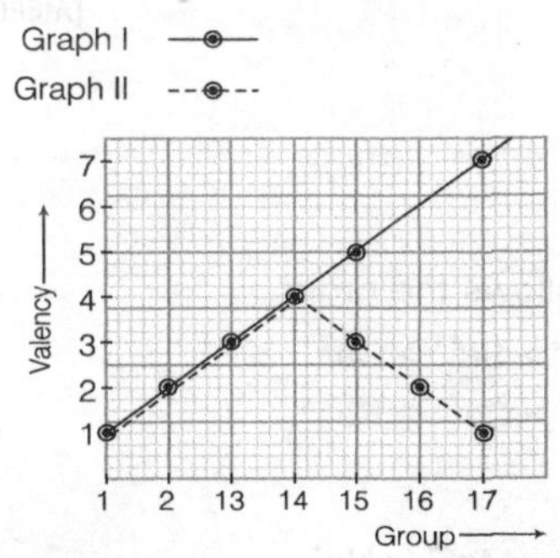

Select the correct statement(s).

a. Graph I is for valency w.r.t oxygen
b. Graph II is for valency w.r.t hydrogen
c. Both (a) and (b)
d. None of these

Working Space

14 Different (variable) oxidation states are shown by transition elements. It is due to the fact

a. $(n-1)d$-electrons may be excited to ns orbital
b. ns-electrons may be excited to $(n-1)d$ orbitals
c. $(n-1)d$-electrons may get involved along with ns electrons in bonding
d. None of the above facts is correct

15 Element (X) has (EC)...ns^2np^1 and element (Y) has (EC)...ns^2np^4.
Thus, most probable compound formed from X and Y is

a. X_2Y b. XY_2
c. X_6Y_3 d. X_2Y_3

16 Which has maximum stability?

a. $AsCl_3$ b. $SbCl_3$
c. $BiCl_3$ d. All of these

17 Which reaction is most spontaneous?

a. $Ga^{3+} + 2e^- \longrightarrow Ga^+$ b. $In^{3+} + 2e^- \longrightarrow In^+$
c. $Tl^{3+} + 2e^- \longrightarrow Tl^+$ d. Cannot be predicted

18 Maximum value of oxidation number is shown by

a. Group 6 b. Group 7
c. Group 16 d. Group 18

19 Arrange hydrides of third period in increasing reducing nature.

a. $NaH < MgH_2 < AlH_3 < SiH_4$ b. $NaH < SiH_4 < AlH_3 < MgH_2$
c. $SiH_4 < AlH_3 < MgH_2 < NaH$ d. $NaH = MgH_2 < AlH_3 = SiH_4$

20 Out of BeO, Al_2O_3, SnO_2, PbO_2 and ZnO, oxide(s) soluble in NaOH is/are

a. SnO_2 and PbO_2 b. BeO, SnO_2 and PbO_2
c. PbO_2 d. None of these

21 Ionic hydrides react with water to give

a. a basic solution b. an acidic solution
c. a hydride ion d. protons

22 Reducing nature of hydrides is due to

a. oxidation of H^- to H b. oxidation of H to H^+
c. reduction of H to H^- d. reduction of H^+ to H

23 Which one of the following orders represents the correct sequence of the increasing basic nature of the given oxides? **[AIEEE 2011]**

a. $Al_2O_3 < MgO < Na_2O < K_2O$
b. $MgO < K_2O < Al_2O_3 < Na_2O$
c. $Na_2O < K_2O < MgO < Al_2O_3$
d. $K_2O < Na_2O < Al_2O_3 < MgO$

24 The heat of hydration of Ca^{2+}, Sr^{2+} and Ba^{2+} follows the order

a. $Ca^{2+} > Sr^{2+} > Ba^{2+}$ b. $Ba^{2+} > Sr^{2+} > Ca^{2+}$
c. $Sr^{2+} > Ba^{2+} > Ca^{2+}$ d. $Sr^{2+} > Ca^{2+} > Ba^{2+}$

25 Which is/are correct order of ionic mobility?

a. $Li^+ < Na^+ < K^+$ b. $Al^{3+} < Mg^{2+} < Na^+$
c. Both (a) and (b) d. None of these

Working Space

Statement Type

Direction (Q. Nos. 26-29) *This section is based on Statement I and Statement II. Select the correct answer from the codes given below.*

a. Both Statement I and Statement II are correct and Statement II is the correct explanation of Statement I
b. Both Statement I and Statement II are correct but Statement II is not the correct explanation of Statement I
c. Statement I is correct but Statement II is incorrect
d. Statement II is correct but Statement I is incorrect

26 **Statement I** Free energy decreases for the following change

$$Ga^{+} \longrightarrow Ga^{3+} + 2e^{-};\ \Delta G^{\circ} < 0$$

Statement II Stability of $Ga^{+} < In^{+} < Tl^{+}$ and that of $Ga^{3+} > In^{3+} > Tl^{3+}$

27 **Statement I** A positive ion (cation) is always smaller than the corresponding atom.

Statement II Greater the positive charge on the ion, smaller the size of cation.

28 **Statement I** There is high jump in the value of radius from F (72 pm) to Ne (160 pm).

Statement II Fluorine is an electronegative atom and Ne is a noble gas.

29 **Statement I** Zr (160 pm) and Hf (159 pm) have equal size.

Statement II The *d*- and *f*-orbitals do not shield the nuclear charge very effectively and thus, there is significant reduction in the size of the ions after *d*- or *f*-orbitals have been filled completely (lanthanide contraction).

Matching List Type

Direction (Q. No. 30) *Choices for the correct combination of elements from Column I and Column II are given as options (a), (b), (c) and (d), out of which one is correct.*

30 Match the oxides given in Column I with their types provided in Column II and mark the correct option from the codes given below.

	Column I		Column II
i.	Na_2O	p.	Neutral
ii.	Al_2O_3	q.	Amphoteric
iii.	Cl_2O_7	r.	Acidic
iv.	CO	s.	Basic

Codes

	i	ii	iii	iv
a.	s	q	r	p
b.	s	r	p	q
c.	s	q	r	r
d.	s	r	q	p

DPP-3 Periodic Trends-II (IE, Electron Affinity and Electronegativity)

Only One Option Correct Type

Direction (Q. Nos. 1-23) *This section contains 23 multiple choice questions. Each question has four choices (a), (b), (c) and (d), out of which ONLY ONE option is correct.*

Working Space

1 The dominant factor in determining the IE of the elements on moving down the group 1 (alkali metals) is their

a. atomic radius
b. effective nuclear charge
c. Both (a) and (b)
d. None of the above

2 The first ionisation potential of Na is 5.1 eV. The value of electron gain enthalpy of Na^+ will be **[JEE Main 2013]**

a. -2.55 eV
b. -5.1 eV
c. -10.2 eV
d. $+2.17$ eV

3 The first ionisation potentials of Na, Mg, Al and Si are in order

a. Na > Mg > Al > Si
b. Na < Mg < Al < Si
c. Na < Mg > Al < Si
d. Na > Al > Si > Mg

4 Among the following elements (whose electronic configurations are given below), the one having the highest ionisation energy is

a. [Ne] $3s^2 3p^1$
b. [Ne] $3s^2 3p^3$
c. [Ne] $3s^2 3p^2$
d. [Ar] $3d^{10} 4s^2 4p^3$

5 Electronegativity of an element is E_1 and ionisation potential is E_2. Hence, electron affinity value is

a. $2E_1 - E_2$
b. $E_1 - E_2$
c. $E_1 - 2E_2$
d. $(E_1 + E_2)/2$

6 For an element, IE values are

$(IE)_1 = 738 \text{ kJ mol}^{-1}$ $\quad (IE)_2 = 1450 \text{ kJ mol}^{-1}$

$(IE)_3 = 7.7 \times 10^3 \text{ kJ mol}^{-1}$ $\quad (IE)_4 = 1.1 \times 10^4 \text{ kJ mol}^{-1}$

This element belongs to

a. alkali metals
b. alkaline earth metals
c. chalcogens
d. halogens

7 Consider the following ionisation processes

$$M(g) \longrightarrow M^+(g) + e^-, \ \Delta H^\circ = 100 \text{ eV}$$

$$M(g) \longrightarrow M^{2+}(g) + 2e^-, \ \Delta H^\circ = 250 \text{ eV}$$

Select the incorrect statement.

a. $(IE)_1$ of $M(g)$ is 100 eV
b. $(IE)_1$ of $M^+(g)$ is 150 eV
c. $(IE)_2$ of $M(g)$ is 250 eV
d. $(IE)_2$ of $M(g)$ is 150 eV

8 Select the correct statement(s).

a. $(IE)_1$ of the corresponding elements of $3d$ and $4d$-series are almost similar
b. $(IE)_1$ of $5d$-series elements are smaller than that of $3d$ and $4d$-series elements
c. Both (a) and (b)
d. None of the above

9 Higher values of ionisation energies of the $5d$-transition elements are consistent with the

a. relatively smaller effective nuclear charge
b. relatively smaller size of their atoms
c. relatively smaller penetration
d. All of the above factors

10 Following the transition elements, $(IE)_1$ drops abruptly in Ga, In and Tl. This is due to

a. decrease in effective nuclear charge
b. increase in atomic radius
c. removal of an electron from the singly occupied *np*-orbitals of higher energy than the *ns*-orbitals of Zn, Cd and Hg
d. None of the above

11 The set representing the correct order of first ionisation potential is

a. K > Na > Li
b. Be > Mg > Ca
c. B > C > N
d. Ge > Si > C

12 Which is incorrect statement?

a. In solid state, O^{2-} is stabilised by neighbouring cations
b. Formation of O^{2-} from O^- is unfavourable in the gas phase
c. Electron affinity of O > S
d. All of the above are incorrect statements

13 Select the pair of elements with most negative and least negative electron affinity.

a. F, Cl
b. Cl, F
c. Cl, S
d. Cl, P

14 The correct order of second ionisation potential of carbon, nitrogen, oxygen and fluorine is

a. C > N > O > F
b. O > N > F > C
c. O > F > N > C
d. F > O > N > C

15 The element with the highest first ionisation potential is

a. boron
b. carbon
c. nitrogen
d. oxygen

16 The incorrect statement among the following is

a. The first ionisation potential of Al is less than the first ionisation potential of Mg
b. The second ionisation potential of Mg is greater than the second ionisation potential of Na
c. The first ionisation potential of Na is less than the first ionisation potential of Mg
d. The third ionisation potential of Mg is greater than the third ionisation potential of Na

17 Which of the following represents the correct increasing order of first ionisation enthalpy (IE) for Ca, Ba, S, Se and Ar? **[JEE Main 2013]**

a. Ca < S < Ba < Se < Ar
b. S < Se < Ca < Ba < Ar
c. Ba < Ca < Se < S < Ar
d. Ca < Ba < S < Se < Ar

Working Space

18 Consider the following ionisation processes with enthalpy changes :

$$Be(g) \xrightarrow{\Delta H_1} Be^{+}(g) \xrightarrow{\Delta H_2} Be^{2+}(g) \xrightarrow{\Delta H_5} Be^{3+}(g)$$

$$B(g) \xrightarrow{\Delta H_3} B^{+}(g) \xrightarrow{\Delta H_4} B^{2+}(g) \xrightarrow{\Delta H_6} B^{3+}(g)$$

For the above cases,

a. $\Delta H_1 > \Delta H_3$
b. $\Delta H_4 > \Delta H_2$
c. $\Delta H_5 >> \Delta H_6$
d. All of these

19 Which represents alkali metal based on $(IE)_1$ and $(IE)_2$ values in kJ mol^{-1}?

	$(IE)_1$	$(IE)_2$
a. *P*	100	110
b. *Q*	110	120
c. *R*	195	500
d. *S*	200	250

20 The second ionisation energy is always higher than the first ionisation energy because the

a. ion becomes more stable attaining an octet or duplet configurations
b. electron is more tightly bound to the nucleus in an ion
c. electron is attracted more by the core electrons
d. None of the above is the correct explanation

21 Which pair represents incorrect first IE?

a. Be > B
b. N > O
c. Li > Na
d. He > He^{+}

22 Fluorine is more electronegative than nitrogen. The best explanation is that

a. the valence electrons in F are on the average, a little closer to the nucleus than in N
b. the charge on a F nucleus is + 9, while that on N nucleus is +7
c. the valence electrons in F and N are in different shells and thus, they differ in energy
d. the electronegativity increases along a period

23 Electronegativities of two elements *X* and *Y* are given

X : 1.2 units

Y : 3.4 units

Thus, type of bond between *X* and *Y* in *XY* is

a. covalent
b. ionic
c. polar covalent
d. coordinate covalent

Statement Type

***Direction** (Q. Nos. 24-28) This section is based on Statement I and Statement II. Select the correct answer from the codes given below.*

a. Both Statement I and Statement II are correct and Statement II is the correct explanation of Statement I
b. Both Statement I and Statement II are correct but Statement II is not the correct explanation of Statement I
c. Statement I is correct but Statement II is incorrect
d. Statement II is correct but Statement I is incorrect

24 **Statement I** The first IE of Be is greater than that of B.

Statement II 2*p*-orbital is lower in energy than 2*s*-orbital.

Working Space

25 **Statement I** F-atom has less electron affinity than Cl-atom.

Statement II Additional electrons are repelled more effectively by 3*p* electrons in Cl-atom than by 2*p* electrons in F-atom.

26 **Statement I** The highest oxidation state of the element is exhibited in its oxide and fluoride.

Statement II Electronegativity increases along a period from left to right.

27 **Statement I** Li and Mg show diagonal relationship.

Statement II Li_2CO_3 and $MgCO_3$ are stable to heat.

28 **Statement I** Be and Al show diagonal relationship.

Statement II Be_2C and Al_4C_3 are decomposed by water forming alkanes.

Matching List Type

Direction *(Q. Nos. 29 and 30) Choices for the correct combination of elements from Column I and Column II are given as options (a), (b), (c) and (d), out of which one is correct.*

29 Match the elements in Column I with their properties in Column II and select the answer from the codes given.

	Column I		Column II
i.	Li	p.	Highest electron gain enthalpy
ii.	Ca	q.	Most electropositive element
iii.	He	r.	Strongest reducing agent
iv.	Cl	s.	Highest ionisation energy

Codes

	i	ii	iii	iv		i	ii	iii	iv
a.	r	q	s	p	b.	q	r	s	p
c.	q	r	p	s	d.	p	q	r	s

30 Match the properties in Column I with the parameters (that they are dependent) in Column II and select the answer from the codes given.

	Column I		Column II
i.	Electron affinity	p.	Radius
ii.	Ionisation potential	q.	Z_{eff}
iii.	Electronegativity	r.	Half-filled or fully filled
iv.	Hydration energy	s.	Screening constant

Codes

	i	ii	iii	iv		i	ii	iii	iv
a.	p,r	q,r	p,s	s	b.	s,q	r,p	q,p	p
c.	p, r	r	s	q	d.	p,q,r,s	p,q,r,s	p,q,r,s	p,q

SET - 1

Revisal Problems for JEE Main

Only One Option Correct Type

Direction *(Q. Nos. 1-34) This section contains 34 multiple choice questions. Each question has four choices (a), (b), (c) and (d), out of which ONLY ONE option is correct.*

Working Space

1 Select the sets of species with approximately same size.

a. Fe, Co Ni
b. Zr, Hf
c. Nb, Ta
d. All of these

2 Decreasing order of ionic radii of $Se^{2-}, I^-, Br^-, O^{2-}$ and F^- is

a. $Se^{2-}, I^-, Br^-, O^{2-}, F^-$
b. $I^-, Se^{2-}, O^{2-}, Br^-, F^-$
c. $Se^{2-}, I^-, Br^-, F^-, O^{2-}$
d. $I^-, Se^{2-}, Br^-, O^{2-}, F^-$

3 For an atom *X*, measured values of different radii are 100 pm, 112 pm and 90 pm. These values respectively are

a. crystal, van der Waals' and covalent
b. crystal, covalent and van der Waals'
c. covalent, van der Waals' and crystal
d. van der Waals', covalent and crystal

4 Select the incorrect variation of gradation of size.

a. Cr= Mn= Fe
b. Ni < Cu < Zn
c. Se < As < Ge
d. Ti < Zr < Hf

5 The first element of a group in many ways differs from the other heavier members of the group. This is due to

a. the small size
b. the high electronegativity and high ionisation potential
c. the unavailability of *d*-orbitals
d. All of the above

6 Select the statement(s) that is/are incorrect for the periodic classification of elements.

a. The properties of elements are the periodic functions of their atomic numbers
b. Non-metallic elements are lesser in number than metallic elements
c. The first ionisation energies of elements along a period do not vary in a regular manner with increase in atomic number
d. For transition elements, the *d*-subshells are filled with electrons monotonically with increase in atomic number

7 Select the correct statement.

a. Elements of groups 11 and 12 are of transition family
b. Elements of group 5 do not contain metalloids
c. Elements of groups 1and 2 are normal elements
d. Elements of group 14 are neither strongly electronegative nor strongly electropositive

8 General electronic configuration of lanthanides is

a. $(n-2)f^{1\text{-}14}\ (n-1)s^2p^6\ d^{0\text{-}1}\ ns^2$
b. $(n-2)f^{1\text{-}14}\ (n-1)d^{10}\ ns^2$
c. $(n-2)f^{0\text{-}14}\ (n-1)d^{10}\ ns^2$
d. $(n-2)d^{0\text{-}1}\ (n-1)f^{1\text{-}4}\ ns^2$

9 In which block will copper be placed if Aufbau rule is not followed and filling of electrons takes place in the following sequence 1s, 2s, 2p, 3s, 3p, 3d, 4s, 4p, 4d, 4f...?

a. *s* b. *p*
c. *d* d. *f*

10 In the long form of the periodic table, the valence shell electronic configuration of $5s^2 5p^4$ corresponds to the element present in **[JEE Main 2015]**

	Group	Period
a.	16	6
b.	17	5
c.	16	5
d.	17	6

11 The element with the electronic configuration as $[Ar]\, 3d^{10} 4s^2\, 4p^3$ represents a

a. metal b. non-metal
c. metalloid d. transition element

12 With respect to oxygen, maximum valency is shown by

a. halogen family b. oxygen family
c. nitrogen family d. boron family

13 Which is the best oxidising agent?

a. Ge^{4+} b. Pb^{4+}
c. Sn^{4+} d. Sn^{2+}

14 Cu^{2+} has a stronger polarising power than that of Ca^{2+} because

a. Cu^{2+} ion is smaller than Ca^{2+}
b. Ca^{2+} ion has inert gas configuration whereas, Cu^{2+} ion does not
c. Copper shows variable valency, calcium does not
d. Cu^{2+} ion is smaller than Ca^{2+} ion and the *d*-electrons in Cu^{2+} ion shield the nucleus poorly

15 The high oxidising power of fluorine is due to

a. high electron affinity
b. high ionising energy
c. Both (a) and (b)
d. None of the above

16 For Ni and Pt, different IE values in kJ mol^{-1} are given below

	$(IE)_1 + (IE)_2$	$(IE)_3 + (IE)_4$
Ni	2.49	8.80
Pt	2.66	6.70

Hence,

a. nickel (II) compounds tend to be thermodynamically more stable than platinum (II)
b. platinum (IV) compounds tend to be more stable than nickel (IV)
c. Both (a) and (b) are correct
d. None of the above is correct

17 Standard heat of formation (ΔH_f°) of different oxides of third period are given below

Oxide	ΔH_f° (kJ mol^{-1})
Na_2O	– 416
MgO	– 602
Al_2O_3	– 1676
SiO_2	– 911
P_4O_{10}	– 2984
SO_3	– 395

Most stable oxide is

a. Na_2O b. MgO
c. P_4O_{10} d. SO_3

18 A, B and C are the elements of the third short period. Given,

Oxide of A is ionic.

Oxide of B is amphoteric.

Oxide of C is giant molecule.

Thus, A, B and C will have atomic number in the order

a. $A < B < C$ b. $C < B < A$
c. $A < C < B$ d. $B < A < C$

19 Mixture containing aqueous Li^+, Na^+ and K^+ ions are electrolysed. Reaction taking place at cathode is

$$M^+ (aq) + e^- \longrightarrow M$$

Cations are discharged at the cathode in the order

a. Li^+, Na^+, K^+ b. K^+, Na^+, Li^+
c. Li^+, K^+, Na^+ d. Na^+, K^+, Li^+

20 Select the correct statement(s).

a. Along a period from left to right, acidic nature of oxides increases
b. Along a group from top to bottom, basic nature of oxides increases
c. Oxide of hydrogen is amphiprotic
d. All of the above are correct statements

21 Given,

Bond energy of H—H bond = 104.2 kcal mol^{-1}

Bond energy of F—F bond = 36.6 kcal mol^{-1}

Bond energy of H—F bond = 134.6 kcal mol^{-1}

Electronegativity of H on Pauling's scale = 2.05

Thus, electronegativity of fluorine is

a. 3.60 b. 5.65
c. 1.55 d. − 3.60

22 Ionic character of a bond (H—X) varies with the difference in electronegativity values as shown below.

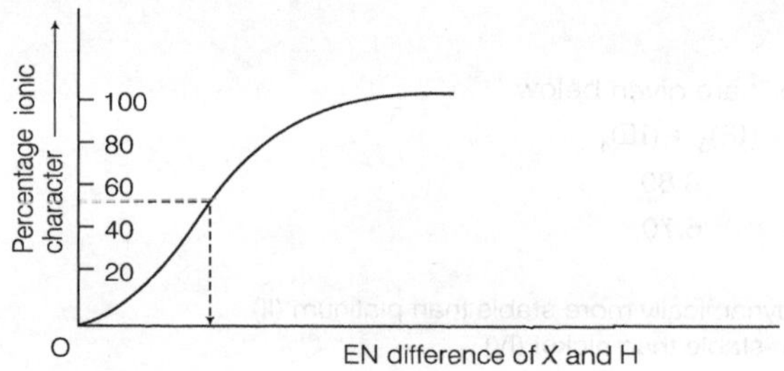

50% ionic character of HX is when EN difference is

a. 1.4 b. 1.5
c. 1.6 d. 1.7

23 Successive (IE) values of three elements of third period are given below

Elements	$(IE)_1$	$(IE)_2$	$(IE)_3$
X	513	4562	6920
Y	738	1451	7733
Z	1521	2666	3931

Atomic numbers of X, Y and Z respectively are

a. 11, 12, 13 b. 11, 12, 14
c. 11, 12, 15 d. 11, 12, 18

24 $(IE)_1$ of Cs is 376 kJ mol^{-1}. Maximum number of Cs^+ ions that can be produced per joule of energy absorbed by a sample of gaseous Cs-atom is

a. 1.60×10^{18}
b. 6.01×10^{23}
c. 6.02×10^{23}
d. 1.60×10^{15}

25 $\frac{N_0}{2}$ atoms of $X(g)$ are converted into $X^+(g)$ by energy ΔH_1°.

$\frac{N_0}{2}$ atoms of $X(g)$ are converted into $X^+(g)$ and $\frac{N_0}{2}$ atoms of $X(g)$ are converted into $X^-(g)$ by energy ΔH_2°.

Thus, IE and EA of $X(g)$ are

	(IE)	**(EA)**
a.	ΔH_1°	$\Delta H_1^\circ + \Delta H_2^\circ$
b.	$\frac{\Delta H_1^\circ}{N_0}$	$\frac{\Delta H_2^\circ - \Delta H_1^\circ}{N_0}$
c.	$\frac{\Delta H_1^\circ}{2N_0}$	$\frac{\Delta H_2^\circ - \Delta H_1^\circ}{2N_0}$
d.	$\frac{2\Delta H_1^\circ}{N_0}$	$\frac{2(\Delta H_2^\circ - \Delta H_1^\circ)}{N_0}$

26 Which has maximum polarising power?

a. Na^+
b. Mg^{2+}
c. Al^{3+}
d. Si^{4+}

27 If each orbital can take maximum of three electrons (elements), then total number of elements in second and fourth periods respectively are

a. 8, 18
b. 9, 18
c. 18, 27
d. 12, 27

28 There is repetition of properties after 2, 8, 8, 18, 18 and 32 elements if they are arranged in increasing atomic numbers. These numbers are called

a. magic numbers
b. atomic numbers
c. mass numbers
d. quantum numbers

29 In the absence of Aufbau rule and also assume that each orbital can take maximum of three electrons, then number of elements in different periods are

Period	3	4	5
a.	18	32	50
b.	18	18	32
c.	27	27	48
d.	27	48	75

30 There is repetition of properties in the periodic table after 2, 8, 8, 18, 18 and 32 elements. This sequence is based on

a. Aufbau rule
b. Pauli's exclusion principle
c. Hund's rule
d. Avogadro's law

31 What is the group number of the divalent cation with 24 electrons?

a. 5
b. 6
c. 7
d. 8

32 The general electronic configuration of the outermost and penultimate shell for a d-block element is given by

$$(n-1)\, s^2p^6d^x ns^y$$

For a divalent cation of an element with $n = 4$, $x = 6$ and $y = 2$, number of protons is

a. 24
b. 25
c. 26
d. 27

33 Assume total number of 112 elements, then how many periods would be required in the periodic table without separate position of lanthanides and actinides, also in the absence of Aufbau rule?

a. 5 b. 6 c. 7 d. 8

34 As predicted by Johann Dobereiner's Triads, elements in triads are found with sequence of three similar elements, when the middle element has a mass equal to the average of the least and most massive.

Elements	Mass number
A	—
B	23
C	39

What is mass number of *A*?

a. 7 b. 8
c. 31 d. 17

Statement Type

Direction (Q. Nos. 35-43) *This section is based on Statement I and Statement II. Select the correct answer from the codes given below.*

a. Both Statement I and Statement II are correct and Statement II is the correct explanation of Statement I
b. Both Statement I and Statement II are correct but Statement II is not the correct explanation of Statement I
c. Statement I is correct but Statement II is incorrect
d. Statement II is correct but Statement I is incorrect

35 **Statement I** Na^{+}, Mg^{2+}, F^{-} and O^{2-} are isoelectronic and have same electronic configuration.

Statement II Isoelectronic species with different number of protons have been placed in different groups.

36 **Statement I** Atomic radius of Ga ($Z = 31$) is less than that of Al ($Z = 13$).

Statement II The presence of additional ten electrons in $3d$-orbitals in Ga offer only poor screening effect for the outer electrons from the increased nuclear charge in Ga.

37 **Statement I** PbO_2 is a good oxidising agent.

Statement II PbO_2 is oxidised to PbO.

38 **Statement I** PCl_5 and NCl_5 are stable compounds.

Statement II In PCl_5 formation, phosphorus uses *ns* electron in excited state.

39 **Statement I** Isoelectronic species (containing equal number of electrons) will have different size.

Statement II N^{3-}, O^{2-}, F^{-}, Ne, Na^{+}, Mg^{2+}, Al^{3+} and Si^{4+} are isoelectronic.

40 **Statement I** For the processes in gaseous phase $O^{-} + e^{-} \longrightarrow O^{2-}; \Delta G° > 0$.

Statement II In solids, O^{2-} is stabilised by neighbouring cations.

41

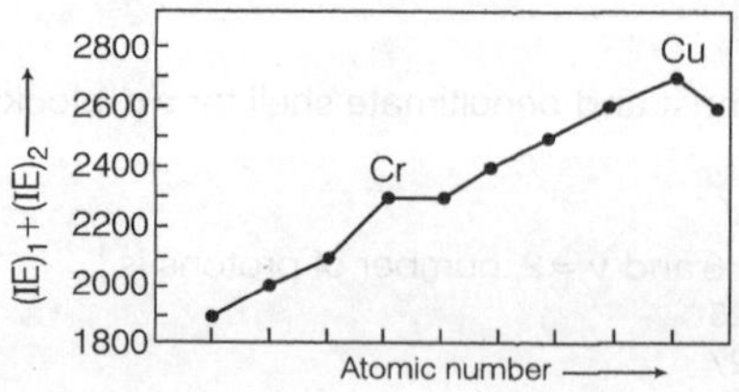

Statement I Higher than the expected values for the sum of $(IE)_1$ and $(IE)_2$ are observed for chromium and copper.

Statement II Chromium has [Ar] $3d^5 4s^1$ and copper has [Ar] $3d^{10} 4s^1$ configuration.

42 **Statement I** Fluorine is more electronegative than nitrogen.

Statement II Effective nuclear charge in F is larger than that of N.

43 **Statement I** Lithium resembles magnesium.

Statement II In the periodic table, they lie diagonally.

Matching List Type

Direction *(Q. Nos. 44-48) Choices for the correct combination of elements from Column I and Column II are given as options (a), (b), (c) and (d), out of which one is correct.*

44 Match the electronic configuration of the elements in Column I with their nature/properties in Column II and select the answer from the codes given.

	Column I		Column II
i.	[Ar] $3d^{10}4s^{2}4p^{6}5s^{1}$	p.	Paramagnetic
ii.	[Ne] $3s^{2}3p^{6}$	q.	Diamagnetic
iii.	[Ar] $4s^{2}3d^{6}$	r.	*p*-block element
iv.	[Xe] $4f^{14}5d^{10}6s^{2}6p^{3}$	s.	*d*-block element
		t.	Zero group (18)
		u.	4th period element

Codes

	i	ii	iii	iv
a.	p	q,r,t	p,s,u	p,r
b.	p	q,r, t	s,u	r
c.	p	r, t	u	p
d.	p,u	r,t	p,s,u	p

45 If all the rules of filling orbitals are not considered, match the elements in Column I with their groups in Column II and select the answer from the codes given.

	Column I		Column II
i.	Na ($Z = 11$)	p.	1
ii.	S ($Z = 16$)	q.	10
iii.	Ca ($Z = 20$)	r.	15
iv.	As ($Z = 33$)	s.	16

Codes

	i	ii	iii	iv
a.	p	q	r	s
b.	s	p	p	q
c.	p	s	q	r
d.	s	r	q	p

46 Match the EC of the elements in Column I with their IE values in Column II and select the answer from the codes given.

	Column I		Column II (kJ mol^{-1})
i.	ns^2	p.	4000
ii.	ns^2np^1	q.	1500
iii.	ns^2np^3	r.	1400
iv.	ns^2np^6	s.	1200

Codes

	i	ii	iii	iv
a.	p	q	r	s
b.	p	r	q	s
c.	s	p	r	q
d.	r	s	q	p

47 Match the EC of the elements in Column I with their groups in Column II and select the answer from the codes given.

	Column I		Column II
i.	[Ar] $3d^7 4s^2$	p.	3
ii.	[Ar] $3d^5 4s^1$	q.	4
iii.	[Kr] $4d^2 5s^2$	r.	6
iv.	[Xe] $4f^3 6s^2$	s.	9

Codes

	i	ii	iii	iv
a.	p	q	r	s
b.	s	r	q	p
c.	p	s	r	q
d.	p	q	s	q

48 Match the processes in Column I with the characteristics (they have) in Column II and select the answer from the codes given.

	Column I		Column II
i.	$S + e^- \longrightarrow S^-$	p.	Exothermic process
ii.	$O^- + e^- \longrightarrow O^{2-}$	q.	Endothermic process
iii.	$Mg \longrightarrow Mg^{2+} + 2e^-$	r.	Inert gas configuration is attained
iv.	$N \longrightarrow N^+ + e^-$	s.	Half-filled configuration is attained
v.	$O \longrightarrow O^+ + e^-$	t.	Half-filled configuration is lost

Codes

	i	ii	iii	iv	v
a.	p	q,r	q,r	q,t	q,s
b.	p,r	q,r	q,r	s	q,s
c.	p,r	q,r	q,r	q,t	q,s
d.	p	q	r	q,t	q,s

JEE Main & AIEEE Archive

(Compilation of Last 13 Years' Questions)

1 Which of the following represents the correct order of increasing first ionisation enthalpy for Ca, Ba, S, Se and Ar? **(JEE Main 2013)**

a. Ca < S < Ba < Se < Ar
b. S < Se < Ca < Ba < Ar
c. Ba < Ca < Se < S < Ar
d. Ca < Ba < S < Se < Ar

2 The first ionisation potential of Na is 5.1 eV. The value of electron gain enthalpy of Na^+ will be **(JEE Main 2013)**

a. – 2.55 eV
b. – 5.1 eV
c. –10.2 eV
d. + 2.55 eV

3 The increasing order of the ionic radii of the given isoelectronic species is **(AIEEE 2012)**

a. Cl^-, Ca^{2+}, K^+, S^{2-}
b. S^{2-}, Cl^-, Ca^{2+}, K^+
c. Ca^{2+}, K^+, Cl^-, S^{2-}
d. K^+, S^{2-}, Ca^{2+}, Cl^-

4 Which one of the following orders represents the correct sequence of the increasing basic nature of the given oxides? **(AIEEE 2011)**

a. $Al_2O_3 < MgO < Na_2O < K_2O$
b. $MgO < K_2O < Al_2O_3 < Na_2O$
c. $Na_2O < K_2O < MgO < Al_2O_3$
d. $K_2O < Na_2O < Al_2O_3 < MgO$

5 The correct order of electron gain enthalpy with negative sign of F, Cl, Br and I, having atomic number 9, 17, 35 and 53 respectively, is **(AIEEE 2011)**

a. I > Br > Cl > F
b. F > Cl > Br > I
c. Cl > F > Br > I
d. Br > Cl > I > F

6 The correct sequence which shows decreasing order of the ionic radii of the elements is **(AIEEE 2010)**

a. $Al^{3+} > Mg^{2+} > Na^+ > F^- > O^{2-}$
b. $Na^+ > Mg^{2+} > Al^{3+} > O^{2-} > F^-$
c. $Na^+ > F^- > Mg^{2+} > O^{2-} > Al^{3+}$
d. $O^{2-} > F^- > Na^+ > Mg^{2+} > Al^{3+}$

7 The set representing the correct order of ionic radius is **(AIEEE 2009)**

a. $Li^+ > Be^{2+} > Na^+ > Mg^{2+}$
b. $Na^+ > Li^+ > Mg^{2+} > Be^{2+}$
c. $Li^+ > Na^+ > Mg^{2+} > Be^{2+}$
d. $Mg^{2+} > Be^{2+} > Li^+ > Na^+$

8 The charge/size ratio of a cation determines its polarising power. Which one of the following sequences represents the increasing order of the polarising power of the cationic species, K^+, Ca^{2+}, Mg^{2+}, Be^{2+}? **(AIEEE 2007)**

a. $Mg^{2+} < Be^{2+} < K^+ < Ca^{2+}$
b. $Be^{2+} < K^+ < Ca^{2+} < Mg^{2+}$
c. $K^+ < Ca^{2+} < Mg^{2+} < Be^{2+}$
d. $Ca^{2+} < Mg^{2+} < Be^{2+} < K^+$

9 Which one of the following sets of ions represents a collection of isoelectronic species? **(AIEEE 2006)**

a. K^+, Cl^-, Ca^{2+}, Sc^{3+}
b. Ba^{2+}, Sr^{2+}, K^+, S^{2-}
c. N^{3-}, O^{2-}, F^-, S^{2-}
d. Li^+, Na^+, Mg^{2+}, Ca^{2+}

10 The ionic mobility of alkali metal ions in aqueous solution is maximum for **(AIEEE 2006)**

a. K^+
b. Rb^+
c. Li^+
d. Na^+

11 The increasing order of the first ionisation enthalpies of the elements B, P, S and F (lowest first) is **(AIEEE 2006)**

a. F < S < P < B
b. P < S < B < F
c. B < P < S < F
d. B < S < P < F

12 Following statements regarding the periodic trends of chemical reactivity of the alkali metals and the halogens are given. Which of these statements gives the correct picture? **(AIEEE 2006)**

a. The reactivity decreases in the alkali metals but increases in the halogens with increase in atomic number down the group
b. In both the alkali metals and the halogens, the chemical reactivity decreases with increase in atomic number down the group
c. Chemical reactivity increases with increase in atomic number down the group in both the alkali metals and halogens
d. In alkali metals, the reactivity increases but in the halogens, it decreases with increase in atomic number down the group

13 In which of the following arrangements, the order is not according to the property indicated against it? **(AIEEE 2005)**

a. Li < Na < K < Rb (Increasing metallic radius)
b. I < Br < F < Cl (Increasing electron gain enthalpy) (with negative sign)
c. B < C < N < O (Increasing first ionisation enthalpy)
d. $Al^{3+} < Mg^{2+} < Na^+ < F^-$ (Increasing ionic size)

14 Based on lattice energy and other considerations, which one of the following alkali metal chlorides is expected to have the highest melting point? **(AIEEE 2005)**

a. RbCl b. KCl
c. NaCl d. LiCl

15 Which one of the following ions has the highest value of ionic radius? **(AIEEE 2004)**

a. Li^+ b. B^{3+}
c. O^{2-} d. F^-

16 The formation of the oxide ion O^{2-} (*g*) requires first an exothermic and then an endothermic step as shown below

$$O(g)+e^- = O^-(g); \quad \Delta H^\circ = -142 \text{ kJ mol}^{-1}$$

$$O^-(g)+e^- = O^{2-}(g); \quad \Delta H^\circ = 844 \text{ kJ mol}^{-1}$$

This is because **(AIEEE 2004)**

a. oxygen is more electronegative
b. oxygen has high electron affinity
c. O^- ion will tend to resist the addition of another electron
d. O^- ion has comparatively larger size than oxygen atom

17 Which among the following factors is the most important in making fluorine the strongest oxidising agent? **(AIEEE 2004)**

a. Electron affinity
b. Ionisation enthalpy
c. Hydration enthalpy
d. Bond dissociation energy

18 Among Al_2O_3, SiO_2, P_2O_3 and SO_2, the correct order of acidic strength is **(AIEEE 2004)**

a. $SO_2 < P_2O_3 < SiO_2 < Al_2O_3$
b. $SiO_2 < SO_2 < Al_2O_3 < P_2O_3$
c. $Al_2O_3 < SiO_2 < SO_2 < P_2O_3$
d. $Al_2O_3 < SiO_2 < P_2O_3 < SO_2$

19 According to the periodic law of elements, the variation in properties of elements is related to their **(AIEEE 2003)**

a. atomic masses
b. nuclear masses
c. atomic numbers
d. nuclear neutron-proton number ratios

20 The radius of La^{3+} (atomic number : La = 57) is 1.06 Å. Which one of the following given values will be closest to the radius of Lu^{3+} (atomic number : Lu = 71)? **(AIEEE 2003)**

a. 1.60 Å b. 1.40 Å c. 1.06 Å d. 0.85 Å

21 The atomic numbers of vanadium (V), chromium (Cr), manganese (Mn) and iron (Fe) are, respectively 23, 24, 25 and 26. Which one of these may be expected to have the highest second ionisation enthalpy? **(AIEEE 2003)**

a. V b. Cr c. Mn d. Fe

22 Ce^{3+}, La^{3+}, Pm^{3+} and Yb^{3+} have ionic radii in the increasing order as **(AIEEE 2002)**

a. $La^{3+} < Ce^{3+} < Pm^{3+} < Yb^{3+}$
b. $Yb^{3+} < Pm^{3+} < Ce^{3+} < La^{3+}$
c. $La^{3+} < Ce^{3+} < Pm^{3+} < Yb^{3+}$
d. $Yb^{3+} < Pm^{3+} < La^{3+} < Ce^{3+}$

2. Hydrogen

DPP-1 Hydrogen and Its Isotopes

Only One Option Correct Type

Working Space

Direction *(Q. Nos. 1-31) This section contains 31 multiple choice questions. Each question has four choices (a), (b), (c) and (d), out of which ONLY ONE is correct.*

1 Which statement is not correct for hydrogen?

a. It has a very high ionisation potential
b. It is always collected at cathode
c. It can form bonds in $+1$ as well as -1 oxidation state
d. It has approximately same electronegativity as that of halogens

2 H^+, D^+ and T^+ differ in all except in

a. number of electrons
b. number of neutrons
c. ionic mass
d. All of the above

3 Average atomic mass of a mixture of two isotopes of hydrogen ${}_1^1H$ and ${}_1^2H$ is 1.00014. Thus, percentage of ${}_1^2H$ is

a. 1.4% b. 0.14%
c. 0.014% d. 0.0014%

4 H_2 reacts much faster with Cl_2 than D_2 because

a. rate of diffusion of H_2 is larger than that of D_2
b. energy of dissociation of H_2 is lower than that of D_2
c. Both (a) and (b)
d. None of the above

5 High purity H_2 is formed by

a. electrolysis of acidified water using platinum electrodes
b. electrolysis of warm aqueous $Ba(OH)_2$ solution between nickel electrodes
c. electrolysis of brine solution using amalgam electrodes
d. reaction of steam on hydrocarbon or coke

6 Coal gasification reaction is

a. $CH_4(g) + H_2O(g) \xrightarrow[Ni]{1270\ K} CO(g) + 3H_2(g)$

b. $C(s) + H_2O(g) \xrightarrow{1270\ K} CO(g) + H_2(g)$

c. $CO(g) + H_2O(g) \xrightarrow[Co]{673\ K} CO(g) + H_2(g)$

d. None of the above

7 Atomic hydrogen can be used as a/an

a. reducing agent b. oxidising agent
c. source of welding d. All of these

8 Consider the following statements about the figure,

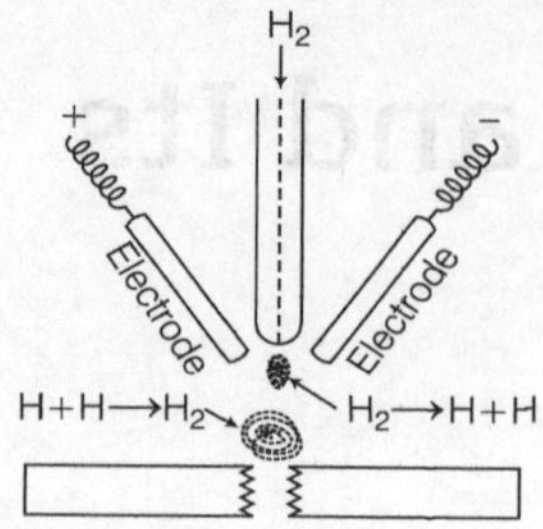

I. It produces atomic H.
II. It underlines the principle of making "atomic hydrogen welding torch".
III. Atomic hydrogen forms metallic hydride with metals.
IV. Transition metals catalyse the reaction.

$$2H \longrightarrow H_2, \ \Delta H = -436 \text{ kJ mol}^{-1}$$

Select the correct statements.

a. II, III and IV
b. Both I and II
c. Both II and III
d. All of these

9 Which is the source of nascent hydrogen for reduction?

a. $Cl_2 + H_2O$
b. $SO_2 + H_2O$
c. Both (a) and (b)
d. None of these

10 In which case(s), yellowish solution of $FeCl_3(aq)$ turns colourless?

a. On passing H_2 gas into it
b. On adding zinc pieces into it
c. Both (a) and (b)
d. None of these

11 Hydroformylation reaction is used in the synthesis of

a. formaldehyde
b. formic acid
c. methyl alcohol
d. ethyl alcohol

12 Cracking reactions are used in the refining of petroleum and H_2 is the byproduct of cracking of propane along with

a. methane
b. ethene
c. propene
d. All of these

13 In water-gas shift reaction, H_2 is formed in excess and CO is oxidised to CO_2

$$CO + H_2 + H_2O \xrightarrow[450\text{-}550 \text{ K}]{Fe_2O_3} CO_2 + 2H_2$$

Mixture of CO_2 and H_2 can be separated by

a. dissolving into water
b. dissolving into lime water
c. diffusion
d. All of the above

14 Henry Cavendish isolated pure H_2 for the first time based on which of the following reaction?

a. $2NaOH + Zn \longrightarrow Na_2ZnO_2 + H_2$
b. $\underset{\text{Dilute}}{H_2SO_4} + Zn \longrightarrow ZnSO_4 + H_2$
c. $2H_2O \xrightarrow{\text{Current}} 2H_2 + O_2$
d. $BeH_2 \xrightarrow{\text{Heat}} Be + H_2$

Working Space

15 Out of the following metals, which will give H_2 on reaction with NaOH and dil. HCl both?

I. Zn II. Mg
III. Al IV. Be

a. I, II, III and IV
b. I, III and IV
c. Both II and IV
d. Both I and III

16 Which gives colour with $K_3[Fe(CN)_6]$?

I. $FeCl_3$ aqueous solution + Zn $\longrightarrow$ Product *X*
II. $FeCl_3$ aqueous solution + H_2 gas $\longrightarrow$ Product *Y*

a. Only *X* b. Only *Y*
c. Both (a) and (b) d. None of these

17 Which is a source of nascent hydrogen out of I to IV?

I. Zn + dil. HCl
II. CH_3OH + Na
III. Electrolysis of H_2O
IV. Silent electric discharge

a. Both I and II b. Both II and III
c. I, II, III and IV d. Only IV

18 Which of the following compounds reacts easily with water producing H_2?

a. PH_3 b. B_2H_6
c. CH_4 d. H_2S

19 Steam is passed over red hot carbon and the gaseous products cooled and passed through as shown

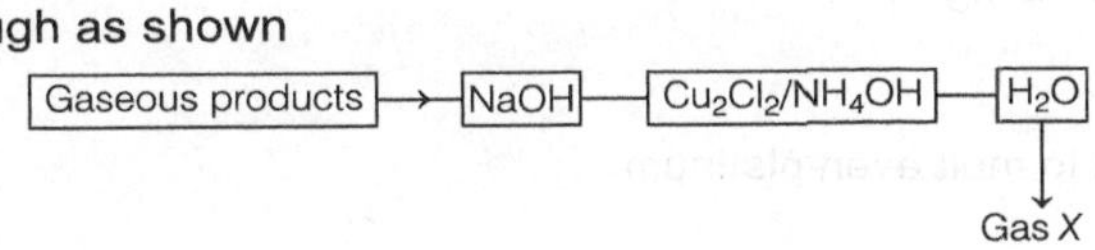

Gas *X* coming out at the end is

a. CO_2
b. a mixture of hydrocarbons
c. $CO + H_2$
d. H_2

20 H_2 gas is liberated at cathode and anode both by electrolysis of the following aqueous solution except in

a. NaH b. HCOONa
c. NaCl d. LiH

21 H, D and T have nuclear spin quantum numbers as

	H	D	T
a.	$\frac{1}{2}$	1	$\frac{1}{2}$
b.	$\frac{1}{2}$	$\frac{1}{2}$	1
c.	$\frac{1}{2}$	1	$\frac{3}{2}$
d.	$\frac{1}{2}$	$\frac{1}{2}$	$\frac{1}{2}$

22 Adsorbed hydrogen by palladium is known as

a. nascent b. atomic
c. heavy d. occluded

Working Space

23 Following reaction takes place at

$$4{}_1^1H \longrightarrow {}_2^4He + 2{}_1^0e$$

a. high temperature, high pressure
b. high temperature, low pressure
c. low temperature, high pressure
d. low temperature, low pressure

24 Hydrogen gas is produced by the reaction

a. $Na_2O_2 + HCl$
b. $Mg + H_2O(steam)$
c. $BaO_2 + HCl$
d. $H_2S_2O_8 + H_2O$

25 The acid not suitable for the preparation of H_2 by the action of metals is

a. dil. HCl
b. CH_3COOH
c. dil. HNO_3
d. dil. H_2SO_4

26 *Ortho* and *para* hydrogens have

a. identical chemical properties but different physical properties
b. identical chemical and physical properties
c. identical physical properties but different chemical properties
d. different chemical and physical properties

27 In context with the industrial preparation of hydrogen from water gas ($CO + H_2$), which of the following is the correct statement? **[AIEEE 2008]**

a. CO and H_2 are fractionally separated using differences in their densities
b. CO is removed by the absorption in aqueous Cu_2Cl_2 solution
c. H_2 is removed through occlusion with Pd
d. CO is oxidised to CO_2 with steam in the presence of a catalyst followed by absorption of CO_2 in alkali

28 Select the correct statement(s) about lime light.

a. Oxy-hydrogen flame
b. Used in welding
c. Temperature of lime light is sufficient to melt even platinum
d. All of the above

29 A device that converts energy of combustion of fuels like H_2 and CH_4, directly into electrical energy is known as **[JEE Main 2015]**

a. electrolytic cell
b. dynamo
c. Ni-Cd cell
d. fuel cell

30 In fuel cells, for use in space-crafts, electrical energy is generated by the reaction of H_2 and O_2 without the evolution of heat.
Select the correct statement(s).

a. Reaction is called cold combustion
b. In acidic or alkaline medium, same types of reactions occur
c. EMF of the cell is 1.23 V
d. All of the above

31 Hydrogen as a fuel has many advantages over the conventional fossil fuels and electrical power. Its advantage(s) is/are

a. It is pollution free and thus is a green fuel
b. It releases greater energy per unit weight compared to gasoline and other fuels
c. Hydrogen can be transported in the form of solid as metal hydride
d. All of the above

Working Space

Statement Type

Direction (Q. Nos. 32-34) *This section is based on Statement I and Statement II. Select the correct answer from the codes given below.*

a. Both Statement I and Statement II are correct and Statement II is the correct explanation of Statement I
b. Both Statement I and Statement II are correct but Statement II is not the correct explanation of Statement I
c. Statement I is correct but Statement II is incorrect
d. Statement II is correct but Statement I is incorrect

32 **Statement I** Intermolecular forces between H_2 molecules are very weak.

Statement II H_2 molecule is non-polar.

33 **Statement I** H_2 molecule is thermally stable with exceptionally high dissociation energy (436 kJ mol^{-1}).

Statement II Stability is attained when H atoms approach each other in the formation of (H — H) bond as shown below.

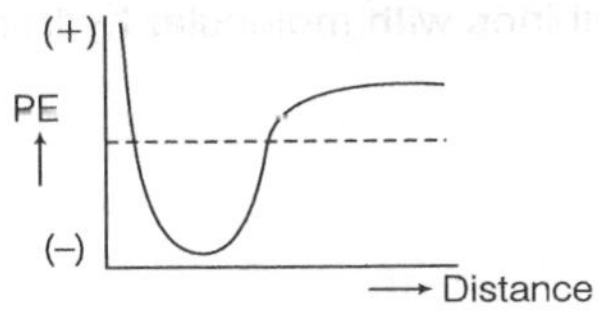

34 **Statement I** Abundance of H_2 in the earth's atmosphere is very small.

Statement II The earth's gravitational field is too small to hold so light an element.

Matching List Type

Direction (Q. No. 35) *Choices for the correct combination of elements from Column I and Column II are given as options (a), (b), (c) and (d), out of which one is correct.*

35 Match the isotopes of hydrogen (and its compounds) in Column I with their properties given in Column II and mark the correct option from the codes given below.

	Column I		Column II
i.	^1_1H	p.	Radioactive
ii.	^2_1H	q.	Moderator in reactor
iii.	^3_1H	r.	0.014%
iv.	D_2O	s.	Stable isotopes

Codes

	i	ii	iii	iv
a.	s	r, s	p	q
b.	s	r	p	q
c.	q	p	r	s
d.	p	r	s	q

DPP-2 Hydrides

Only One Option Correct Type

Direction *(Q. Nos. 1-19) This section contains 19 multiple choice questions. Each question has four choices (a), (b), (c) and (d), out of which ONLY ONE is correct.*

Working Space

1 Select the ionic hydride out of

a. NaH, CaH_2, BaH_2
b. BH_3, AlH_3, SiH_4
c. MnH_2, ZrH_2, LaH_3
d. HF, HCl, HBr, HI

2 Select the correct statement(s) about ionic hydrides.

a. They are formed when electropositive metal combines with molecular hydrogen at high temperature
b. They conduct electricity in fused state
c. They have high melting points
d. All of the above

3 Consider the following reaction,

$$H^-(aq) + H_2O(l) \longrightarrow OH^-(aq) + H_2(g)$$

H^- is a

a. Lewis base
b. Lowry-Bronsted base
c. Lewis acid
d. Lowry-Bronsted acid

4 H^- is a Lewis base in which of the following reaction?

a. $NH_3 + H^+ \longrightarrow NH_4^+$
b. $AlH_3 + H^- \longrightarrow AlH_4^-$
c. $H^- + H_2O \longrightarrow H_2 + OH^-$
d. In all of these

5 Energetic stability of the following hydrides is in the order

$$\underset{\text{I}}{BeH_2}, \underset{\text{II}}{MgH_2}, \underset{\text{III}}{CaH_2}, \underset{\text{IV}}{SrH_2}, \underset{\text{V}}{BaH_2}$$

a. I < II < III < IV < V
b. V < IV < III < II < I
c. I < II > III < IV > V
d. II < I > III < IV > V

6 Which of the following is not a reducing agent?

a. H_2S
b. H_2O
c. H_2Se
d. H_2Te

7 Mass of one atom is 6.66×10^{-23}g. Its percentage in a hydride is 95.24. Thus, hydride is

a. *MH*
b. MH_2
c. MH_3
d. MH_4

8 Hydride ion is a

a. strong conjugate acid of H_2
b. strong conjugate base of H_2
c. strong conjugate acid of H^+
d. strong conjugate base of H^-

9 Saline hydrides are known to react with water violently producing fire which can be put off by

a. CO_2
b. sand
c. H_2O
d. None of these

Working Space

10 Select the correct statement(s).

a. Hydrides are instant source of H_2
b. Saline hydrides remove traces of moisture from organic compounds
c. Metallic hydrides can be useful for hydrogen storage
d. All of the above

11 Electrolysis of aqueous NaCl and NaH differs in

a. formation of basic solution at the cathode by NaCl only
b. formation of basic solution at the cathode by NaH only
c. formation of H_2 gas at the cathode as well as at the anode by NaH only
d. formation of H_2 gas at the cathode as well as at the anode by NaCl only

12 Select the correct statement(s).

a. Hydride ion is larger than any of the halide ions except iodide
b. Hydride ions are reducing agents
c. Boranes are electron deficient hydrides
d. All of the above

13 Select the correct statement(s) about metallic hydrides.

a. They are called interstitial hydrides
b. They are non-stoichiometric, being deficient in hydrogen
c. They are poor conductors of electricity, exhibit less paramagnetism and have hydrogen as atom and not as molecule
d. All of the above

14 The set of elements which could form stable, covalent and hydrogen bond compounds with hydrogen is

a. nitrogen, oxygen and fluorine
b. lithium, sodium and potassium
c. sulphur, selenium and tellurium
d. chlorine, bromine and iodine

15 Among the following, hydride is

a. calamine
b. nitrolim
c. hydrolith
d. minium

16 In which of the following compounds, H is covalent bonded?

a. BaH_2
b. CaH_2
c. NaH
d. SiH_4

17 Consider the following reactions,

I. $AlH_3 + H^- \longrightarrow AlH_4^-$

II. $H_2O + H^- \longrightarrow H_2 + OH^-$

What is correct nature of H^- in I and II?

	I	II
a.	Lewis acid	Lewis base
b.	Lewis base	Bronsted base
c.	Lewis acid	Bronsted acid
d.	Lewis base	Lewis base

18 Electron-rich hydrides have excess electrons which are present as lone pairs. Number of lone pairs are incorrectly matched in

a. PH_3 -1
b. H_2O - 2
c. HF - 3
d. CH_4 - 4

19 LiH reacts with Al_2Cl_6 and B_2H_6 respectively. Products formed are

a. AlH_3 and LiB_2H_7
b. AlH_3 and $LiBH_4$
c. $LiAlH_4$ and $LiBH_4$
d. $LiAlH_4$ and $LiBH_3$

Statement Type

Working Space

Direction *(Q. Nos. 20-24) This section is based on Statement I and Statement II. Select the correct answer from the codes given below.*

a. Both Statement I and Statement II are correct and Statement II is the correct explanation of Statement I
b. Both Statement I and Statement II are correct but Statement II is not the correct explanation of Statement I
c. Statement I is correct but Statement II is incorrect
d. Statement II is correct but Statement I is incorrect

20 **Statement I** Hydrides of group 14 (CH_4, SiH_4, GeH_4 ...) have weak van der Waals' forces of attraction.

Statement II Boiling point of hydrides increases on going down the group.

21 **Statement I** Boiling points of the group 15, 16 and 17 hydrides are higher than that of the group 14 hydrides.

Statement II There is the formation of permanent dipole moment resulting in H-bonding.

22 **Statement I** Ionic hydrides are readymade source of hydrogen.

Statement II CaH_2 is a covalent hydride.

23 **Statement I** Electron deficient hydrides are electron-pair acceptor.

Statement II NH_3 is electron-pair donor.

24 **Statement I** Phosphorus with outer electronic configuration $3s^2 3p^3$ forms PH_5.

Statement II H_2 has high dissociation energy.

Matching List Type

Direction *(Q. No. 25) Choices for the correct combination of elements from Column I and Column II are given as options (a), (b), (c) and (d), out of which one is correct.*

25 Match the hydrides in Column I with their nature (may be more than one) given in Column II and mark the correct option from the codes given below.

	Column I		Column II
i.	B_2H_6	p.	Lewis acid
ii.	CH_4	q.	Lewis base
iii.	AsH_3	r.	Instant source of H_2
iv.	LiH	s.	Electron-deficient
		t.	Electron-rich
		u.	Electron-precise
		v.	Lowry-Bronsted base

Codes

	i	ii	iii	iv		i	ii	iii	iv
a.	p,s	u	q,t	r,v	b.	p	u	q,t	r,v
c.	s	p	q,t	r,v	d.	p,s	q	u	r,v

DPP-3 Soft, Hard and Heavy Water

Only One Option Correct Type

Working Space

Direction *(Q. Nos. 1-26) This section contains 26 multiple choice questions. Each question has four choices (a), (b), (c) and (d), out of which ONLY ONE is correct.*

1 Which is used as the green chemistry solvent?

a. H_2O b. D_2O
c. T_2O d. CH_3CH_2OH

2 Heavy water is

a. H_2O^{18} b. H^1H^2O
c. H^2H^2O d. that does not form lather with soap

3 Temporary hardness of water can be removed by

a. $CaCO_3$ b. Na_2CO_3
c. CaO d. HCl

4 When ordinary water is electrolysed several times, product formed is

a. enriched in H_2O b. enriched in D_2O ($D = {}^2_1H$)
c. enriched in HDO d. enriched in T_2O ($T = {}^3_1H$)

5 Hard water can be purified by passing through synthetic resins in two stages to remove insoluble cations and anions.

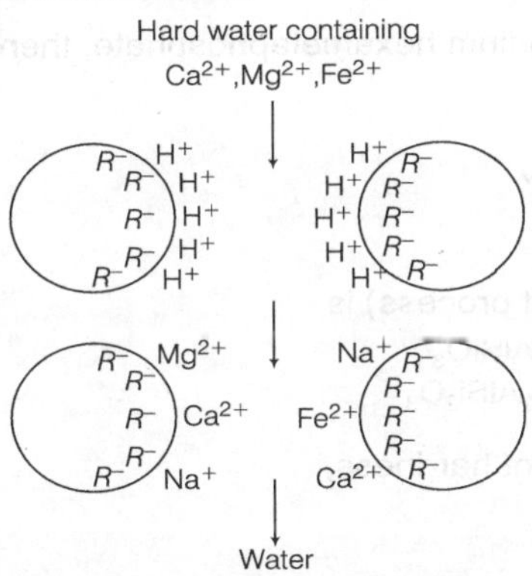

Consider the following statements about the above figure,

I. This is cation exchange resin.
II. This is anion exchange resin.
III. H^+ ions of resin are replaced by cations present in hard water.
IV. Water coming out of the exchange resin is acidic.

Select the correct statements.

a. Both I and III b. I to IV except I
c. I to IV except II d. All of these

Working Space

6 Water is said to play a vital role in the biosphere due to higher values of

I. specific heat
II. thermal conductivity
III. surface tension
IV. dipole moment
V. dielectric constant
VI. viscosity

Which is not the correct choice?

a. Both I and II
b. Both I and IV
c. Both III and V
d. Only VI

7 Select the correct statement(s) about structure of ice and liquid water.

a. Coordination number of water in liquid water is four
b. Ice has highly ordered three-dimensional hydrogen-bonded structure
c. In ice, each oxygen atom is surrounded tetrahedrally by four other oxygen atoms at a distance of 276 pm
d. All of the above

8 In H_2O, angle θ and bond length x respectively are

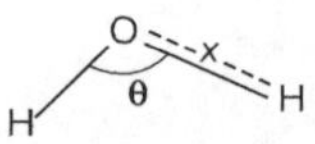

a. $109°28'$, 95.7 pm
b. $104.5°$, 95.7 pm
c. $107°$, 95.7 pm
d. $104.5°$, 100 pm

9 Formation of lather and scum with soap indicates that water used is

a. soft, hard
b. hard, soft
c. hard, hard
d. soft, soft

10 In Clark's method, hard water is treated with $Ca(OH)_2$ to remove temporary hardness in the form of

a. $CaCO_3$ and $MgCO_3$
b. $CaCO_3$ and $Mg(HCO_3)_2$
c. $CaCO_3$ and $Mg(OH)_2$
d. $Ca(HCO_3)_2$ and $Mg(OH)_2$

11 When permanent hard water is treated with sodium hexametaphosphate, then

a. soluble complex is formed
b. method is called Calgon's method
c. Ca^{2+} and Mg^{2+} ions remain soluble in complex
d. All of the above

12 Zeolite used in ion exchange method (Permutit process) is

a. $NaAlSiO_4$
b. $NaAlSiO_3$
c. $Na_3AlSi_2O_4$
d. $Na_2AlSi_2O_3$

13 In ion exchange method of removing permanent hardness,

a. Ca^{2+} or Mg^{2+} ions are replaced by Al^{3+}
b. Ca^{2+} or Mg^{2+} ions are replaced by Na^+
c. Al^{3+} ions are replaced by Ca^{2+}
d. Ca^{2+} or Mg^{2+} ions are replaced by Na^+ and Al^{3+}

14 Synthetic resin method uses cation exchangers. In this method,

a. water insoluble resin RSO_3Na is used
b. Ca^{2+} and Mg^{2+} ions of hard water are replaced by Na^+ ions
c. exhausted resin can be regenerated by adding aqueous NaCl
d. All of the above

Working Space

15 Cl_2 dissolves in water forming HCl and HOCl

$$Cl_2(g) + H_2O(l) \longrightarrow HOCl(aq) + HCl(aq)$$

In this reaction, select the incorrect statement.

a. Cl_2 disproportionates to HOCl and HCl
b. HOCl is oxidising agent, bleaching agent and antibacterial agent
c. H_2O is oxidising agent as well as reducing agent
d. Cl_2 is oxidising agent as well as reducing agent

16 Which is not the hydrolysis reaction?

a. $NaCl + H_2O \rightleftharpoons NaOH + HCl$
b. $CH_3COONa + H_2O \rightleftharpoons CH_3COOH + NaOH$
c. $CuSO_4 + 2H_2O \rightleftharpoons Cu(OH)_2 + H_2SO_4$
d. $CH_3COOC_2H_5 + H_2O \rightleftharpoons CH_3COOH + C_2H_5OH$

17 Excessive H-bonding between water molecules leads to

a. high boiling point and freezing point
b. high heat of vaporisation
c. high heat of fusion
d. All of the above

18 For H_2O, $K_w = 1.0 \times 10^{-14}$ and for D_2O, $K_w = 3.0 \times 10^{-15}$

Electrolysis of H_2O and D_2O separately produces H_2 and D_2. Which is liberated at faster rate?

a. H_2 at cathode
b. D_2 at cathode
c. O_2 at anode
d. All at equal rate

19 Which of the following processes will produce hard water? **[AIEEE 2003]**

a. Saturation of water with $CaCO_3$
b. Saturation of water with $MgCO_3$
c. Saturation of water with $CaSO_4$
d. Addition of Na_2SO_4 to water

20 Select the correct statement(s).

a. The layer of ice on the surface of river in winter acts as a thermal insulator between the water below and the air above
b. The fish and the other organisms are enabled to survive long periods of freezing weather due to the fact that the ice is lighter than water
c. Both (a) and (b) are correct
d. None of the above

21 Which of the following disproportionates when treated with H_2O?

a. SO_3
b. F_2
c. N_2
d. Cl_2

22 D_2O has maximum density at

a. 4°C
b. 11.6°C
c. 0°C
d. 0 K

23 Permanent hardness in water cannot be cured by **[JEE Main 2015]**

a. boiling
b. ion-exchange method
c. Calgon's method
d. treatment with washing soda

24 The molecular formula of a commercial resin used for exchanging ions in water softening is $C_8H_7SO_3Na$ (mol. wt. 206). What would be the maximum uptake of Ca^{2+} ions by the resin expressed in mole per gram resin? **[JEE Main 2015]**

a. 1/103
b. 1/206
c. 2/309
d. $\frac{1}{412}$

Working Space

25 25.00 mL of hard water having hardness due to $CaSO_4$ is passed through anion exchange resin and water coming out of resin required 20 mL of 0.001 M H_2SO_4 for its titration. Thus, hardness of water in ppm of $CaSO_4$ is

a. 108.8
b. 54.4
c. 217.6
d. 27.2

26 The hardness of water sample containing 0.002 mol L^{-1} $MgSO_4$, expressed as ppm of $MgSO_4$, is

a. 120
b. 240
c. 60
d. 480

Statement Type

Direction *(Q. Nos. 27-29) This section is based on Statement I and Statement II. Select the correct answer from the codes given below.*

a. Both Statement I and Statement II are correct and Statement II is the correct explanation of Statement I
b. Both Statement I and Statement II are correct but Statement II is not the correct explanation of Statement I
c. Statement I is correct but Statement II is incorrect
d. Statement II is correct but Statement I is incorrect

27 **Statement I** Boiling point of H_2O is abnormally high compared to boiling point of H_2S.

Statement II H_2O is liquid while H_2S is gas.

28 **Statement I** Boiling point of H—F is higher than boiling point of H_2O.

Statement II H—F bond is more polar than H—O bond due to more electronegativity of F atom.

29 **Statement I** Ice floats on water.

Statement II Density of water is maximum at 4°C.

Matching List Type

Direction *(Q. No. 30) Choices for the correct combination of elements from Column I and Column II are given as options (a), (b), (c) and (d), out of which one is correct.*

30 Match the reactions in Column I with nature of water in Column II and mark the correct option from the codes given below.

	Column I		Column II
i.	$NH_3 + H_2O \longrightarrow NH_4OH$	p.	Photochemical reaction
ii.	$CH_3COOH + H_2O \longrightarrow CH_3COO^- + H_3O^+$	q.	Oxidising nature of H_2O
iii.	$H_2O + H_2O \longrightarrow H_3O^+ + OH^-$	r.	Amphiprotic nature of H_2O
iv.	$3Fe + 4H_2O \xrightarrow{\Delta} Fe_3O_4 + 4H_2$	s.	Reducing nature of H_2O
v.	$2F_2 + 2H_2O \longrightarrow 4HF + O_2$	t.	Acidic nature of H_2O
vi.	$6CO_2 + 12H_2O \longrightarrow C_6H_{12}O_6 + 6H_2O + 6O_2$	u.	Basic nature of H_2O

Codes

	i	ii	iii	iv	v	vi
a.	t	u	r	q	s	p
b.	t	u	r	q	s	p,s
c.	u	t	r	q,s	s	p,s
d.	r	p	q	r,s	s	u

DPP-4 Hydrogen Peroxide

Subjective Question

Working Space

Direction (Q. No. 1) *This question is subjective in nature, need to be solved completely on notebook.*

1 Express concentration of '*x* volume' H_2O_2 in terms of

i. density
ii. $g\ L^{-1}$
iii. percentage
iv. normality
v. molarity

Only One Option Correct Type

Direction (Q. Nos. 2-27) *This section contains 26 multiple choice questions. Each question has four choices (a), (b), (c) and (d), out of which ONLY ONE is correct.*

2 H_2O_2 cannot be prepared by

a. $BaO_2 + H_2SO_4 \longrightarrow$
b. $Na_2O_2 + H_2O \longrightarrow$
c. $KO_2 + H_2O \longrightarrow$
d. $PbO_2 + H_2SO_4 \longrightarrow$

3 Reducing action of H_2O_2 is not shown in

a. $H_2O_2 \longrightarrow 2H^+ + O_2 + 2e^-$
b. $H_2O_2 + 2H^+ + 2e^- \longrightarrow 2H_2O$
c. $O_3 + H_2O_2 \longrightarrow 2O_2 + H_2O$
d. $[Fe(CN)_6]^{3-} + H_2O_2 + 2OH^- \longrightarrow [Fe(CN)_6]^{4-} + 2H_2O + O_2$

4 H_2O_2 is obtained by electrolysis of H_2SO_4 or NH_4HSO_4 at high current density using platinum electrodes. Consider the following statements about this method,

I. Intermediates are $S_2O_8^{2-}$, SO_5^{2-} which are hydrolysed to H_2O_2.
II. H_2O_2 is obtained at anode.
III. H_2 is obtained at cathode.
IV. H_2O_2 is obtained at cathode and anode but H_2 is obtained at cathode.

The correct statements are

a. I, II and III
b. I, III and IV
c. Both I and IV
d. Both II and III

5 Consider the following redox reaction,

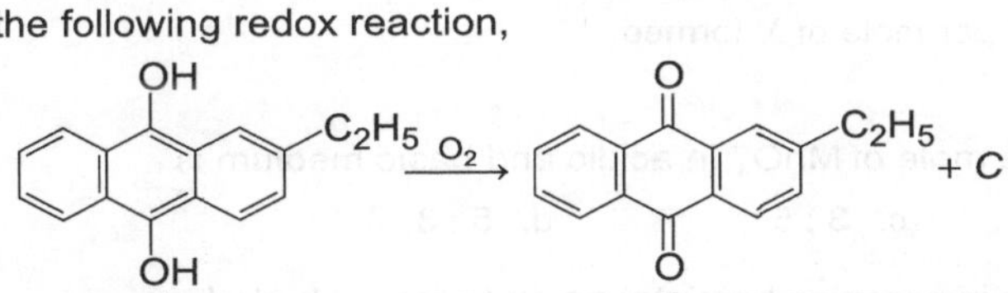

A
Ethyl anthraquinol

B

Correct statement is

a. Product *C* is H_2O_2 obtained by the reduction of O_2
b. *B* can be recycled to *A* by reduction with Pd/H_2 or Raney Ni/H_2
c. H_2O_2 obtained is concentrated under the reduced pressure
d. All of the above

Working Space

6 Partial oxidation of 2-propanol forms

a. propanone and water
b. propanone and hydrogen peroxide
c. propanoic acid and water
d. propanone and carbon dioxide

7 Stabiliser added to H_2O_2 is

a. 8-hydroquinoline
b. sodium stannate
c. Both (a) and (b)
d. None of these

8 H_2O_2 changes to H_2O when $[H^+] = 1\,M$, $E^\circ = 1.77\,V$. Thus, H_2O_2 is

a. a strong oxidising agent
b. a strong reducing agent
c. both oxidising and reducing agent
d. neither reducing nor oxidising agent

9 Select the correct statement about the reaction of MnO_4^- with H_2O_2. It is reduced to

a. Mn^{2+} in acidic medium and to MnO_2 in basic medium
b. MnO_2 in acidic medium and to Mn^{2+} in basic medium
c. Mn^{2+} in acidic as well as in basic medium
d. MnO_2 in acidic as well as in basic medium

10 MnO_2 (black) changes to a colourless (in dilute solution) by H_2O_2 in acidic medium due to the formation of

a. MnO_4^-
b. MnO_4^{2-}
c. Mn_2O_3
d. Mn^{2+}

11 Consider the following reaction,

$$PbO_2 + H{-}\overset{18}{O}{-}\overset{18}{O}{-}H \longrightarrow PbO + H_2O + O_2$$

Oxygen atoms of H_2O_2 have been labelled $\overset{18}{O}$. In the products, $\overset{18}{O}$ goes with

a. PbO
b. H_2O
c. O_2
d. All of the products

12 Coinage metal-oxides can be reduced to metals by

a. H_2
b. CO
c. H_2O_2
d. All of these

13 Oil paintings is blackened by atmospheric H_2S due to the formation of *X*, H_2O_2 changes *X* to colourless (white) *Y*. *X* and *Y* respectively are

a. PbS and $PbSO_3$
b. PbS and $PbSO_4$
c. PbS and PbO
d. PbS and PbO_2

14 H_2O_2 oxidises I^- to product *X* in acidic medium. Product *X*

a. can be estimated by hypo
b. gives blue colour with starch
c. is formed by lose of two electrons per mole of *X* formed
d. All of the above

15 Molar ratio of H_2O_2 oxidised by one mole of MnO_4^- in acidic and basic medium is

a. 1 : 1
b. 1 : 2
c. 3 : 5
d. 5 : 3

16 Following types of radical reactions in organic chemistry occur to form dihydrol

$$\begin{array}{c} CH_2 \\ \| \\ CH_2 \end{array} + 2OH \longrightarrow \begin{array}{c} CH_2OH \\ | \\ CH_2OH \end{array}$$

Free OH radicals are formed by

a. glycerol
b. H_2O_2/Fe^{3+}
c. H_2O_2/Fe^{2+}
d. MnO_2/OH^-

Working Space

17 H_2O_2 ionises as, $H_2O_2 \rightleftharpoons H^+ + HO_2^-$

At 25°C, pH of H_2O_2 is found to be 6.0, thus auto-protolysis constant of water is

a. 8.0 b. 10^{-8} c. 10^{-6} d. 10^{-12}

18 From the following statements regarding H_2O_2, choose the incorrect statement. **[JEE Main 2015]**

a. It can act only as an oxidising agent
b. It decomposes on exposure to light
c. It has to be kept in plastic or wax lined glass bottles in dark
d. It has to be kept away from dust

19 How many mL of perhydrol is required to produce sufficient O_2 which can convert 2 L of SO_2 gas to SO_3 gas?

a. 10 mL b. 5 mL c. 20 mL d. 30 mL

20 Consider the following reaction,

$$H_2O_2 + O_3 \longrightarrow 2O_2 + H_2O$$

Magnetic behaviour of

a. H_2O_2 changes
b. O_3 changes
c. Both H_2O_2 and O_3 changes
d. None of the two changes

21 Select the correct statement(s).

a. H_2O_2 is used to counteract Cl_2
b. Na_2O_2 and KO_2 are CO_2 absorber and O_2 giver
c. H_2O_2 neutralises $Ba(OH)_2$
d. All of the above

22 10 mL of H_2O_2 solution on treatment with KI and titration of liberated I_2 required 10 mL of 1N hypo solution. Thus, H_2O_2 is

a. 1N
b. "5.6" volume
c. 17 g L^{-1}
d. All are correct factors

23 In the given Fig., representing structure of H_2O_2, bond angles α, β and (O—O) bond length x are

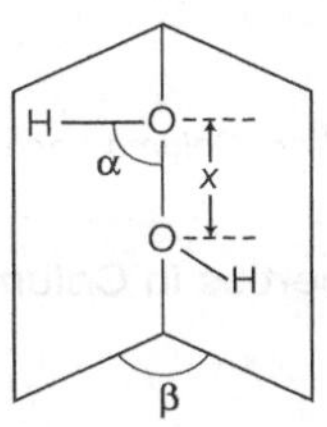

	$\angle\alpha$	$\angle\beta$	x (pm)
a.	94° 48′	111° 30′	148
b.	104° 30′	111° 30′	160
c.	94° 38′	111° 30′	160
d.	111° 30′	111° 30′	130

24 Select the incorrect statement.

a. $[Fe(CN)_6]^{4-}$ is oxidised to $[Fe(CN)_6]^{3-}$ in acidic medium and $[Fe(CN)_6]^{3-}$ is reduced to $[Fe(CN)_6]^{4-}$ in basic medium by H_2O_2
b. Mn^{2+} is oxidised to MnO_4^- in basic medium by H_2O_2
c. Cr^{3+} is oxidised to CrO_4^{2-} in basic medium by H_2O_2
d. Equivalent weight of H_2O_2 in oxidising and reducing properties is same

25 Which of the following reactions is accompanied by the emission of light in the dark (chemiluminescence)?

a. $2HCHO + H_2O_2 \longrightarrow 2HCOOH + H_2$
b. $Na_2CO_3 + H_2O_2 \longrightarrow Na_2O_2 + H_2O + CO_2$
c. $AsO_3^{3-} + H_2O_2 \longrightarrow AsO_4^{3-} + H_2O$
d. $NO_2^- + H_2O_2 \longrightarrow NO_3^- + H_2O$

Working Space

26 H_2O_2 behaves as

a. an acid
b. an oxidising agent
c. a reducing agent
d. All of these

27 $2H_2O_2 \longrightarrow 2H_2O + O_2$

This represents

a. oxidation of H_2O_2
b. reduction of H_2O_2
c. disproportionation of H_2O_2
d. acidic nature of H_2O_2

Statement Type

Direction *(Q. Nos. 28 and 29) This section is based on Statement I and Statement II. Select the correct answer from the codes given below.*

a. Both Statement I and Statement II are correct and Statement II is the correct explanation of Statement I
b. Both Statement I and Statement II are correct but Statement II is not the correct explanation of Statement I
c. Statement I is correct but Statement II is incorrect
d. Statement II is correct but Statement I is incorrect

28 **Statement I** When a few drops of blood are added to aqueous H_2O_2, it (H_2O_2) decomposes rapidly.

Statement II The decomposition of H_2O_2 is catalysed by catalase, an enzyme present in blood.

29 **Statement I** In alkaline medium at room temperature, H_2O_2 exists as HO_2^- ion.

Statement II Ionisation constant of H_2O_2 at room temperature is

$$\frac{[H^+][HO_2^-]}{[H_2O_2]} = 2\times10^{-12}$$

Matching List Type

Direction *(Q. No. 30) Choices for the correct combination of elements from Column I and Column II are given as options (a), (b), (c) and (d), out of which one is correct.*

30 Match the species in Column I with corresponding properties in Column II and select the answer from the codes given.

	Column I		Column II
i.	H_2O	p.	Reducing agent
ii.	H_2O_2	q.	Oxidising agent
iii.	Aqueous Na_2O	r.	Paramagnetic
iv.	Aqueous KO_2	s.	Diamagnetic
v.	Aqueous CaH_2	t.	Turns red litmus blue
		u.	Turns blue litmus red

Codes

	i	ii	iii	iv	v
a.	p,s	p,q,s,u	s,t	q,r,t	p,s,t
b.	p,s	p,s,u	s,t	r,t	p,s,t
c.	p	p,q	s,t	q,r	p,t
d.	p	p,q,r	s,t	q,t	s,t

SET - 1

Revisal Problems
for JEE Main

Only One Option Correct Type

Working Space

Direction *(Q. Nos. 1-26) This section contains 26 multiple choice questions. Each question has four choices (a), (b), (c) and (d), out of which ONLY ONE option is correct.*

1 Very pure H_2 (99.9%) can be made by which of the following processes? **[AIEEE 2012]**

a. Reaction of methane with steam
b. Mixing natural hydrocarbons of high molecular weight
c. Electrolysis of water
d. Reaction of ionic hydrides with water

2 H_2 is manufactured by allowing steam to react with red hot coke at about 1250 K

$$C + H_2O \xrightarrow{1250\text{ K}} CO + H_2$$

Select the correct statement(s).
a. Mixture of CO and H_2 is called water gas
b. Mixture of CO and H_2 is called synthesis gas
c. This is a redox reaction
d. All of the above

3 The correct decreasing order of basic strength is
a. $AsH_3 > SbH_3 > PH_3 > NH_3$
b. $SbH_3 > AsH_3 > PH_3 > NH_3$
c. $NH_3 > PH_3 > AsH_3 > SbH_3$
d. $PH_3 > AsH_3 > SbH_3 > NH_3$

4 H_2O_2 is marked "22.4 volume". How much of it (in mL) are required to oxidise 3.4 g H_2S gas?

a. 100 mL
b. 50 mL
c. 10 mL
d. 5 mL

5 When H_2O_2 is added to a mixture of $K_2Cr_2O_7$ and conc. H_2SO_4,

a. CrO_4^{2-} is formed
b. Cr^{3+} is formed
c. CrO_5 is formed
d. CrO_3 is formed

6 Consider the following reactions,

I. $SO_3^{2-} + H^+ + Cr_2O_7^{2-} \longrightarrow X$

II. $SO_3^{2-} + H_2O_2 + H^+ \longrightarrow Y \xrightarrow{Cr_2O_7^{2-}} Z$

Thus,
a. X and Z both are green
b. X is green, Y is colourless and it does not react with $Cr_2O_7^{2-}$
c. X, Y and Z all are colourless
d. X is yellow, Y is colourless and Z is green

7 1 mole of IO_4^- is reduced to a new compound by 1 mole of H_2O_2. New compound is

a. IO_2^-
b. IO^-
c. IO_3^-
d. I_2

8 Perhydrol is not
a. 30% H_2O_2 solution
b. 30 volume H_2O_2 solution
c. "100 volume" H_2O_2 solution
d. 8.93 molar H_2O_2 solution

Working Space

9 100 mL of a water sample contains hardness equivalent to 25 mL of 0.08 N $MgSO_4$. Thus, hardness in ppm of $CaCO_3$ is

a. 200
b. 2000
c. 500
d. 1000

10 A sample of hard water has hardness expressed as 80 ppm Ca^{2+}. This sample is passed through a cation exchange resin column. Thus, pH of water coming out of exchange resin is

a. 2.70
b. 2.40
c. 3.00
d. 4.00

11 100 mL of sample of hard water is passed through a column of the cation exchange resin H_2R. The water coming out of the column requires 20 mL of 0.025 M NaOH for the titration. Thus, hardness of water expressed as ppm of Ca^{2+} ion is

a. 100
b. 50
c. 40
d. 20

12 Exhausted zeolite can be put into fresh operation by treating it with

a. Na_2SO_4
b. NaCl
c. $AlCl_3$
d. All of these

13 In Calgon's method of removing permanent hardness, Ca^{2+} and Mg^{2+} remain soluble due to the formation of complex

a. $Na_2[Na_2CaP_6O_{18}]$
b. $Na_4[CaP_6O_{18}]$
c. $Ca[NaPO_3]_4$
d. None of these

14 Pure de-mineralised (de-ionised) water free from all soluble mineral salts is obtained by passing hard water through

a. cation exchange resin which removes dissolved cations
b. anion exchange resin which removes dissolved anions
c. Both (a) and (b)
d. None of the above

15 There are three isotopes of hydrogen and three naturally occurring isotopes of oxygen (O^{16}, O^{17}, O^{18}). How many types of water are possible?

a. 18
b. 16
c. 9
d. 8

16 Select the incorrect statement.

a. H^+ can exist as $H_9O_4^+$ in water
b. H_2 is thermally stable
c. Ionisation of CH_3COOH is slower than that of CH_3COOD
d. Kinetic isotopic effect is observed when there is retardation in the rate when H_2O is replaced by D_2O

17 Product of the following reaction is a reducing agent

$$LiH + Al_2Cl_6 \longrightarrow$$

which can reduce the given reaction

a. $CH_3CHO \longrightarrow CH_3CH_2OH$
b. $CH_3COCH_3 \longrightarrow (CH_3)_2CHOH$
c. $CH_3COOH \longrightarrow CH_3CH_2OH$
d. All of the above

18 How many kilograms of CaH_2 is needed to fill a 100 L tank with compressed H_2 gas at 150 atm pressure and at 298 K?

a. 613.1 kg
b. 0.042 kg
c. 12.875 kg
d. 6.131 kg

19 Which is not the correct order of properties as indicated?

a. $BeH_2 < CaH_2 < TiH_2$, Electrical conductivity
b. $LiH < NaH < CsH$; Ionic character
c. $F—F < H—H < D—D$; Bond dissociation energy
d. $NaH < MgH_2 < AlH_3$; Reducing nature

Working Space

20 Kinetic isotopic effect is not observed in which of the following by replacing H by D?

a. 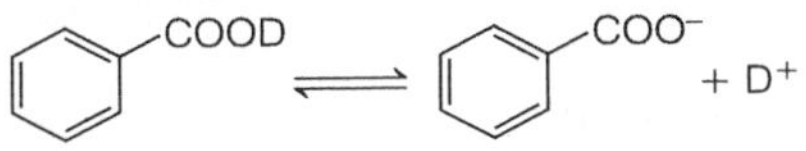

b. 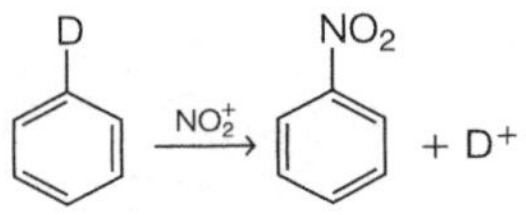

c. $CH_3-\overset{\overset{\large O}{\|}}{C}-CH_3 \xrightarrow{D^+} CH_3-\overset{\overset{\large OD}{|}}{C}=CH_2 + H^+$

d. In all of the above cases

21 Variation of various properties of H_2, D_2 and T_2 are shown graphically:

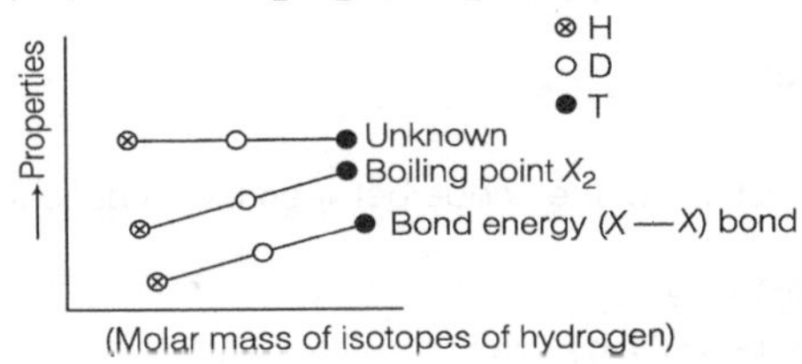

Unknown property is constant. This is

a. critical temperature
b. freezing point
c. bond length
d. latent heat of fusion

22 Select the correct fact(s) about *ortho*- and *para*-hydrogen.

a. Total spin of *ortho*-hydrogen is 1 while that of *para*-hydrogen is zero
b. Ratio of *ortho* to *para* increases with increasing temperature and at 300 K and above, ratio is 3 : 1
c. 100% of *ortho* hydrogen can't be obtained
d. All of the above facts are correct

23 A thermochemical, non-electrolytic method of preparation of hydrogen is based on cycle

I. $3FeCl_2 + 4H_2O \xrightarrow{775\text{ K}} Fe_3O_4 + 6HCl + H_2$

II. $2Fe_3O_4 + 3Cl_2 + 12HCl \xrightarrow{375\text{ K}} 6FeCl_3 + 6H_2O + O_2$

III. $2FeCl_3 \xrightarrow{575\text{ K}} 2FeCl_2 + Cl_2$

Net electrolytic reaction is

a. $2H_2O \longrightarrow 2H_2 + O_2$
b. $H_2O \longrightarrow H_2 + \frac{1}{2}O_2$
c. $H_2O + Cl_2 \longrightarrow 2HCl + \frac{1}{2}O_2$
d. None of the above

24 Atomic hydrogen is formed in the following reaction

$$H_2(g) \xrightarrow{\text{Tungsten}} 2H(g)$$

It is a powerful reducing agent. Which reaction does not show the reducing nature of hydrogen?

a. $HgO + 2H \longrightarrow Hg + H_2O$
b. $CuO + 2H \longrightarrow Cu + H_2O$
c. $AgNO_3 + H \longrightarrow Ag + HNO_3$
d. $Na + H \longrightarrow NaH$

25 Water gas shift reaction is

a. $C + H_2O \xrightarrow{1250\text{ K}} CO + H_2$
b. $CH_4 + H_2O \xrightarrow{\text{Ni, 920-1170 K}} CO + 3H_2$
c. $CO + H_2O + H_2 \xrightarrow{Fe_2O_3,\ 450\text{-}550\text{ K}} CO_2 + 2H_2$
d. $2H_2O \xrightarrow{\text{Current}} 2H_2 + O_2$

Working Space

26 Which one of the following statements is incorrect with regard to *ortho* and *para* isomers of hydrogen?

a. They are nuclear spin isomers
b. The *ortho* isomer has zero nuclear spin whereas the *para* isomer has one nuclear spin
c. The *para* isomer is favoured at low temperature
d. The thermal conductivity of the *para* isomer is 50% greater than that of the *ortho* isomer

Statement Type

Direction (Q. Nos. 27-37) ***This section is based on Statement I and Statement II. Select the correct answer from the codes given below.***

a. Both Statement I and Statement II are correct and Statement II is the correct explanation of Statement I
b. Both Statement I and Statement II are correct but Statement II is not the correct explanation of Statement I
c. Statement I is correct but Statement II is incorrect
d. Statement II is correct but Statement I is incorrect

27 **Statement I** During electrolysis, there is a competition at the anode between the oxidation of chloride ion and oxidation of water.

$$2Cl^-(aq) \longrightarrow Cl_2(g) + 2e^-$$
$$2H_2O(l) \longrightarrow O_2(g) + 4H^+(aq) + 4e^-$$

Statement II In brine solution, water is reduced as compared to Na^+.

28 **Statement I** It is customary to place hydrogen in group 1 (along with alkali metals) as well as in group 17 (along with halogens).

Statement II In view of the anomalous behaviour of hydrogen, it is difficult to assign any definite position to hydrogen in the periodic table.

29 **Statement I** Hydrogen is placed in group 1 (alkali metals).

Statement II Alkali metals and hydrogen both form X^- and attain inert gas configuration.

30 **Statement I** Hydrogen resembles halogen.

Statement II Both gain one electron and attains inert gas configuration.

31 **Statement I** *d*- and *f*- block elements form non-stoichiometric hydrides.

Statement II These elements have holes in their lattices and hydrogen atoms occupy these holes.

32 **Statement I** Metallic hydrides can be useful for hydrogen storage.

Statement II Adsorbed H_2 on free valencies can be regenerated by heating.

33 **Statement I** D_2O resembles H_2O in many respects, but D_2O is toxic in nature.

Statement II Kinetic isotopic effect is observed in the ionisation of a weak acid.

34 **Statement I** Hydrate copper (II) sulphate is $CuSO_4 \cdot 5H_2O$.

Statement II Coordination number of Cu(II) is four, one unit of H_2O joined to SO_4^{2-} by H-bonding.

35 **Statement I** Oxygen atom in water is sp^3-hybridised with tetrahedral structure.

Statement II Lone pair-lone pair repulsion distorts its tetrahedral angle to 104.5°.

36 **Statement I** Boiling point of H_2O_2 is higher than that of H_2O.

Statement II H_2O_2 is more hydrogen bonded than H_2O.

37 **Statement I** Hydrogen occurs in a diatomic form rather than in a monoatomic form under normal condition.

Statement II Bond order of H_2 is 1 and formation of H_2 from H-atoms is an exothermic change.

Working Space

Matching List Type

Direction *(Q. Nos. 38-40) Choices for the correct combination of elements from Column I and Column II are given as options (a), (b), (c) and (d), out of which one is correct.*

38 Match the reactions in Column I with the type of products in Column II and select the answer from the codes given below.

	Column I		Column II
i.	$2\,^{2}_{1}D \longrightarrow$	p.	$^{3}_{1}T + ^{12}_{6}C$
ii.	$^{6}_{3}Li + ^{1}_{0}n \longrightarrow$	q.	$^{3}_{1}T + ^{1}_{1}H$
iii.	$^{14}_{7}N + ^{1}_{0}n \longrightarrow$	r.	$^{3}_{1}T + ^{4}_{2}He$
iv.	$^{3}_{1}T \longrightarrow$	s.	$^{3}_{2}He + ^{0}_{-1}e$

Codes

	i	ii	iii	iv
a.	q	r	p	s
b.	q	r	s	p
c.	p	r	q	s
d.	p	q	r	s

39 Match the hydrides in Column I with the types in Column II and select the answer from the codes given below.

	Column I		Column II
i.	$NiH_{0.6}$	p.	Stoichiometric
ii.	KH	q.	Ionic
iii.	CoH_2	r.	Non-stoichiometric
iv.	SiH_4	s.	Covalent

Codes

	i	ii	iii	iv
a.	p	q	s	r
b.	r	q	p	s
c.	r	p	q	s
d.	q	p	r	s

40 Match the species in Column I with the corresponding properties in Column II and select the answer from the codes given below.

	Column I		Column II
i.	LiH	p.	Reducing agent
ii.	H_2O	q.	Lewis base
iii.	H_3O^+	r.	Conjugate base
iv.	BH_3	s.	Lewis acid
v.	H	t.	Lowry-Bronsted acid
		u.	Lowry-Bronsted base

Codes

	i	ii	iii	iv	v
a.	p,q	t,u	p,q, s	p,s	p,q,r
b.	p,q,r	p,q,t,u	q,r,t	s	p
c.	p,q	p,q	q,s	p,s	p,q
d.	q	p,q	q,s	s	p

JEE Archive

(Compilation of Last 13 Years' Questions of JEE Main, AIEEE, JEE Advanced and IIT JEE)

1 A device that converts energy of combustion of fuels like H_2 and CH_4, directly into electrical energy is known as **(JEE Main 2015)**

a. electrolytic cell b. dynamo
c. Ni-Cd cell d. fuel cell

2 Permanent hardness in water cannot be cured by **(JEE Main 2015)**

a. boiling
b. ion-exchange method
c. Calgon's method
d. treatment with washing soda

3 The molecular formula of a commercial resin used for exchanging ions in water softening is $C_8H_7SO_3Na$ (mol. wt. 206). What would be the maximum uptake of Ca^{2+} ions by the resin expressed in mole per gram resin? **(JEE Main 2015)**

a. 1/103 b. 1/206 c. 2/309 d. $\frac{1}{412}$

4 From the following statements regarding H_2O_2, choose the incorrect statement. **(JEE Main 2015)**

a. It can act only as an oxidising agent
b. It decomposes on exposure to light
c. It has to be stored in plastic or wax lined glass bottles in dark
d. It has to be kept away from dust

5 In which of the following reactions, H_2O_2 acts as a reducing agent? **(JEE Main 2014)**

I. $H_2O_2 + 2H^+ + 2e^- \longrightarrow 2H_2O$
II. $H_2O_2 - 2e^- \longrightarrow O_2 + 2H^+$
III. $H_2O_2 + 2e^- \longrightarrow 2OH^-$
IV. $H_2O_2 + 2OH^- - 2e^- \longrightarrow O_2 + 2H_2O$

a. I and II b. III and IV
c. I and III d. II and IV

6 Very pure hydrogen (99.9) can be made by which of the following processes? **(AIEEE 2012)**

a. Reaction of methane with steam
b. Mixing natural hydrocarbons of high molecular weight
c. Electrolysis of water
d. Reaction of salts like hydrides with water

7 In context with the industrial preparation of hydrogen from water gas $(CO + H_2)$, which of the following is the correct statement? **(AIEEE 2008)**

a. CO and H_2 are fractionally separated using differences in their densities
b. CO is removed by absorption in aqueous Cu_2Cl_2 solution
c. H_2 is removed through occlusion with Pd
d. CO is oxidised to CO_2 with steam in the presence of a catalyst followed by the absorption of CO_2 in alkali

8 Hydrogen peroxide in its reaction with KIO_4 and NH_2OH respectively, is acting as **(2014 Adv., Only One Option Correct Type)**

a. reducing agent, oxidising agent
b. reducing agent, reducing agent
c. oxidising agent, oxidising agent
d. oxidising agent, reducing agent

9 The reagent(s) used for softening the temporary hardness of water is(are) **(IIT JEE 2010, One or More than One Options Correct Type)**

a. $Ca_3(PO_4)_2$
b. $Ca(OH)_2$
c. Na_2CO_3
d. NaOCl

3. Alkali Metals (Group 1)

DPP
DAILY PRACTICE PROBLEMS

DPP-1 Periodicity in Group 1

Only One Option Correct Type

Working Space

Direction (Q. Nos. 1-12) This section contains 12 multiple choice questions. Each question has four choices (a), (b), (c) and (d), out of which ONLY ONE is correct.

1 Atomic radius of Na, K, Rb and Cs is mainly dependent on

a. effective nuclear charge
b. number of shells
c. Both (a) and (b)
d. None of these

2 Select the correct variation of thermal stabilities of compounds of group 1.

a. $LiOH < NaOH < KOH < RbOH$
b. $LiF > NaF > KF > RbF$
c. $LiNO_3 < NaNO_3 < KNO_3 < RbNO_3$
d. All are correct variation

3 Arrange carbonates of alkali metals in increasing melting points.

a. $Li_2CO_3 < Na_2CO_3 < K_2CO_3 < Rb_2CO_3 < Cs_2CO_3$
b. $Cs_2CO_3 < Rb_2CO_3 < K_2CO_3 < Na_2CO_3 < Li_2CO_3$
c. $Na_2CO_3 < Li_2CO_3 < K_2CO_3 < Rb_2CO_3 < Cs_2CO_3$
d. None of the above is correct

4 Name the alkali metal which floats on water without any apparent reaction with it.

a. Li
b. Na
c. K
d. Cs

5 The reactivity of alkali metals increases on moving down the group. Slow reactivity of Li is due to

a. kinetic factor
b. thermodynamic factor
c. Both (a) and (b)
d. None of these

6 Solubility of CsF, CsCl, CsBr and CsI is in order

a. $CsF < CsCl < CsBr < CsI$
b. $CsF > CsCl > CsBr > CsI$
c. $CsI > CsF > CsCl > CsBr$
d. $CsF > CsI > CsCl > CsBr$

7 Difference in lattice energy between NaCl and KCl is $67\ kJ\ mol^{-1}$ and difference in $\Delta G^\circ_{hydration}$ for Na^+ and K^+ is $71\ kJ\ mol^{-1}$. Thus,

a. KCl is more soluble than NaCl
b. NaCl is more soluble than KCl
c. KCl is precipitated while NaCl is highly soluble
d. Both are equally soluble

8 Which has maximum solubility in water?

a. Li_2CO_3
b. Na_2CO_3
c. K_2CO_3
d. Rb_2CO_3

Working Space

9 Which decomposes to CO_2 and oxide on heating?

a. Li_2CO_3
b. Na_2CO_3
c. K_2CO_3
d. All of these

10 Ionic conductance of chlorides of group 1 is in order

a. LiCl < NaCl < KCl < RbCl < CsCl
b. LiCl > NaCl > KCl > RbCl > CsCl
c. NaCl > LiCl > RbCl > KCl > CsCl
d. KCl > NaCl > RbCl > LiCl > CsCl

11 What is lattice energy *U* (energy required to break of lattice into gaseous ions) of the salt *MX* based on following Born-Haber cycle?
(*M* = alkali metal, *X* = halide)

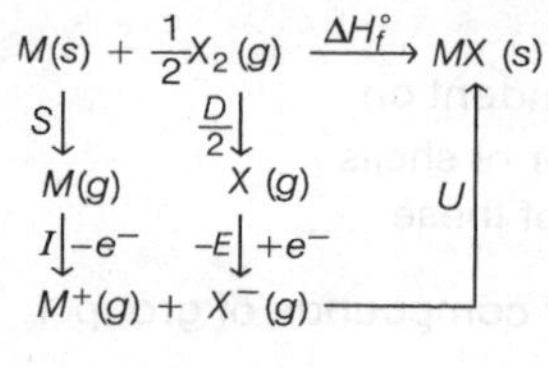

a. $U = -\Delta H_f^\circ + S + I + \frac{D}{2} - E$

b. $U = -\Delta H_f^\circ + S + I + \frac{D}{2} + E$

c. $U = \Delta H_f^\circ - S - I - \frac{D}{2} + E$

d. $U = -\Delta H_f^\circ - S + I + \frac{D}{2} - E$

12 The products obtained on heating $LiNO_3$ will be **[AIEEE 2011]**

a. $LiNO_2 + O_2$
b. $Li_2O + NO_2 + O_2$
c. $Li_3N + O_2$
d. $Li_2O + OH + O_2$

Statement Type

Direction *(Q. Nos. 13-20) This section is based on Statement I and Statement II. Select the correct answer from the codes given below.*

a. Both Statement I and Statement II are correct and Statement II is the correct explanation of Statement I
b. Both Statement I and Statement II are correct but Statement II is not the correct explanation of Statement I
c. Statement I is correct but Statement II is incorrect
d. Statement II is correct but Statement I is incorrect

13 **Statement I** Lithium compounds are covalent in nature.

Statement II Lithium ion has maximum polarising power among alkali metal ions.

14 **Statement I** Li_2CO_3 is thermally unstable.

Statement II Li and Mg show diagonal relationship.

15 **Statement I** Li_3N is formed but K_3N is not formed.

Statement II Li_3N is decomposed by H_2O forming NH_3.

16 **Statement I** The variation in lattice energy for fluorides of alkali metals is greater than the variation in lattice energy for chlorides, bromides and iodides.

Statement II Lattice energy is the driving force for the formation of an ionic compound and its stability.

Working Space

17 **Statement I** LiCl can be extracted into ether, NaCl cannot be.

Statement II Li resembles Mg.

18 **Statement I** Alkali metals are soft and can be cut with a knife.

Statement II They have low IE.

19 **Statement I** Alkali metals are good conductors of heat and electricity.

Statement II Valence electrons are tightly bound to nucleus.

20 **Statement I** Lithium, the first element of group 1 shows considerable difference from rest of the group.

Statement II Li^+ has high charge density in its compounds.

One or More than One Options Correct Type

Direction *(Q. Nos. 21-25) This section contains 5 multiple choice questions. Each question has four choices (a), (b), (c) and (d), out of which ONE or MORE THAN ONE are correct.*

21 Select the correct statement(s).

a. Na_2CO_3 is thermally stable while Ag_2CO_3 decomposes into Ag, CO_2 and O_2
b. Ag^+ forms complexes, Na^+ does not
c. NaCl is water soluble, AgCl is insoluble
d. NaCl and AgCl both give colour in flame when ignited

22 Select the correct statement(s).

a. NaOH does not form hydrated salt
b. Stability of peroxides and superoxides of alkali metals increases with increase in size of the metal ion
c. Stability of MO_2 and M_2O is due to stabilisation of larger anions by larger cations through lattice energy effects
d. The low solubility of LiF is due to its lattice energy whereas low solubility of CsI is due to smaller hydration energy

23 Alkali metals (group 1) and coinage metals (group 11) have ns^1 configuration of valence shell. They differ in

a. electronic configuration of monovalent cation
b. reactivity with H_2O
c. solubility of sulphates, halides and sulphides
d. electropositive nature

24 Block of which of the following elements would change if Aufbau rule is not followed?

a. Li b. Na
c. K d. Rb

25 Select the correct statement (s).

a. All of the alkali halides, MX (X = F, Cl, Br, I) have negative standard enthalpy of formation (ΔH_f°)
b. ΔH_f° for fluorides becomes less negative as we go down the group, while reverse is true for Cl^-, Br^- and I^-
c. For a given alkali metal, ΔH_f° always becomes less negative as we go from F^- to I^-
d. ΔH_f° can be derived from Born-Haber cycle

Comprehension Type

Working Space

Direction *(Q. Nos. 26-30) This section contains 2 paragraphs, each describing theory, experiments, data, etc. Five questions related to the paragraphs have been given. Each question has only one correct answer among the four given options (a), (b), (c) and (d).*

Passage I

Given,

Hydration energy $\Delta H^\circ_{hydration}$ (kJ mol^{-1})		Lattice energy ΔH_U (kJ mol^{-1})	
Li^+	− 499	LiCl	− 840
Na^+	− 390	NaCl	− 776
K^+	− 305	KCl	− 703
Cl^-	− 382		

26 Which has maximum heat of hydration?

a. LiCl
b. NaCl
c. KCl
d. KCl and LiCl equally

27 Which salt can be used to control humidity?

a. LiCl
b. NaCl
c. KCl
d. None of these

28 $\Delta H^\circ_{solution}$ (standard heat of solution) is most exothermic in case of

a. KCl
b. NaCl
c. LiCl
d. All are endothermic

Passage II

Lattice energies depend on the polarising power of cation and anion.

29 Arrange the fluorides of Li to Cs in the increasing order of lattice energy.

a. LiF < NaF < KF < RbF < CsF
b. CsF < RbF < KF < NaF < LiF
c. KF < RbF < NaF < CsF < LiF
d. None of the above is correct

30 Arrange LiF, LiCl, LiBr, LiI in the increasing order of lattice energy.

a. LiF < LiCl < LiBr < LiI
b. LiI < LiBr < LiCl < LiF
c. LiCl < LiBr < LiI < LiF
d. LiI < LiCl < LiF < LiBr

One Integer Value Correct Type

Direction *(Q. No. 31) This section contains 1 question. When worked out will result in an integer from 0 to 9 (both inclusive).*

31 Standard hydration energy of Na^+ ion = −390 kJ mol^{-1} that of Cl^- ion = −380 kJ mol^{-1} and lattice energy of NaCl = −776 kJ mol^{-1}. Thus, heat of solution of NaCl is = x kJ mol^{-1} What is the value of x?

DPP-2 Compounds of Alkali Metals

Only One Option Correct Type

Working Space

Direction *(Q. Nos. 1-14) This section contains 14 multiple choice questions. Each question has four choices (a), (b), (c) and (d), out of which ONLY ONE is correct.*

1 CsI_3 ionises as

a. CsI, I_2 b. Cs^+, I^-, I_2
c. Cs^+, I_3^- d. CsI_2, I^-

2 Which is used as a mild antiseptic for skin infections?

a. $NaHCO_3$ b. C_6H_5COONa
c. CH_3COONa d. $Na_2B_4O_7$

3 Which is used in the treatment of manic-depressive disorders?

a. Na_2CO_3 b. Li_2CO_3
c. K_2CO_3 d. $MgCO_3$

4 Nitride of *s*-block elements can be

a. Na_3N b. Li_3N, Mg_3N_2
c. Li_3N, Na_3N d. Li_3N, Mg_3N_2, Na_3N

5 HCO_3^- forms dimeric and infinite chain type structures

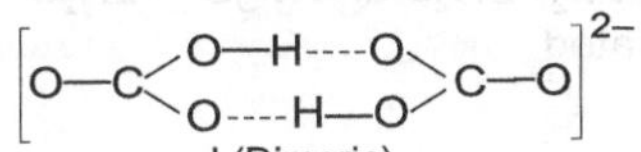

I (Dimeric)

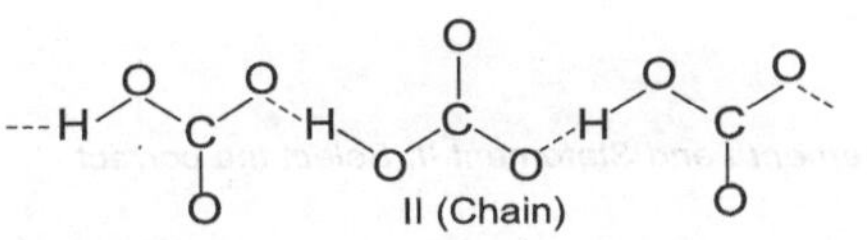

II (Chain)

Select the correct statement

a. I is in $NaHCO_3$, II is in $KHCO_3$ b. I is in $KHCO_3$, II is in $NaHCO_3$
c. I and II are for $NaHCO_3$ d. I and II are of $KHCO_3$ and $NaHCO_3$

6 Na_2CO_3 is obtained by

a. Castner-Kellner method b. Solvay (ammonia-soda) process
c. Nelson method d. None of these

7 Consider the following reactions to prepare K_2CO_3,

$$KCl + H_2SO_4 \longrightarrow KHSO_4 + HCl$$
$$KHSO_4 + KCl \longrightarrow K_2SO_4 + HCl$$

Select the correct statement(s).

a. Reactions are part of Le-blanc process to prepare K_2CO_3
b. K_2SO_4 is converted to K_2CO_3 by heating with $CaCO_3$ and carbon
c. Both (a) and (b)
d. None of the above is correct

Working Space

8 Intermediate formed when CO_2 is passed into a solution of KCl and $MgCO_3 \cdot 3H_2O$ at 290 K in Precht's process (also called Magnesia process) is

a. $KCl \cdot MgCO_3 \cdot 3H_2O$
b. $KHCO_3 \cdot MgCO_3 \cdot 3H_2O$
c. $KHCO_3 \cdot MgCO_3 \cdot 4H_2O$
d. $KHCO_3 \cdot MgCl_2 \cdot 4H_2O$

9 Which of the following effloresces and gives white precipitate with $BaCl_2$ solution?

a. $Na_2SO_4 \cdot 10H_2O$ b. $Na_2CO_3 \cdot 10H_2O$ c. Both (a) and (b) d. None of these

10 Gun powder is a mixture of

a. potassium nitrate and ammonium nitrate
b. potassium nitrate, sulphur and potassium chlorate
c. potassium nitrate, sulphur and carbon
d. potassium nitrate, sulphur and potassium chloride

11 Sodium and potassium ions differ

a. quantitatively in their ability to penetrate cell membranes
b. in their transport mechanism
c. in their efficiency to activate enzymes
d. All of the above

12 Sodium ions are found primarily on the outside of cells being located in

a. blood plasma b. interstitial fluid c. Both (a) and (b) d. None of these

13 Sodium ions participitate in

a. transmission of nerve signals
b. regulating the flow of water across cell membranes
c. transport of sugars and amino acids into cells
d. all of the above

14 Out of LiCl and $MgCl_2$,

a. Both are soluble in ethanol
b. Both are deliquescent and form hydrates $LiCl_2 \cdot 2H_2O$ and $MgCl_2 \cdot 2H_2O$
c. Both show exothermic reaction when hydrated
d. All of the above are correct statements

Statement Type

***Direction** (Q. Nos. 15-19) This section is based on Statement I and Statement II. Select the correct answer from the codes given below.*

a. Both Statement I and Statement II are correct and Statement II is the correct explanation of Statement I
b. Both Statement I and Statement II are correct but Statement II is not the correct explanation of Statement I
c. Statement I is correct but Statement II is incorrect
d. Statement II is correct but Statement I is incorrect

15 **Statement I** Alkali metals dissolve in liquid ammonia to give blue solution.

Statement II Alkali metals in liquid ammonia give solvated species of the type $[M(NH_3)_n]^+$, where M is alkali metal. **[AIEEE 2007]**

16 **Statement I** The Solvay (ammonia-soda) process is used to produce Na_2CO_3 synthetically, but K_2CO_3 cannot be.

Statement II $KHCO_3$ is highly soluble in water.

17 **Statement I** KO_2 is used in space and submarines.

Statement II It absorbs CO_2 and releases O_2.

Working Space

18 **Statement I** Sodium is some 30 times as abundant as is K in the oceans.

Statement II Potassium is more strongly bound to the complex silicates and aluminosilicates in soils.

19 **Statement I** K_2O_3 (potassium sesquioxide) is diamagnetic.

Statement II K_2O_3 is a peroxodisuperoxide.

One or More than One Options Correct Type

Direction (Q. Nos. 20-24) This section contains 5 multiple choice questions. Each question has four choices (a), (b), (c) and (d), out of which ONE or MORE THAN ONE are correct.

20 Highly pure dilute solution of sodium in liquid ammonia

a. shows blue colour
b. exhibits electrical conductivity
c. produces sodium amide
d. produces hydrogen gas

21 Sodium nitrate decomposes above 800°C to give **[AIEEE 2007]**

a. N_2
b. O_2
c. NO_2
d. Na_2O

22 The compound(s) formed upon combustion of sodium metal in excess of air is (are) **[AIEEE 2007]**

a. Na_2O_2
b. Na_2O
c. NaO_2
d. $NaOH$

23 The pair(s) of reagents that yield paramagnetic species is/are **[IITJEE 2011]**

a. Na and excess of NH_3
b. K and excess of O_2
c. Cu and dilute HNO_3
d. O_2 and 2-ethyl anthraquinol

24 Cell as shown in the Fig. is used to manufacture NaOH. Select the correct statement(s) about this.

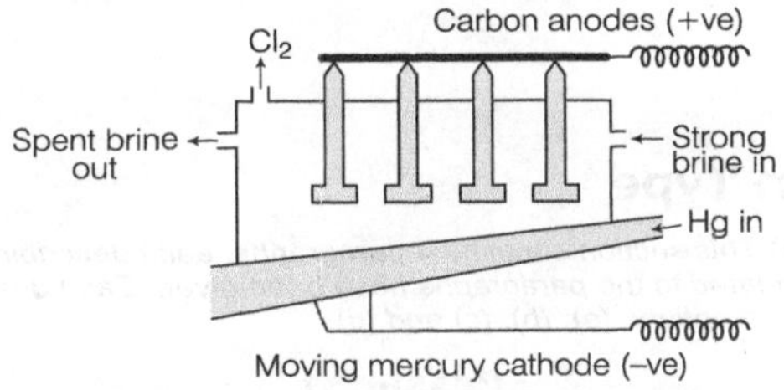

a. It is called Castner-Kellner cell
b. It is called Nelson cell
c. Sodium formed combines with Hg cathode to give sodium amalgam
d. Side reactions may cause corrosion of carbon anodes

Matching List Type

Direction (Q. Nos. 25-27) Choices for the correct combination of elements from Column I and Column II are given as options (a), (b), (c) and (d), out of which one is correct.

25 Match the different forms of sodium carbonate in Column I with their common names in Column II and select the answer from the codes given below.

	Column I		Column II
i.	$Na_2CO_3 \cdot 10H_2O$	p.	Sesquicarbonate
ii.	$Na_2CO_3 \cdot 7H_2O$	q.	Soda-ash
iii.	$Na_2CO_3 \cdot H_2O$	r.	Crystal carbonate
iv.	$Na_2CO_3 \cdot NaHCO_3 \cdot 2H_2O$	s.	Heavy ash
v.	Na_2CO_3	t.	Washing soda

Codes

	i	ii	iii	iv	v
a.	t	r	s	p	q
b.	t	s	p	q	r
c.	p	q	s	t	r
d.	r	q	p	s	t

Working Space

26 Match the species in Column I with their uses in Column II and select the answer from the codes given.

	Column I		Column II
i.	Liquid sodium metal	p.	Life supporting system
ii.	Potassium stearate	q.	Explosive
iii.	Potassium nitrate	r.	Coolant in nuclear reactor
iv.	Potassium superoxide	s.	Soft soap

Codes

	i	ii	iii	iv		i	ii	iii	iv
a.	r	s	q	p	b.	r	p	q	s
c.	s	r	q	p	d.	p	s	q	r

27 Match the compounds in Column I with their uses in Column II and select the answer from the codes given.

	Column I		Column II
i.	Na_2CO_3	p.	Alum formation
ii.	Na_2SiO_3	q.	Glass
iii.	NaOH	r.	Bleach
iv.	NaOCl	s.	SO_2 absorber
v.	K_2SO_4	t.	Detergent

Codes

	i	ii	iii	iv	v
a.	t	r	s	q	p
b.	t	q	s	r	p
c.	q	p	s	t	r
d.	s	q	p	r	t

Comprehension Type

Direction *(Q. Nos. 28-36) This section contains 4 paragraphs, each describing theory, experiments, data, etc. Nine questions related to the paragraphs have been given. Each question has only one correct answer among the four given options (a), (b), (c) and (d).*

Passage I

Sodium forms two oxides *X* and *Y*.

28 Oxide *X* turns red litmus blue. Thus, it is

a. Na_2O b. Na_2O_2
c. NaO_2 d. None of these

29 Oxide *Y* decolourises red litmus. Thus, it is

a. Na_2O b. Na_2O_2
c. NaO_2 d. None of these

Passage II

Sodium is obtained by the electrolysis of fused (molten) NaCl.

30 NaCl has high melting point. To lower the melting point, sodium is obtained by taking a mixture of

a. 40% NaCl and 60% $CaCl_2$
b. 50% NaCl and 50% $CaCl_2$
c. 60% NaCl and 40% $CaCl_2$
d. 80% NaCl and 20% $CaCl_2$

Working Space

31 Electrolysis of fused NaCl and $CaCl_2$ is carried out in

a. Nelson cell
b. Downs cell
c. Castner-Kellner cell
d. None of these

32 Valuable by-product is

a. Cl_2 and collects at anode
b. Cl_2 and collects at cathode
c. NaClO and collects at cathode
d. H_2 and collects at cathode

Passage III

$$A + NH_4Cl \longrightarrow B + NaCl$$
$$B \xrightarrow{\Delta} H_2O + C$$
$$A + Na \longrightarrow D + C$$
$$D + H_2O \longrightarrow E$$

B is decomposed by *E* forming a gas that gives white fumes with HCl.

33 *A*, *B* and *C* respectively are

a. $NaNO_2$, NH_4NO_2, N_2O
b. $NaNO_3$, NH_4NO_3, N_2O
c. $NaNO_2$, NH_4NO_2, N_2
d. $NaNO_2$, NH_4NO_3, H_2O

34 *D* and *E* respectively are

a. Na_2O, NaOH
b. Na_2O_2, H_2O_2
c. Na_2O_2, NaOH
d. Na_2O, H_2O_2

Passage IV

A binary salt KO_2 reacts with S forming *X*. Aqueous solution of $Al_2(SO_4)_3$ reacts with *X* forming *Y*.

35 *X* and *Y* respectively are

a. K_2SO_4, $K_2SO_4 \cdot Al_2(SO_4)_3 \cdot 24H_2O$
b. K_2SO_3, $K_2SO_3 \cdot Al_2(SO_4)_3 \cdot 24H_2O$
c. $K_2S_2O_3$, $K_2SO_4 \cdot Al_2(SO_4)_3 \cdot 24H_2O$
d. K_2SO_2, $K_2SO_2 \cdot Al_2(SO_4)_3 \cdot 24H_2O$

36 *X* on heating with carbon forms

a. K_2S
b. K_2SO_3
c. K_2CO_3
d. $K_2C_2O_4$

One Integer Value Correct Type

***Direction** (Q. Nos. 37 and 38) This section contains 2 questions. When worked out will result in an integer from 0 to 9 (both inclusive).*

37 If aqueous solution of Na_2SO_4 (*A*) is cooled below 32°C, $Na_2SO_4 \cdot xH_2O$ is formed. If *A* is cooled below 12°C, $Na_2SO_4 \cdot yH_2O$ is formed. What is the value of $(x - y)$?

38 Mixture containing equal moles of Li_2CO_3 and Na_2CO_3 form 6 moles CO_2. Na_2CO_3 in mixture is ...... moles.

SET - 1

Revisal Problems for JEE Main

Only One Option Correct Type

Direction *(Q. Nos. 1-30) This section contains 30 multiple choice questions. Each question has four choices (a), (b), (c) and (d), out of which ONLY ONE is correct.*

Working Space

1 The metallic lustre exhibited by sodium metal is explained by

a. diffusion of sodium ion
b. oscillation of loose electrons
c. excitation of free protons
d. existence of body-centred cubic lattice

2 Which is the bridge element of group 1 (alkali metals) and group 11 (coinage metals)?

a. Li b. Na
c. Cu d. Ag

3 All alkali metals form following compounds except

a. amide, MNH_2 b. superoxide (like KO_2)
c. ionic salt like hydride, MH d. basic oxide

4 Li_2CO_3 is

a. added to bauxite in electrolytic production of Al
b. used to toughen glass
c. used as medicine as affects the balance between Na^+ and K^+, and Ca^{2+} and Mg^{2+}
d. All of the above are correct

5 Which oxide undergoes redox reaction to M_2O_2 and metal M on heating?

a. Li_2O b. Na_2O
c. Both (a) and (b) d. None of these

6 When NO is passed into KOH solution at 400°C, product(s) is(are)

a. N_2 b. KNO_2
c. Both (a) and (b) d. None of these

7 Molten sodium chloride conducts electricity due to the presence of

a. free electrons b. ions
c. sodium atom d. chlorine atom

8 CO_2 gas along with solid Y is obtained when sodium salt X is heated. X is again obtained when CO_2 gas is passed into aqueous solution Y. Thus, X and Y are

a. Na_2CO_3, Na_2O b. Na_2CO_3, NaOH
c. $NaHCO_3$, Na_2CO_3 d. Na_2CO_3, $NaHCO_3$

9 A colourless solid X on heating evolved CO_2 and also gave a white residue soluble in water. Residue also gave CO_2 when treated with dilute acid. Thus, X is

a. Na_2CO_3 b. $CaCO_3$
c. $Ca(HCO_3)_2$ d. $NaHCO_3$

10 Oxone is

a. Na_2O b. Na_2O_2
c. KO_2 d. K_2O_2

11 In the laboratory bottle in which NaOH is stored, there is white deposit of

a. Na_2CO_3 b. Na_2SiO_3
c. Both (a) and (b) d. None of these

Working Space

12 Alkali metals catch fire in water. Which of the following methods would you use to extinguish fire?

a. Water
b. Nitrogen
c. Carbon dioxide
d. Asbestos blanket

13 The molar conductivity of solution of sodium in ether decreases as the concentration is raised. This is due to

a. solvated electrons
b. formation of Na^+ ions
c. formation of Na^- ions
d. formation of ion pair Na^+ and Na^-

14 In some foam-type fire extinguishers, $NaHCO_3$ is one of the reactants. Other suitable reactant can be

a. NaCl
b. CH_3COONa
c. NH_4Cl
d. $Al_2(SO_4)_3$

15 $NaNO_3$ is converted to KNO_3 by following reaction

$$NaNO_3(aq) + KCl(aq) \rightleftharpoons KNO_3(aq) + NaCl\,(aq)$$

In the above equilibrium, which is least soluble in hot water?

a. $NaNO_3$
b. KCl
c. KNO_3
d. NaCl

16 Na^+ ions with combination of K^+ ions

a. activate enzymes
b. participate in the oxidation of glucose to produce ATP
c. are responsible for transmission of nerve signals
d. regulate flow of water across cell membrane

17 Out of the following

I. Na_2CO_3
II. $NaHCO_3$
III. Starch
IV. $Ca(H_2PO_4)_2$
V. $NaAl(SO_4)_2$
VI. $Mg(HCO_3)_2$

which are not present in baking powder?

a. I and III
b. I, III and VI
c. I and VI
d. I, III, V and VI

18 In Solvay process, which side-product(s) can be recycled to further use?

a. CO_2, NH_3, $NaHCO_3$
b. CO_2, NH_4Cl, CaO
c. NH_4Cl, CO_2
d. CO_2

19 What is the best description of the change that occurs when $Na_2O(s)$ is dissolved in water? **[AIEEE 2011]**

a. Oxidation number of sodium decreases
b. Oxide ion accepts a shared pair of electrons
c. Oxide ion donates a pair of electrons
d. Oxidation number of oxygen increases

20 Which of the following on thermal decomposition yields a basic as well as acidic oxide? **[AIEEE 2012]**

a. $NaNO_3$
b. $KClO_3$
c. $CaCO_3$
d. NH_4NO_3

21 A solution of sodium metal in liquid ammonia is strongly reducing agent due to the presence of

a. sodium atoms
b. sodium hydride
c. sodium amide
d. solvated electrons

22 Molecular formula of Glauber's salt is

a. $MgSO_4 \cdot 7H_2O$
b. $CuSO_4 \cdot 5H_2O$
c. $FeSO_4 \cdot 7H_2O$
d. $Na_2SO_4 \cdot 10\,H_2O$

23 Operation of (Na^+- K^+) pump is biological in each and every cell of all animals in which

a. Na^+ ions from intramolecular fluids are transferred to extracellular fluids with the help of carrier-proteins
b. K^+ ions from extracellular fluids are transferred to intramolecular fluids
c. Both (a) and (b) are correct
d. None of the above is correct

Working Space

24 Which is not isolated in solid state?

a. $KHCO_3$
b. $NaHCO_3$
c. $LiHCO_3$
d. Li_2CO_3

25 A mixture contains 1 mole each of $NaHCO_3$, Na_2CO_3 and Li_2CO_3. CO_2 collected at NTP would be

a. 3.0 moles
b. 2.0 moles
c. 1.5 moles
d. 1.0 mole

26 Nitrogen dioxide cannot be obtained by heating

a. KNO_3
b. $Pb(NO_3)_2$
c. $Cu(NO_3)_2$
d. $AgNO_3$

27 Out of LiF, LiCl, LiBr and LiI, maximum solubility in water is of

a. LiF
b. LiCl
c. LiBr
d. LiI

28 Solubility of fluorides of alkali metals in water increases in the order

a. LiF < NaF < KF < RbF < CsF
b. CsF < NaF < KF < RbF < LiF
c. CsF < RbF < KF < NaF < LiF
d. None is the correct alternate

29 The increasing stability of the peroxide (M_2O_2) or superoxide (MO_2) as the size of the alkali metal ion increases is due to the stabilisation of the

a. larger anion by smaller cation
b. larger anion by larger cation
c. smaller anion by larger cation
d. smaller anion by smaller cation

30 For the following reaction

$$LiI + KF \longrightarrow LiF + KI$$

a. $\Delta G^\circ < 0$
b. $\Delta G^\circ > 0$
c. $\Delta S^\circ < 0$
d. $\Delta S^\circ > 0$

Statement Type

Direction ***(Q. Nos. 31-35) This section is based on Statement I and Statement II. Select the correct answer from the codes given below.***

a. Both Statement I and Statement II are correct and Statement II is the correct explanation of Statement I
b. Both Statement I and Statement II are correct but Statement II is not the correct explanation of Statement I
c. Statement I is correct but Statement II is incorrect
d. Statement II is correct but Statement I is incorrect

31 **Statement I** The crystalline salts of alkaline earth metals contain more water of crystallisation than the corresponding alkali metal salts.

Statement II Enthalpy of hydration of alkali metal salts are larger than that of alkaline earth metal salts.

32 **Statement I** The reaction

$$2Na(g) \rightleftharpoons Na^+(g) + Na^-(g) \text{ is spontaneous.}$$

Statement II This is an endothermic by 438 kJ mol^{-1}.

33 **Statement I** LiCl is predominantly a covalent compound.

Statement II Electronegativity difference between Li and Cl is too small.

34 **Statement I** All the alkali metals give characteristic colours in Bunsen flame.

Statement II Colour emission by these elements is the basis of analytical methods for precise estimation of these elements.

35 **Statement I** Lithium salts are not involved in alums.

Statement II Alum is an anhydrous double salt.

SET - 2

Revisal Problems
for JEE Advanced

Working Space

Only One Option Correct Type

Direction *(Q. Nos. 1-6) This section contains 6 multiple choice questions. Each question has four choices (a), (b), (c) and (d), out of which ONLY ONE is correct.*

1 Which metal is used in photo voltaic cell?

a. Li b. Na
c. Cs d. K

2 NaOH can be stored without any effect in

a. Al vessel b. Zn vessel
c. Glass vessel d. Poly vessel

3 To an acidified dichromate solution, a pinch of Na_2O_2 and ether is added and shaken. There is the formation of

a. deep violet colour
b. red colour changing to green
c. O_2 gas
d. bluish green precipitate

4 Aqueous Na_2CO_3 is saturated with SO_2 gas. Product formed is

a. $NaHSO_4$ b. $NaHSO_3$
c. Na_2SO_3 d. Na_2SO_4

5 Li and Na are placed in dry air. We get

a. $NaOH, Na_2O, Li_2O$
b. Na_2O, Li_2O
c. $Na_2O, Li_2O, Li_3N, NH_3$
d. Na_2O, Li_3N, Li_2O

6 Which species can be placed together?

a. $NaHCO_3$ with NaOH
b. $NaHCO_3$ with NH_4Cl
c. Na_2CO_3 with $NaHCO_3$
d. Na_2CO_3 with NH_4Cl

Statement Type

Direction *(Q. Nos. 7-10) This section is based on Statement I and Statement II. Select the correct answer from the codes given below.*

a. Both Statement I and Statement II are correct and Statement II is the correct explanation of Statement I
b. Both Statement I and Statement II are correct but Statement II is not the correct explanation of Statement I
c. Statement I is correct but Statement II is incorrect
d. Statement II is correct but Statement I is incorrect

7 **Statement I** $NaHCO_3$ is also added in small amount to table salt (NaCl).

Statement II NaCl is often contaminated with small amount of $MgCl_2$.

8 **Statement I** In the following reaction, $NH_4HCO_3 + KCl \longrightarrow KHCO_3 + NH_4Cl$
$KHCO_3$ is separated by crystallisation.

Statement II $KHCO_3$ is highly soluble in water.

Working Space

9 Consider the following graph,

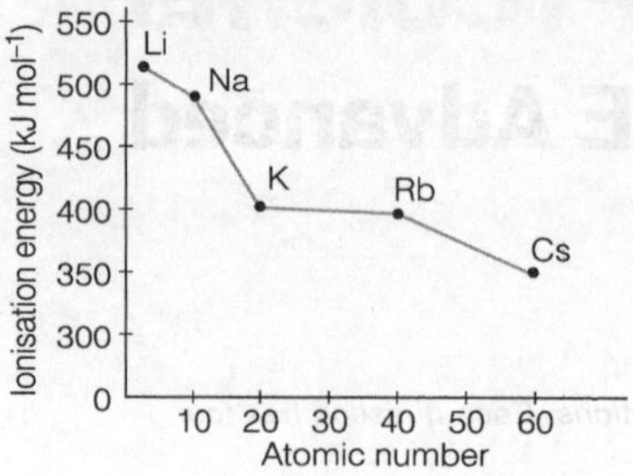

Statement I First ionisation energies (IE) of group 1 elements do not vary regularly.

Statement II Group 1 elements are electropositive elements.

10 Consider the following graph between atomic number and ionic radius,

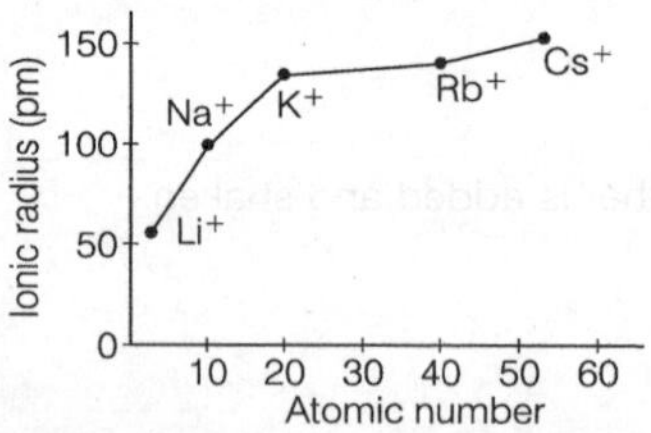

Statement I In alkali metals (group 1), ionic radius increases rapidly upto K^+, then increases slowly after K^+.

Statement II Valence shell electrons are screened and screening effect is in order $s > p > d > f$.

One or More than One Options Correct Type

Direction *(Q. Nos. 11-21) This section contains 11 multiple choice questions. Each question has four choices (a), (b), (c) and (d), out of which ONE or MORE THAN ONE are correct.*

11 Select the correct statement(s) about sesquioxide of alkali metals.

a. They are dark coloured paramagnetic
b. Thermal decomposition of $MO_2 (M = K^+, Rb^+, Cs^+)$ form sesquioxide
c. They are peroxide disuperoxide
d. They are diamagnetic

12 This cell is used to manufacture NaOH. Select the correct statement(s) about this.

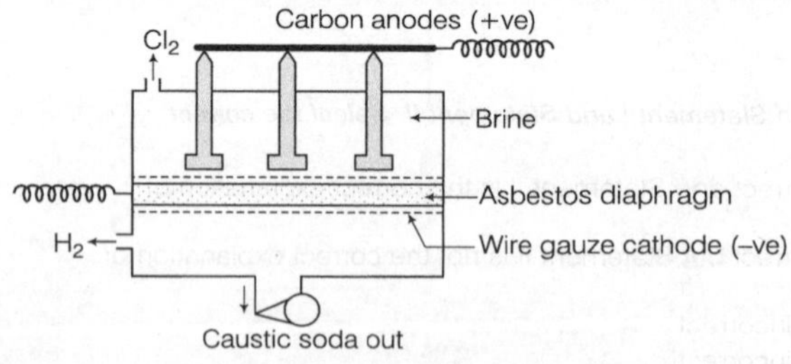

a. This is called Castner-Kellner cell
b. This is called Nelson cell
c. NaOH obtained is 11% concentration which is further concentrated by steam evaporation
d. H_2 and Cl_2 in the cell may combine violently

13 Select the species in which the precipitate formed on adding NaOH further dissolves in excess of NaOH.

a. $ZnCl_2$
b. $AlCl_3$
c. $FeCl_3$
d. $CrCl_3$

Working Space

14 Na_2O_2 can oxidise which of the following?

a. Al
b. Cr^{3+}
c. $Mn(OH)_2$
d. CO

15 In which of the following, O_2 is also one of the products when Na_2O_2 reacts with

a. CO_2
b. dil. H_2SO_4
c. hot HCl
d. CO

16 Select the correct statement(s).

a. $(IE)_1$, first ionisation energy, measures the tendency for a gaseous atom to ionise
b. E°_{red}, standard reduction potential, measures the tendency for a metal atom to go from solid state to the ion in aqueous solution
c. $Li(g)$ is the most difficult to ionise to $Li^+(g)$
d. $Li(s)$ is the easiest to oxidise to $Li^+(aq)$

17 Sodium sulphate is soluble in water whereas barium sulphate is sparingly soluble because

a. the hydration energy of sodium sulphate is more than its lattice energy
b. the lattice energy of barium sulphate is more than its hydration energy
c. the lattice energy has no role to play in solubility
d. the hydration energy of sodium sulphate is less than its lattice energy

18 Select the correct statement(s) about KO_2.

a. It finds use in life support system
b. Its acidic and aqueous solution has bleaching properties
c. It is paramagnetic in nature
d. Bond order of the anion part is 2.0

19 Select the correct statement(s).

a. Li_2CO_3 is only sparingly soluble in water and no $LiHCO_3$ has been isolated
b. K_2CO_3 cannot be made by a method similar to the ammonia soda process
c. Li_2CO_3 and $MgCO_3$ both are thermally stable
d. $Na_2CO_3 \cdot NaHCO_3 \cdot 2H_2O$ is a mineral called trona

20 Select the correct alternate(s).

a. Solid $LiNO_3$ and $NaNO_3$ are deliquescent
b. KNO_3 is used in preference to $NaNO_3$ in gun powder
c. $NaNO_3$ is used in preference to KNO_3 in gun powder
d. NH_4HCO_3 exists in solid state

21 Which is/are a pair of paramagnetic species?

a. NO_2, KO_2
b. NO, KO_2
c. N_2O_4, K_2O
d. N_2O_2, K_2O_2

Matching List Type

***Direction** (Q. Nos. 22 and 23) Choices for the correct combination of elements from Column I and Column II are given as options (a), (b), (c) and (d), out of which one is correct.*

22 Match the compounds in Column I with their properties in Column II and mark the correct option from the codes given.

	Column I		Column II
i.	KO_2	p.	Paramagnetic
ii.	K_2O_2	q.	Diamagnetic
iii.	K_2O	r.	Aqueous alkaline solution
iv.	KO_3	s.	CO_2 absorber and O_2 releasor
v.	K_2O_3	t.	Coloured

Codes

	i	ii	iii	iv	v
a.	p,r,s,t	q,r,s	q,r	q,r	p,r,s,t
b.	p,r	q,r	q,r	q,r,s	p,r,t
c.	p,r,s,t	q,r	q,r	q,r,s	s,t
d.	p,r,s,t	q,r	p,q,r	q,r	p,r,s,t

Working Space

23 Match the compounds in Column I with given properties in Column II and mark the correct option from the codes given.

	Column I		Column II
i.	Li_3N	p.	Paramagnetic
ii.	LiCl	q.	Ether soluble
iii.	KO_2	r.	Humidity control
iv.	Rb_2O_3	s.	Automobile airbags
		t.	Coloured compound

Codes

	i	ii	iii	iv
a.	s,t	q,r	p,t	p,t
b.	s,t	q,t	p,r	p,t
c.	q,r	p,t	q,r	p,t
d.	s,t	q,t	p,r	p,s

Comprehension Type

Direction (Q. Nos. 24-32) ***This section contains 4 paragraphs, each describing theory, experiments, data, etc. Nine questions related to the paragraphs have been given. Each question has only one correct answer among the four given options (a), (b), (c) and (d).***

Passage I

A binary salt *X* of an alkali metal on heating with sulphur gives *Y*.
Y with $BaCl_2$ solution gives white precipitate *Z*.
Z is insoluble in concentrated HCl.
14.2 g of *X* gave 23.3 g of *Z* $[Ba = 137, S = 32, O = 16]$.

24 *X* is

a. KO_2
b. paramagnetic in nature
c. used in submarines
d. All of these are correct

25 *Y* and *Z* respectively are

a. K_2SO_3 and $BaSO_3$
b. K_2CO_3 and $BaCO_3$
c. K_2SO_4 and $BaSO_4$
d. $K_2S_2O_3$ and $BaSO_4$

Passage II

Electrolysis of aqueous NaCl gives NaOH at cathode and Cl_2 at anode.

26 If cathode and anode are not separated, side products may be formed at room temperature are

a. NaCl and $NaClO_3$
b. NaCl and NaClO
c. Cl_2 and NaCl
d. Cl_2 and H_2

27 After electrolysis, there is change in pH from

a. 7.0 to 9.0
b. 7.0 to 5.0
c. 7.5 to 9.0
d. 5.0 to 7.0

Passage III

Sodium forms two oxides *P* and *Q* having 25.81% and 41.03% oxygen respectively.

28 *P* on heating with MnO_2 forms

a. Na_2MnO_4 (green)
b. $NaMnO_4$ (purple)
c. $Mn(OH)_2$ (buff coloured ppt)
d. No reaction

29 *Q* on heating with $Cr(OH)_3$ forms

a. Na_2CrO_4
b. $Na_2Cr_2O_7$
c. CrO_3
d. CrO_5

Working Space

Passage IV

An improved combination of baking powder contains 40% starch, 30% $NaHCO_3$, 20% $Na Al(SO_4)_2$ and 10% $Ca(H_2PO_4)_2$.

30 Rise in bread and cakes is due to ...... when $NaHCO_3$ decomposes between 50° and 100°C.

a. H_2O
b. CO_2
c. Na_2CO_3
d. None of these

31 $Ca(H_2PO_4)_2$ behaves as

a. an acid in moist condition and decomposes $NaHCO_3$ to CO_2
b. a base in moist condition and gives H_3PO_4
c. a binder
d. a filler

32 $NaAl(SO_4)_2$ in baking powder

a. makes the decomposition of $NaHCO_3$ faster
b. slows the decomposition of $NaHCO_3$
c. acts as filler
d. makes the bread rise

One Integer Value Correct Type

Direction *(Q. Nos. 33-35) This section contains 3 questions. When worked out will result in an integer from 0 to 9 (both inclusive).*

33 How many of the following undergo disproportionation in NaOH medium?
$CO_2, SO_2, NO_2, S, Cl_2, P, B, Al, Zn, CO$

34 Sodium superoxide is not formed directly from sodium and oxygen, but by reaction of sodium peroxide with oxygen. How much sodium superoxide is formed from 6.382 g sodium peroxide?

35 KO_2 (potassium superoxide) is paramagnetic due to the presence of $(x-1)$ unpaired electrons. What is the value of x?

JEE Archive

(Compilation of Last 13 Years' Questions of JEE Main, AIEEE, JEE Advanced and IIT JEE)

1 The products obtained on heating $LiNO_3$ will be **(AIEEE 2011)**

a. $LiNO_2 + O_2$ b. $Li_2O + NO_2 + O_2$
c. $Li_3N + O_2$ d. $Li_2O + OH + O_2$

2 What is the best description of the change that occurs when $Na_2O(s)$ is dissolved in water? **(AIEEE 2011)**

a. Oxidation number of sodium decreases
b. Oxide ion accepts a shared pair of electrons
c. Oxide ion donates a pair of electrons
d. Oxidation number of oxygen increases

3 The metallic sodium dissolves in liquid ammonia to form a deep blue coloured solution. The deep blue colour is due to the formation of **(AIEEE 2002)**

a. solvated electron, $[e^-(NH_3)_x]$
b. solvated atomic sodium, $[Na(NH_3)_y]$
c. $(Na^+ + Na^-)$
d. $NaNH_2 + H_2$

4 KO_2 is used in space and submarines because it **(AIEEE 2002)**

a. absorbs CO_2 and increases O_2 concentration
b. absorbs moisture
c. absorbs CO_2
d. produces ozone

5 The pair(s) of reagents that yield paramagnetic species is/are **(IIT JEE 2011, One or More than One Options Correct Type)**

a. Na and excess of NH_3 b. K and excess of O_2
c. Cu and dilute HNO_3 d. O_2 and 2-ethyl anthraquinol

6 The compound(s) formed upon combustion of sodium metal in excess air is (are) **(IIT JEE 2007, One or More than One Options Correct Type)**

a. Na_2O_2 b. Na_2O
c. NaO_2 d. NaOH

7 **Statement I** Alkali metals dissolve in liquid ammonia to give blue solutions.
Statement II Alkali metals in liquid ammonia give solvated species of the type $[M(NH_3)_n]^+$ (M = alkali metals). **(IIT JEE 2007, Statement Type)**

a. Both Statement I and Statement II are correct and Statement II is the correct explanation of Statement I
b. Both Statement I and Statement II are correct but Statement II is not the correct explanation of Statement I
c. Statement I is correct but Statement II is incorrect
d. Statement II is correct but Statement I is incorrect

4. Alkaline Earth Metals (Group 2)

DPP-1 Periodicity in Group 2

Only One Option Correct Type

Working Space

Direction (Q. Nos. 1-10) *This section contains 10 multiple choice questions. Each question has four choices (a), (b), (c) and (d), out of which ONLY ONE is correct.*

1 The following compounds have been arranged in order of their increasing thermal stabilities. Identify the correct order.

K_2CO_3(I) $MgCO_3$(II) $CaCO_3$(III) $BeCO_3$(IV)

a. I < II < III < IV
b. IV < II < III < I
c. IV < II < I < III
d. II < IV < III < I

2 Which is not the correct variation of the property indicated of chlorides of alkaline earth metals?

a. $BeCl_2 < MgCl_2 < CaCl_2 < BaCl_2$ (Covalent nature)
b. $BeCl_2 < MgCl_2 < CaCl_2 < BaCl_2$ (Melting point)
c. $BaCl_2 < CaCl_2 < MgCl_2 < BeCl_2$ (Lattice energy)
d. $BaCl_2 < CaCl_2 < MgCl_2 < BeCl_2$ (Thermal stability)

3 Basic nature of MgO(I), SrO (II), K_2O (III), Cs_2O (IV) and Na_2O(V) will be in order

a. I < II < III < IV < V
b. I < II < V < III < IV
c. IV < III < V < II < I
d. None is correct

4 Be and Al exhibit many properties which are similar. But two elements differ in **[AIEEE 2004]**

a. exhibiting maximum covalency in compounds
b. forming polymeric hydrides
c. forming covalent halides
d. exhibiting amphoteric nature in their oxides

5 The solubilities of carbonates decrease down the magnesium group due to decrease in **[AIEEE 2003]**

a. lattice energies of solids
b. hydration energies of cations
c. inter-ionic attractions
d. entropy of solution formation

6 Group 2 (alkaline earth elements) and group 12 (zinc family) have ns^2 configuration of valence shell. Thus,

a. both form M^{2+} ions with inert gas configuration
b. BeO and ZnO are amphoteric
c. sulphides of both groups are soluble
d. solubility of hydroxide increases as we go down the group

7 F^-, Na^+, Mg^{2+}, Al^{3+} and Si^{4+} are isoelectronic. Their ionic sizes are in order

a. $F^- > Na^+ > Mg^{2+} > Al^{3+} > Si^{4+}$
b. $F^- < Na^+ < Mg^{2+} < Al^{3+} < Si^{4+}$
c. $Na^+ < Mg^{2+} < Al^{3+} < Si^{4+} < F^-$
d. None of the above

Working Space

8 Alkaline earth metals (group 2) and zinc family (group 12) resemble in which of the following?

a. $MgSO_4 \cdot 7H_2O$ is isomorphous with $ZnSO_4 \cdot 7H_2O$
b. Cations of both are precipitated by H_2S in acidic medium
c. Both (a) and (b)
d. None of the above

9 When moving along a group 2 from Be to Ba, following properties increase except

a. solubility of hydroxide in H_2O
b. hydration energy
c. thermal stability of carbonates
d. ionic radius

10 Following are the ionisation potentials of
(I_1) 899 kJ mol^{-1}, (I_2) 1757 kJ mol^{-1}, (I_3) 15000 kJ mol^{-1}

a. Na
b. K
c. Be
d. Ne

Statement Type

Direction *(Q. Nos. 11-15) This section is based on Statement I and Statement II. Select the correct answer from the codes given below.*

a. Both Statement I and Statement II are correct and Statement II is the correct explanation of Statement I
b. Both Statement I and Statement II are correct but Statement II is not the correct explanation of Statement I
c. Statement I is correct but Statement II is incorrect
d. Statement II is correct but Statement I is incorrect

11 **Statement I** In the preparation of BaO by thermal decomposition of barium carbonate, carbon is added to reaction mixture.
Statement II Lattice energy of $BaCO_3$ is very high.

12 **Statement I** Beryllium and aluminium both react with HNO_3 forming NO_2.
Statement II Beryllium and aluminium show diagonal relationship.

13 **Statement I** Ca, Sr and Ba react rapidly with H_2O.
Statement II They have high E°_{red} values similar to alkali metals.

$$M^{2+} + 2e^- \longrightarrow M, E^\circ_{red}$$

14 **Statement I** Alkaline earth elements are denser than alkali metals.
Statement II Alkali metals have one and alkaline earth metals have two valence electrons.

15 **Statement I** Thermal stability of hydroxides of group 2 increases moving down the group from Be to Ba.
Statement II $Be(OH)_2$ is amphoteric.

One or More than One Options Correct Type

Direction *(Q. Nos. 16-18) This section contains 3 multiple choice questions. Each question has four choices (a), (b), (c) and (d), out of which ONE or MORE THAN ONE are correct.*

16 Based on data, Na $\xrightarrow{I_1}$ Na^+ $\xrightarrow{I_2}$ Na^{2+},

$$Mg \xrightarrow{I_1'} Mg^+ \xrightarrow{I_2'} Mg^{2+}, \quad Al \xrightarrow{I_1''} Al^+ \xrightarrow{I_2''} Al^{2+}$$

Which comparison is/are correct?

a. $I_2 < I_2'$
b. $I_1 < I_1' < I_2''$
c. $I_2 < I_2' < I_2''$
d. $I_1 < I_1'' < I_1'$

Working Space

17 Be and Al resemble in which of the following properties?

a. Both are amphoteric
b. Oxides are amphoteric
c. $BeCl_2$ and $AlCl_3$ exist in polymeric form and are Lewis acids
d. Salts are hydrated

18 Which is/are correct variation of the indicated properties?

a. $BeSO_4 < MgSO_4 < BaSO_4$ (Solubility)
b. $Be(OH)_2 < Mg(OH)_2 < Ba(OH)_2$ (Solubility)
c. $BeCl_2 < MgCl_2 < BaCl_2$ (Ionic nature)
d. $BeCl_2 < LiCl < MgCl_2 < NaCl$ (Ionic nature)

Matching List Type

***Direction** (Q. No. 19) Choices for the correct combination of elements from Column I and Column II are given as options (a), (b), (c) and (d), out of which one is correct.*

19 Match the species in Column I with the property(ies) in Column II and mark the correct option from the codes given below.

	Column I		Column II
i.	BeO	p.	Amphoteric
ii.	Al_2O_3	q.	Diagonal relationship
iii.	MgO	r.	Water insoluble
iv.	BaO	s.	Estimated by EDTA
		t.	Maximum soluble in water

Codes

	i	ii	iii	iv
a.	p,q,r	p,q,r	q,s	t
b.	p,q,s	p,q,t	q,s	s
c.	p,q	p,q	q,s	s,t
d.	q,r	p,q	q,s	s,t

Comprehension Type

***Direction** (Q. Nos. 20 and 21) This section contains a paragraph, each describing theory, experiments, data, etc. Two questions related to the paragraph have been given. Each question has only one correct answer among the four given options (a), (b), (c) and (d).*

Passage

Hydration energies (in kJ mol^{-1})

$$Ca^{2+} = -1577, \quad Mg^{2+} = -1921, Ba^{2+} = -1305, F^{-} = -457$$

Lattice energies (in kJ mol^{-1})

$$MgF_2 = -2906, \quad CaF_2 = -2610, \quad BaF_2 = -2367$$

20 Hydration energy is maximum for

a. MgF_2
b. CaF_2
c. BaF_2
d. Can't be predicted

21 Which is maximum soluble in H_2O?

a. CaF_2
b. MgF_2
c. BaF_2
d. Equal solubility

DPP-2 Compounds of Group 2

Only One Option Correct Type

Working Space

Direction *(Q. Nos. 1-18) This section contains 18 multiple choice questions. Each question has four choices (a), (b), (c) and (d), out of which ONLY ONE is correct.*

1 About 2.5% gypsum ($CaSO_4 \cdot 2H_2O$) is also added to cement as one of the constituents. Its function is to

a. slow down the setting process to increase strength
b. speed up the setting process to increase strength
c. absorb moisture
d. make it anti-rusting

2 Lithopone is a white paint. It contains two components which are

a. $ZnCO_3$ and $BaCO_3$
b. $ZnSO_4$ and $BaSO_4$
c. ZnS and Na_2SO_4
d. ZnS and $BaSO_4$

3 Magnalium is an alloy of

a. magnesium and copper
b. magnesium, aluminium and copper
c. magnesium and aluminium
d. magnesium and alum

4 Mixture of $BaSO_4$ and Na_2CO_3 is heated strongly, then products (residue) are

a. $BaCO_3$ and Na_2SO_4
b. BaO and Na_2SO_4
c. BaO and Na_2O
d. $BaCO_3$ and Na_2SO_4

5 Select the correct statement(s).

a. Presence of $MgCl_2$ in table salt causes it to clump
b. Addition of $NaHCO_3$ to table salt converts $MgCl_2$ to non-hygroscopic $MgCO_3$
c. Both (a) and (b)
d. None of the above

6 What is impurity (as a salt) associated with table salt obtained from sea water?

a. $NaHCO_3$
b. $MgCO_3$
c. $MgCl_2$
d. NaI

7 Magnesium in sea water is enriched by ion-exchange through

a. calcined dolomite
b. carbon reduction
c. Both (a) and (b)
d. None of these

8 Saturated solutions of Na_2CO_3 and $Ca(OH)_2$ are mixed.

Thus, $Ca(OH)_2 + Na_2CO_3 \longrightarrow 2NaOH + CaCO_3$

a. above reaction is spontaneous
b. pH > 14
c. this reaction can be used to prepare NaOH
d. All of the above

Working Space

9 Which of the following salts fumes in moisture?

a. $BeCl_2$ b. $AlCl_3$
c. Both (a) and (b) d. None of these

10 Beryllium chloride as ...... in solid state.

a. $BeCl_2$ b. $(BeCl_2)_4$
c. $(BeCl_2)_2$ d. $(BeCl_2)_3$

11 If strongly ignited magnesium oxide is added to a concentrated solution of magnesium chloride, the slurry so obtained solidifies to a solid mass after some time. Solid mass so obtained is called

a. Sorel cement b. Portland cement
c. Plaster of Paris d. None of these

12 The substance not likely to contain calcium carbonate, is

a. a marble statue b. calcined gypsum
c. sea shells d. dolomite

13 Calcium is obtained by

a. electrolysis of molten $CaCl_2$
b. electrolysis of aqueous $CaCl_2$
c. reduction of $CaCl_2$ with carbon
d. roasting of limestone

14 In mineral, barium ion can be detected by

a. green colour in flame
b. yellow precipitate with CrO_4^{2-} solution
c. white precipitate with SO_4^{2-} ion
d. All of the above

15 Mg and Zn do not resemble in which the of the following property?

a. Oxides are amphoteric
b. Carbonates on heating form metal oxide
c. Widely used as electrodes
d. Used to prevent corrosion

16 When Cl_2 is passed into slaked lime, compound formed is

a. CaO_2Cl_2 b. CaO_2Cl
c. $CaOCl_2$ d. $CaCl_2O_4$

17 Colourless salt $(X) \xrightarrow{\Delta} CO_2$ + Residue (Y)
Residue (Y) is water insoluble. Residue (Y) liberates CO_2 on heating and also on reaction with dilute acid. Colourless salt (X) is

a. $Ca(HCO_3)_2$ b. $CaCO_3$
c. $NaHCO_3$ d. Na_2CO_3

18 Be and Al do not resemble in which of the following property?

a. Both become passive on reaction with HNO_3 due to the formation of oxide layers
b. Their chlorides are Lewis bases
c. Chlorides exist in polymeric form
d. Hydroxides are soluble in acid as well as in base

Statement Type

Working Space

Direction (Q. Nos. 19 and 20) *This section is based on Statement I and Statement II. Select the correct answer from the codes given below.*

a. Both Statement I and Statement II are correct and Statement II is the correct explanation of Statement I
b. Both Statement I and Statement II are correct but Statement II is not the correct explanation of Statement I
c. Statement I is correct but Statement II is incorrect
d. Statement II is correct but Statement I is incorrect

19 **Statement I** $BeCl_2$ can be easily hydrolysed.

Statement II $BeCl_2$ is a Lewis acid.

20 **Statement I** Hydrated beryllium chloride exists as $[Be(H_2O)_2Cl_2]$.

Statement II It is Lowry-Bronsted acid.

One or More than One Options Correct Type

Direction (Q. Nos. 21-25) *This section contains 5 multiple choice questions. Each question has four choices (a), (b), (c) and (d), out of which ONE or MORE THAN ONE are correct.*

21 On heating $MgCl_2 \cdot 6H_2O$, it changes to

a. $MgCl_2$
b. MgO
c. $Mg(OH)_2$
d. $Mg(OH)Cl$

22 Carbides are decomposed by H_2O forming hydrocarbons. Select the correct reaction(s) of carbides with H_2O.

a. $CaC_2 + H_2O \longrightarrow CH_4$
b. $Al_4C_3 + H_2O \longrightarrow CH_4$
c. $Be_2C + H_2O \longrightarrow CH_4$
d. $Mg_2C_3 + H_2O \longrightarrow CH_3C{\equiv}CH$

23 Which of the following is/are correct statement(s)?

a. $Ca_3(PO_4)_2$ is a part of bones
b. $3Ca_3(PO_4)_2 \cdot CaF_2$ is a part of enamel on teeth
c. Ca^{2+} ions are important in blood clotting
d. Ca^{2+} is an important constituent of chlorophyll which initiates the process of photosynthesis in green plants

24 Select the correct statement(s).

a. Mg^{2+} ions are necessary for the activation of phosphate transfer enzymes
b. Mg is present in chlorophyll used in photosynthesis in green plants
c. Mg^{2+} ions are concentrated more in intracellular than in extracellular fluids in animal bodies
d. Mg^{2+} ions are involved in carbohydrate metabolism

25 High alumina cement is made by fusing limestone and bauxite with small amount of SiO_2 and TiO_2 at 1400-1500°C. It is better than Portland cement because

a. it sets much quicker
b. it develops high strength within one day
c. it has good resistance to sea water and dilute mineral acids
d. it is insoluble in water

Working Space

Matching List Type

Direction (Q. Nos. 26 and 27) *Choices for the correct combination of elements from Column I and Column II are given as options (a), (b), (c) and (d), out of which one is correct.*

26 Match the compounds in Column I with their uses in Column II and mark the correct option from the codes given below.

	Column I		Column II
i.	CaC_2	p.	Paint
ii.	$CaCN_2$	q.	Food ripening
iii.	$BaSO_4$	r.	Automobile air-bag
iv.	Mg_3N_2	s.	Fertiliser

Codes

	i	ii	iii	iv
a.	p	q	r	s
b.	s	p	q	r
b.	r	q	p	s
d.	q	s	p	r

27 Match the constituents in Column I with their composition in Column II and mark the correct option from the codes given below.

	Column I		Column II
i.	CaO	p.	2%
ii.	SiO_2	q.	3%
iii.	Al_2O_3	r.	5%
iv.	Fe_2O_3	s.	20%
v.	$CaSO_4 \cdot 2H_2O$	t.	70%

Codes

	i	ii	iii	iv	v
a.	p	q	r	s	t
b.	t	s	r	q	p
c.	t	s	p	q	r
d.	s	t	q	r	p

Comprehension Type

Direction (Q. Nos. 28-31) *This section contains 2 paragraphs, each describing theory, experiments, data, etc. Four questions related to the paragraphs have been given. Each question has only one correct answer among the four given options (a), (b), (c) and (d).*

Passage I

All the group 2 elements except Be form hydrides by direct combination with H_2.

28 Beryllium hydride is formed by

a. $Be + NaOH \longrightarrow$
b. $Be + H_2O \longrightarrow$
c. $BeCl_2 + LiAlH_4 \longrightarrow$
d. $Be + NH_3 \longrightarrow$

29 Which of the following is/are correct fact(s) about beryllium hydride?

a. Polymeric covalent $(BeH_2)_n$
b. Two electrons are shared between three atoms Be • H • Be
c. BeH_2 is electron deficient
d. All of the above

Working Space

Passage II

Anhydrous $BeCl_2$ is not obtained by the reaction of Be on dil. HCl.

30 Which of the following is not the indirect method of preparation of anhydrous $BeCl_2$?

a. $BeO + C + Cl_2 \xrightarrow{700°C} BeCl_2 + CO$

b. $2BeO + CCl_4 \xrightarrow{800°C} 2BeCl_2 + CO_2$

c. $Be(OH)_2 + 2SOCl_2 \xrightarrow{100°C} BeCl_2 + 2SO_2 + 2HCl$

d. All of the above

31 $BeCl_2$ exists as

a. polymeric three centre-two electron bonds

b. linear $BeCl_2$ with sp^2-hybridisation

c. polymeric two centre-three electron bonds

d. polymeric two centre-two electron bonds

One Integer Value Correct Type

***Direction** (Q. Nos. 32-36) This section contains 5 questions. When worked out will result in an integer from 0 to 9 (both inclusive).*

32 How many of the following are amphoteric species?

BeO, MgO, $Mg(HCO_3)_2$, $NaHCO_3$, $Ca(HCO_3)_2$, CaO, ZnO, Al_2O_3, MgH_2PO_4, $Mg_2P_2O_7$

33 General formula of Portland cement constituent is $Ca_9(Al_6O_{18})$. How many (AlO_4) tetrahedra are present in it?

34 One mole of magnesium nitride on reaction with an excess of water gives x moles of NH_3. The value of x is

35 Number of water molecules per formula of epsom salt is

36 Difference in water molecules in Epsom salt and Kieserite is

SET - 1

Revisal Problems
for JEE Main

Only One Option Correct Type

Working Space

Direction *(Q. Nos. 1-19) This section contains 19 multiple choice questions. Each question has four choices (a), (b), (c) and (d), out of which ONLY ONE option is correct.*

1 BeF_2 in the presence of NaF forms a complex. Be remains with

a. cationic part
b. anionic part
c. Both (a) and (b)
d. None of these

2 Anhydrous $MgCl_2$ is formed by

a. $MgCl_2 \cdot 6H_2O \xrightarrow{\Delta}$
b. $MgCl_2 \cdot 6H_2O + SOCl_2 \longrightarrow$
c. $Mg + Cl_2 \xrightarrow{\Delta}$
d. $MgO + Cl_2 \longrightarrow$

3 When plaster of Paris sets in the presence of water, it forms

a. $CaSO_4 \cdot H_2O$
b. $CaSO_4 \cdot \frac{1}{2}H_2O$
c. $2CaSO_4 \cdot H_2O$
d. $CaSO_4 \cdot 2H_2O$

4 Select the correct alternate(s).

a. $MgSO_4$ is deliquescent and is readily soluble in water
b. $MgSO_4$ forms double salt with K_2SO_4 as $K_2SO_4 \cdot MgSO_4 \cdot 6H_2O$ called 'potash magnesis'
c. Potash magnesis is used as a fertiliser
d. All of the above are correct alternates

5 Hardness of water in terms of Ca^{2+} and Mg^{2+} is determined by titration using

a. ethylenediamine
b. ethylenediamine tetraacetic acid
c. glyoxime
d. sodium bicarbonate

6 Setting of cement is

a. exothermic reaction
b. endothermic reaction
c. neither endothermic nor exothermic
d. dehydration reaction

7 Which is used for fixing atmospheric nitrogen?

a. $CaCN_2$
b. Li_3N
c. Mg_3N_2
d. All of these

8 1 mole of a substance (*X*) was treated with excess of water. There were formation of 2 moles of readily combustible gas and a solution which when treated with CO_2 turned milky. The substance (*X*) is

a. Ca
b. CaH_2
c. $Ca(OH)_2$
d. $CaCO_3$

9 Which of the following is used to treat indigestion?

a. $Be(OH)_2$
b. KOH
c. $Mg(OH)_2$
d. $Ca(OH)_2$

10 A metal *M* readily forms water soluble MSO_4. It also forms oxide *MO* which becomes inert on heating. *MO* is insoluble in water but soluble in dil. HCl as well as in NaOH solution. Thus, metal *M* is

a. Mg
b. Ca
c. Ba
d. Be

11 Important constituent of Portland and high alumina cements, is

a. $Ca_3Al_2O_6$
b. $CaSiO_3$
c. $CaSO_4 \cdot 2H_2O$
d. Fe_2O_3

12 The pair whose both species are used in antacid medicines, is

a. $NaHCO_3$ and $Mg(OH)_2$
b. Na_2CO_3 and $Ca(HCO_3)_2$
c. $Ca(HCO_3)_2$ and $Mg(OH)_2$
d. $Ca(OH)_2$ and $NaHCO_3$

Working Space

13 Stalactite and stalagmite formation is related to the property of
a. calcium and barium carbonates to decompose to form oxides
b. calcium and barium bicarbonates to decompose to give carbonates
c. calcium and barium sulphates to give oxides
d. None of the above

14 In which of the following, the hydration energy is higher than the lattice energy?
a. $BaSO_4$ b. $MgSO_4$
c. $CaSO_4$ d. $RaSO_4$

15 Which has maximum IE?
a. He^+ b. H
c. Li^{2+} d. Be^{3+}

16 Thermal stabilities of various compounds have been compared. Which is incorrect comparison?
a. $NaHCO_3 < Ca(HCO_3)_2$ b. $CaCO_3 < Na_2CO_3$
c. $BeCO_3 < Li_2CO_3$ d. $NaHCO_3 < Na_2CO_3$

17 Which one of the following pairs is different from others?
a. Li - Mg b. Be - Al
c. Mg - Na d. B - Si

18 The activity of alkaline earth metals as reducing agent
a. decreases from Be to Ba
b. increases from Be to Ba
c. increases from Be to Ca and decreases from Ca to Ba
d. decreases from Be to Ca and increases from Ca to Ba

19 Which comparison is incorrect for reaction of group 1 and group 2 metals with NH_3?

	Group 1 (Na)	**Group 2** (Mg)
a.	Dissolve in liquid NH_3, no H_2 is evolved	Dissolve in liquid NH_3, no H_2 is evolved
b.	Blue solution due to solvated electrons	Blue solution due to solvated electrons
c.	Evaporation of ammonia from solutions of ammonia yields alkali metals	Evaporation of ammonia gives metals
d.	Concentrated solutions are blue	Concentrated solutions are bronze

Statement Type

***Direction** (Q. Nos. 20 and 21) This section is based on Statement I and Statement II. Select the correct answer from the codes given below.*
a. Both Statement I and Statement II are correct and Statement II is the correct explanation of Statement I
b. Both Statement I and Statement II are correct but Statement II is not the correct explanation of Statement I
c. Statement I is correct but Statement II is incorrect
d. Statement II is correct but Statement I is incorrect

20 **Statement I** The solubility of the halides in water decreases from Mg to Ra, but the fluorides show the reverse trend.

Statement II Lattice energies of fluorides decrease unusually rapidly.

21 **Statement I** $BeSO_4$ and $MgSO_4$ are water soluble.

Statement II The greater hydration energies of Be^{2+} and Mg^{2+} overcome the lattice energy factor.

Working Space

Matching List Type

Direction *(Q. Nos. 22-25) Choices for the correct combination of elements from Column I and Column II are given as options (a), (b), (c) and (d), out of which one is correct.*

22 Match the compounds (in Column I) with the colour (in Column II) which impart on ignition in flame and mark the correct option from the codes given below.

	Column I		Column II
i.	KCl	p.	Brick red
ii.	$CaCl_2$	q.	Crimson red
iii.	$SrCl_2$	r.	Apple green
iv.	$BaCl_2$	s.	Violet

Codes

	i	ii	iii	iv		i	ii	iii	iv
a.	p	q	s	r	b.	r	s	p	q
c.	q	r	p	s	d.	s	p	q	r

23 Match the different forms of calcium sulphate in Column I with their names in Column II and mark the correct option from the codes given below.

	Column I		Column II
i.	$CaSO_4$	p.	Alabaster
ii.	$CaSO_4 \cdot \frac{1}{2}H_2O$	q.	Gypsum
iii.	$CaSO_4 \cdot H_2O$	r.	Plaster of Paris
iv.	$CaSO_4 \cdot 2H_2O$	s.	Anhydrite
v.	$2CaSO_4 \cdot H_2O$		

Codes

	i	ii	iii	iv	v
a.	q	r	p	s	s
b.	s	r	p	q	r
c.	s	r	p	q	q
d.	p	r	s	q	r

24 Match the compounds in Column I with the specific properties in Column II and mark the correct option from the codes given below.

	Column I		Column II
i.	$BeCO_3$	p.	Temporary hardness
ii.	$BaCl_2$	q.	Permanent hardness
iii.	$MgSO_4$	r.	Decomposes readily
iv.	$Ca(HCO_3)_2$	s.	Gives green edged flame on heating

Codes

	i	ii	iii	iv		i	ii	iii	iv
a.	p	s	q	r	b.	r	s	p	q
c.	r	s	q	p	d.	q	s	r	p

25 Match the mineral names in Column I with the chemical formula in Column II and mark the correct option from the codes given below.

	Column I		Column II
i.	Magnesite	p.	$KCl \cdot MgCl_2 \cdot 6H_2O$
ii.	Dolomite	q.	$MgSO_4 \cdot H_2O$
iii.	Kieserite	r.	$MgCO_3$
iv.	Carnallite	s.	$MgCO_3 \cdot CaCO_3$

Codes

	i	ii	iii	iv		i	ii	iii	iv
a.	r	s	q	p	b.	p	q	r	s
c.	r	s	p	q	d.	q	r	p	s

SET - 2

Revisal Problems for JEE Advanced

Only One Option Correct Type

Direction *(Q. Nos. 1-6)* ***This section contains 6 multiple choice questions. Each question has four choices (a), (b), (c) and (d), out of which ONLY ONE option is correct.***

Working Space

1 Identify the correct statement.

a. Gypsum is obtained by heating plaster of Paris
b. Plaster of Paris can be obtained by hydration of gypsum
c. Plaster of Paris contains higher percentage of calcium than gypsum
d. Plaster of Paris is obtained from gypsum by oxidation

2 Conducting power of Be^{2+}, Mg^{2+}, Ca^{2+}, Sr^{2+} and Ba^{2+} is in the order

a. $Be^{2+} > Mg^{2+} > Ca^{2+} < Sr^{2+} < Ba^{2+}$
b. $Be^{2+} > Mg^{2+} > Ca^{2+} > Sr^{2+} > Ba^{2+}$
c. $Be^{2+} < Mg^{2+} < Ca^{2+} < Sr^{2+} < Ba^{2+}$
d. Equal conducting power

3 Which is maximum covalent in nature?

a. $BeCl_2$ b. $MgCl_2$
c. $CaCl_2$ d. $BaCl_2$

4 Lattice energy of group 2 compounds (oxides, carbonates and fluorides)

a. decreases as size of cation increases b. increases as size of cation increases
c. constant for a given type of anion d. All of the above are incorrect

5 Which is the most stable out of the following?

a. $[Be(H_2O)_4]^{2+}$ b. $[Mg(H_2O)_4]^{2+}$
c. $[Ca(H_2O)_4]^{2+}$ d. $[Sr(H_2O)_4]^{2+}$

6 Given, ΔH_f° of $CaCl_2 = -795\,kJ\,mol^{-1}$

ΔH_f° of $CaCl = -188\,kJ\,mol^{-1}$

Which of the following reaction(s) is/are spontaneous?

a. $CaCl_2(s) + Ca(s) \longrightarrow 2CaCl(s)$
b. $2CaCl(s) \longrightarrow CaCl_2(s) + Ca(s)$
c. Both (a) and (b)
d. None of the above

Statement Type

Direction *(Q. Nos. 7-10)* ***This section is based on Statement I and Statement II. Select the correct answer from the codes given below.***

a. Both Statement I and Statement II are correct and Statement II is the correct explanation of Statement I
b. Both Statement I and Statement II are correct but Statement II is not the correct explanation of Statement I
c. Statement I is correct but Statement II is incorrect
d. Statement II is correct but Statement I is incorrect

7 **Statement I** Beryllium is the strongest reducing agent in Group 2 series.
Statement II Standard reduction potential $Be^{2+}(aq) + 2e^- \longrightarrow Be(s)$ is less negative.

8 **Statement I** Beryllium compounds are largely ionic with low melting and boiling point.
Statement II Polarising power of Be^{2+} is larger than that of other alkaline earth metal ions.

Working Space

9 **Statement I** Solubility of sulphates decreases on moving down the group.

Statement II Heat of solution is dependent on hydration and lattice energy. Heat of solution determines solubility of a compound.

10 **Statement I** Alkaline earth metals (except Be) are strong reducing agents.

Statement II They lose their valence electrons easily.

One or More than One Options Correct Type

Direction (Q. Nos. 11-18) *This section contains 8 multiple choice questions. Each question has four choices (a), (b), (c) and (d), out of which ONE or MORE THAN ONE are correct.*

11 Consider the following changes:

$$Be \xrightarrow{I_1} Be^{+} \xrightarrow{I_2} Be^{2+} \xrightarrow{I_3} Be^{3+}$$

$$B \xrightarrow{I_1'} B^{+} \xrightarrow{I_2'} B^{2+} \xrightarrow{I_3'} B^{3+}$$

Select the correct statement(s).

a. $I_1 < I_2 < I_3$
b. $I_1' < I_2' < I_3'$
c. $I_1 > I_1'$ but $I_2 < I_2'$
d. $I_3 >> I_3'$

12 Be and Al show diagonal relationship, hence both have

a. same degree of electronegativity
b. amphoteric nature of oxides
c. approximately same charge/radius ratio
d. low solubility in H_2O

13 Select the correct statement(s).

a. Be and Mg are not decomposed by cold water but Ca, Sr and Ba are decomposed
b. Alkaline earth metals form nitrides at high temperatures and stability decreases down the group
c. Barium forms peroxides on reaction with O_2
d. BeH_2 is electron-deficient hydrides while Ca, Sr and Ba form ionic salt type hydrides

14 Select the correct statement(s) about Group 2 salts.

a. Thermal stability of sulphates and carbonates increases from Be^{2+} to Ba^{2+}
b. $Mg(NO_3)_2$ and $LiNO_3$ differ in the nature of decomposition products
c. The lattice energies of Group 2 salts are much higher than those of Group 1 salts
d. Hydration energies of Group 2 salts are higher than those of Group 1 salts

15 The reagent(s) used for softening the temporary hardness of water is (are)

a. $Ca_3(PO_4)_2$
b. $Ca(OH)_2$
c. Na_2CO_3
d. NaOCl

16 $MgSO_4$ on reaction with NH_4OH and Na_2HPO_4 forms a white crystalline precipitate of formula **[IIT JEE 2006]**

a. $Mg(NH_4)PO_4$
b. $Mg_3(PO_4)_2$
c. $MgCl_2 \cdot MgSO_4$
d. Na_2SO_4

17 NO_2, O_2 and oxide are formed on heating

a. $LiNO_3$
b. $NaNO_3$
c. $Mg(NO_3)_2$
d. $Ca(NO_3)_2$

18 Which of the following salts on decomposition gives acidic and basic oxides?

a. $CaCO_3$
b. Na_2CO_3
c. $LiNO_3$
d. $MgCO_3$

Matching List Type

Working Space

Direction *(Q. Nos. 19-21) Choices for the correct combination of elements from Column I and Column II are given as options (a), (b), (c) and (d), out of which one is correct.*

19 Match the species in Column I with the corresponding property(ies) in Column II and mark the correct option from the codes given below.

	Column I		Column II
i.	Be	p.	Diagonal relationship
ii.	Mg	q.	Maximum standard reduction potential ($E°$)
iii.	Ca	r.	Cation is maximum hydrated
iv.	Sr	s.	Cation has maximum ionic mobility
		t.	Ion is involved in the transmission of electrical impulses in human body

Codes

	i	ii	iii	iv
a.	p,r	p,q	t	s
b.	q,t	t	s	s
c.	t	p	q	r
d.	p,r	p,t	t	q,s

20 Match the commercial names in Column I with the chemical formula in Column II and mark the correct option from the codes given below.

	Column I		Column II
i.	Magnesia	p.	BeO
ii.	Epsom salt	q.	$(Ca(OH)_2$
iii.	Anhydrone	r.	$MgSO_4 \cdot 7H_2O$
iv.	Hydrolith	s.	CaH_2
v.	Beryllia	t.	MgO
vi.	Slaked lime	u.	$Mg(ClO_4)_2$

Codes

	i	ii	iii	iv	v	vi
a.	p	q	r	s	t	u
b.	t	r	u	s	p	q
c.	t	r	s	u	q	p
d.	r	p	q	s	t	u

21 High alumina cement is better than Portland cement. Match the contents of high alumina cement in Column I with their precentages in Column II and mark the correct option from the codes given below.

	Column I		Column II
i.	Al_2O_3	p.	40
ii.	CaO	q.	10
iii.	SiO_2	r.	15
iv.	Fe_2O_3	s.	35

Codes

	i	ii	iii	iv
a.	p	q	r	s
b.	p	r	q	s
c.	p	p	q	q
d.	q	p	p	q

Comprehension Type

Working Space

Direction *(Q. Nos. 22-27) This section contains 3 paragraphs, each describing theory, experiments, data, etc. Six questions related to the paragraphs have been given. Each question has only one correct answer among the four given options (a), (b), (c) and (d).*

Passage I

In water, beryllium salts are extensively hydrolysed.

22 $BeCl_2$ on hydrolysis exists as

a. $Be(OH)_2$ b. $[Be_2(OH)_6]^{2-}$
c. $[Be(OH)_4]^{2-}$ d. None of these is correct

23 If alkali is added to above solution, it breaks down to form

a. $[Be(OH)_4]^{2-}$ b. $[Be(OH)_6]^{4-}$
c. $Be(OH)_2$ d. BeO

Passage II

Calcium burns in nitrogen to produce a white powder (*A*) which dissolves in sufficient water to produce a gas (*B*) and an alkaline solution. The solution on exposure to air produces a thin solid layer of (*C*) on the surface.

24 *A* and *B* respectively are

a. $Ca(NO_2)_2$ and HNO_2 b. Ca_3N_2 and NH_3
c. CaN and NH_3 d. $Ca(NO_3)_2$ and NH_3

25 Thin solid layer (*C*) formed is of

a. $Ca(OH)_2$ b. $CaSO_4$ c. $CaCO_3$ d. $Ca(HCO_3)_2$

Passage III

Calcium oxide, CaO, is used to remove SO_2 from power-plant exhaust.

26 In what form, CaO removes SO_2?

a. $CaSO_3$ b. $CaSO_4$
c. $Ca(HSO_3)_2$ d. $Ca(HSO_4)_2$

27 Presence of SO_2 from power-plant exhaust can be detected by

a. $Cr_2O_7^{2-}$ b. MnO_4^-
c. Both (a) and (b) d. None of these

One Integer Value Correct Type

Direction *(Q. Nos. 28-34) This section contains 7 questions. When worked out will result in an integer from 0 to 9 (both inclusive).*

28 Coordination number of complexes of beryllium is...

29 Magnesium nitride is represented by Mg_xN_y, where $(x + y)$ is...

30 Magnesium is coordinated to *x* N atoms in chlorophyll. Value of *x* is...

31 In chlorophyll, Mg is coordinated to N-atoms present in rings. How many five membered rings are there in chlorophyll?

32 Magnesium allyide + $H_2O \longrightarrow$ Propyne.
How many total atoms are there in one formula unit of magnesium allyide?

33 96 g of Mg is burnt in air in a closed chamber. On analysis, 25% of Mg is converted into oxide and remaining Mg into other possible product. Residue is dissolved in H_2O and neutralised by H_2SO_4. Number of moles of H_2SO_4 required is...

34 $(Mg)_x(Ba)_y(CO_3)_z$ has following composition

$$Mg : 8.54\%,\ Ba : 48.75\%,\ CO_3 : 42.71\%$$

What is the value of $(x + y + z)$ if its molar mass is $281 g mol^{-1}$?

JEE Archive

(Compilation of Last 10 Years' Questions of JEE Main, AIEEE, JEE Advanced & IIT JEE)

1 Which of the following on thermal decomposition yields a basic as well as acidic oxide? **(AIEEE 2012)**

a. $NaNO_3$ b. $KClO_3$
c. $CaCO_3$ d. NH_4NO_3

2 One mole of magnesium nitride on reaction with an excess of water gives **(AIEEE 2004)**

a. one mole of ammonia
b. one mole of nitric acid
c. two moles of ammonia
d. two moles of nitric acid

3 Beryllium and aluminium exhibit many properties which are similar. But the two elements differ in **(AIEEE 2004)**

a. exhibiting maximum covalency in compounds
b. forming polymeric hydrides
c. forming covalent halides
d. exhibiting amphoteric nature in their oxides

4 The solubilities of carbonates decrease down the magnesium group due to a decrease in **(AIEEE 2003)**

a. lattice energies of solids
b. hydration energies of cations
c. inter ionic attraction
d. entropy of solution formation

5 The substance not likely to contain $CaCO_3$, is **(AIEEE 2003)**

a. a marble statue b. calcined gypsum
c. sea shells d. dolomite

6 In curing cement plasters, water is sprinkled from time to time. This helps in **(AIEEE 2003)**

a. keeping it cool
b. developing interlocking needle-like crystals of hydrated silicates
c. hydrating sand and gravel mixed with cement
d. converting sand into silicic acid

7 A metal *M* forms water soluble MSO_4 and inert *MO*. *MO* in aqueous solution forms insoluble $M(OH)_2$, soluble in NaOH. Metal *M* is **(AIEEE 2002)**

a. Be b. Mg
c. Ca d. Sr

8 Alum is widely used to purify water since **(AIEEE 2002)**

a. it forms complex with clay particles
b. it coagulates the mud particles
c. it exchanges Ca^{2+} and Mg^{2+} ions present in hard water
d. its sulphate ion is water purifier

9 $MgSO_4$ on rection with Na_2HPO_4) forms a white crystalline precipiatate. What is its formula? **(IIT JEE 2006, Only One Option Correct Type)**

a. $Mg(NH_4)PO_4$
b. $Mg_3(PO_4)_2$
c. $MgCl_2 \cdot MgSO_4$
d. $MgSO_4$

5. Group 13 Elements

DPP

DAILY PRACTICE PROBLEMS

DPP-1 Periodicity in Group 13

Only One Option Correct Type

Working Space

Direction (*Q. Nos. 1-9*) *This section contains 9 multiple choice questions. Each question has four choices (a), (b), (c) and (d), out of which ONLY ONE is correct.*

1 Ga(31) would be *d*-block element if

a. Aufbau rule is not followed
b. each orbital has three elements (electrons) instead of two, but other rules followed
c. Both (a) and (b)
d. None of the above

2 Basic nature of oxides of group 13 increases in the order

a. $B_2O_3 < Al_2O_3 < Ga_2O_3 < In_2O_3 < Tl_2O_3$
b. $Tl_2O_3 < In_2O_3 < Ga_2O_3 < Al_2O_3 < B_2O_3$
c. $B_2O_3 < Ga_2O_3 < Al_2O_3 < Tl_2O_3 < In_2O_3$
d. $B_2O_3 \approx Al_2O_3 < Ga_2O_3 \approx In_2O_3 < Tl_2O_3$

3 Stabilities of adducts formed with BX_3 is in the order

$$BX_3 + L(\text{Lewis base}) \longrightarrow (L \rightarrow BX_3)\text{ Adduct}$$

a. $BF_3 < BCl_3 < BBr_3$
b. $BBr_3 < BCl_3 < BF_3$
c. $BCl_3 < BBr_3 < BF_3$
d. $BF_3 < BBr_3 < BCl_3$

4 $B^{3+}(aq)$ ion is not known in aqueous solution because

a. the ionisation energy required would be very large
b. the small ion would be highly polarising
c. Both (a) and (b)
d. None of the above

5 Properties of group 13 (Ga, In, Tl) are quite different from those of group 1 and group 2. It is due to the

a. presence of d^{10} core electrons in Ga and In and d^{10} and f^{14} core electrons in Tl
b. larger size of Ga, In and Tl
c. increase in atomic number
d. variable oxidation number

6 $(IE)_1$ first ionisation energies of group 13 elements have been given in the figure. From this, $(IE)_1$ of Ga and Tl are higher than expected. It is due to

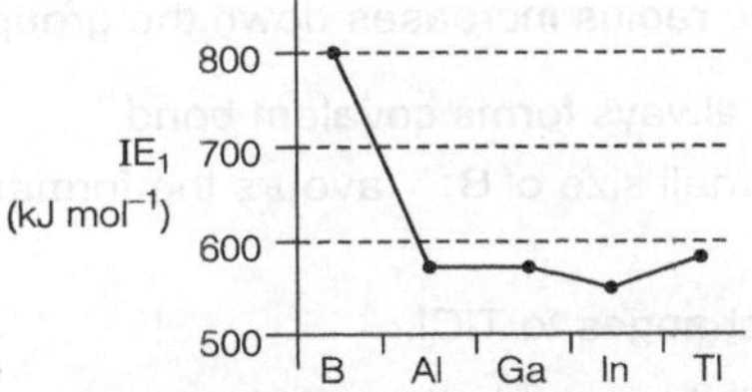

a. poor screening by $3d^{10}$ and $4f^{14}$ core electrons
b. larger radii of Ga and Tl
c. smaller electronegativity of Ga and Tl
d. None of the above

7 Boron cannot form which one of the following anions? **[AIEEE 2011]**

a. BF_6^{3-} b. BH_4^-
c. $[B(OH)_4]^-$ d. BO_2^-

8 The bond dissociation energy of B—F bond in BF_3 is 646 kJ mol^{-1} whereas that of C—F bond in CF_4 is 515 kJ mol^{-1}. The correct reason for higher B—F bond dissociation energy as compared to that of C—F bond is

a. Smaller size of B-atom as compared to that of C-atom
b. Stronger σ-bond between B and F in BF_3 as compared to that between C and F in CF_4
c. Significance *pπ-pπ* interaction between B and F in BF_3 whereas there is no probability of such interaction between C and F in CF_4
d. Lower degree of *pπ-pπ* interaction between B and F in BF_3 than that between C and F

9 Element $M + N_2 \xrightarrow{\Delta} \xrightarrow{H_2O} NH_3$. Element *M* belonging to group 13 can be

a. B or Al b. Ga or Al
c. Al or B or Mg d. Mg or B

Statement Type

Direction *(Q. Nos. 10-13) This section is based on Statement I and Statement II. Select the correct answer from the codes given below.*

a. Both Statement I and Statement II are correct and Statement II is the correct explanation of Statement I
b. Both Statement I and Statement II are correct but Statement II is not the correct explanation of Statement I
c. Statement I is correct but Statement II is incorrect
d. Statement II is correct but Statement I is incorrect

10 **Statement I** Down the group, the electronegativity first decreases from B to Al and then increases marginally from Ga to Tl.

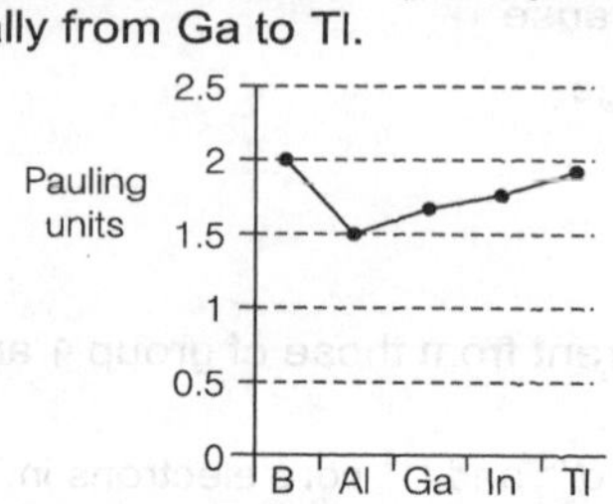

Statement II Presence of *d* and *f*- electrons decreases screening effect causing increase in force of attraction for the additional electron.

11 **Statement I** The atomic radius of gallium (135 pm) is less than that of aluminium (143 pm).
Statement II Atomic radius increases down the group from B to Tl.

12 **Statement I** Boron always forms covalent bond.
Statement II The small size of B^{3+} favours the formation of covalent bond. **[IIT JEE 2007]**

13 **Statement I** $TlCl_3$ changes to TlCl.
Statement II $TlCl_3$ is less stable than TlCl.

Working Space

One or More than One Options Correct Type

Direction (Q. Nos. 14-16) *This section contains 3 multiple choice questions. Each question has four choices (a), (b), (c) and (d), out of which ONE or MORE THAN ONE are correct.*

14 Select the correct statement(s).

a. When AlH_3 changes to AlH_4^-, bond angle increases
b. Stability of hydrides (MH_3) decreases as we move down the group
c. MF_3 is ionic while MCl_3, MBr_3 and MI_3 are predominantly covalent
d. Group 13 metal ions form square planar $[M(H_2O)_4]^{3+}$ (M = Al, Ga ...)

15 Select the correct statement(s).

a. Tl^{3+} salts are better oxidising agents
b. Ga^+ salts are better reducing agents
c. Tl^+ salts are better reducing agents
d. Ga^{3+} salts are better oxidising agents

16 Standard electrode potential values, $E°$ for Al^{3+}/Al is -1.66 V and that of Tl^{3+}/Tl is 1.26 V. Thus,

a. aluminium has high tendency to make $Al^{3+}(aq)$ ions, but Tl^{3+} is not stable in aqueous solution
b. Tl^{3+} salts are powerful oxidising agents
c. Al is more electropositive than Tl
d. Tl is more electropositive than Al

Matching List Type

Direction (Q. No. 17) *Choices for the correct combination of elements from Column I and Column II are given as options (a), (b), (c) and (d), out of which one is correct.*

17 α-Al_2O_3 (corundum) when coloured with metal ion impurities appears as gemstones and are as costly as gems. Match the metal impurity in Column I with the name of gem formed in Column II and select the answer from the codes given below.

	Column I		Column II
i.	Cr^{3+}	p.	Green sapphire
ii.	$Fe^{3+}/Fe^{2+}/Ti^{4+}$	q.	Oriental topaz
iii.	Cr^{3+}/V^{3+}	r.	Oriental emerald
iv.	Fe^{3+}	s.	Sapphire (blue)
v.	Co^{3+}	t.	Ruby

Codes

	i	ii	iii	iv	v
a.	p	q	r	s	t
b.	t	s	r	q	p
c.	t	r	p	q	s
d.	s	t	p	q	r

Comprehension Type

Working Space

Direction (Q. Nos. 18 and 19) *This section contains a paragraph, describing theory, experiments, data, etc. Two questions related to the paragraph have been given. Each question has only one correct answer among the four given options (a), (b), (c) and (d).*

Passage

Effective size of B^{3+} ion may be increased by complexing a number of suitable donor molecules around it.

18 Donor molecules increases effective size of B^{3+} by

a. reducing polarising power of the cation
b. compensating the ionisation energy by the strong ion-dipole interaction
c. Both (a) and (b)
d. None of the above

19 Suitable donor molecule that can increase effective size of B^{3+} is

a. C_5H_5N (pyridine)
b. AlH_3
c. $AlCl_3$
d. $GaCl_3$

One Integer Value Correct Type

Direction (Q. No. 20) *This section contains a question. When worked out will result in an integer from 0 to 9 (both inclusive).*

20 $[Me_3N-BBr_3]$ adduct on reaction with C_5H_5N gives a conducting solution due to the presence of ...*x*... ions. What is the value of *x*?

DPP-2 Compounds of Group 13

Only One Option Correct Type

Working Space

Direction *(Q. Nos. 1-7) This section contains 7 multiple choice questions. Each question has four choices (a), (b), (c) and (d), out of which ONLY ONE is correct.*

1 N_2 gas is passed into heated Mg and B. Product(s) formed is/are

a. Mg_3N_2 and BN
b. Mg_3N_2, BN and Mg_3B_2
c. Only Mg_3N_2
d. BN and Mg_3B_2

2 In the extraction of aluminium, purification of bauxite is done by using

a. Baeyer's process in which bauxite is dissolved in alkali solution
b. Serpeck's process in which bauxite is converted to nitride and then hydrolysed
c. Both (a) and (b)
d. None of the above

3 $2B(OH)_3 + 2NaOH \rightleftharpoons NaBO_2 + Na[B(OH)_4] + 2H_2O$

How can this reaction is made to proceed in forward direction? **[IIT JEE 2006]**

a. Addition of *cis*-1, 2-diol
b. Addition of borax
c. Addition of *trans*-1, 2-diol
d. Addition of Na_2HPO_4

4 At higher concentration, orthoboric acid exists as polymeric metaborate species. In the form of anion, it is

a. $B_2O_4^{2-}$
b. $[B_2O_4(OH)_4]^{6-}$
c. $[B_3O_3(OH)_4]^-$
d. $[B_2O_3(OH)_4]^{2-}$

5 Some of the common properties of boron, silicon and their compounds are given below. Which property(ies) is/are different?

I. Halides are easily hydrolysed.
II. Both form binary salts with metals in which these ions are anions.
III. Oxides are acidic in nature and are weakly acidic.
IV. B is metal and Si is non-metal.
V. Both are semi-conductors and dissolve in alkali.
VI. Both exists as two allotropic forms-amorphous and crystalline.

a. Both II and IV
b. III, IV and V
c. Only IV
d. Both IV and VI

6 Anhydrous aluminium chloride is covalent. Given,

$$Al(s) \longrightarrow Al^{3+}(g) + 3e^-, \Delta H = 5137 \text{ kJ mol}^{-1}$$

$$\Delta H_{\text{hydration}} \text{ for } Al^{3+} = -4665 \text{ kJ mol}^{-1}, \text{ for } Cl^- = -381 \text{ kJ mol}^{-1}$$

Aluminium chloride in aqueous solution exists as

a. $AlCl_3$
b. Al_2Cl_6
c. $[Al(H_2O)_6]^{3+}$ and Cl^-
d. Al^{3+} and Cl^-

Working Space

7 Amorphous boron is obtained by the

a. reduction of B_2O_3 with Na or Mg
b. reduction of BCl_3 with H_2
c. pyrolysis of BI_3
d. thermal decomposition of B_2H_6

Statement Type

Direction (Q. No. 8) *This section is based on Statement I and Statement II. Select the correct answer from the codes given below.*

a. Both Statement I and Statement II are correct and Statement II is the correct explanation of Statement I
b. Both Statement I and Statement II are correct but Statement II is not the correct explanation of Statement I
c. Statement I is correct but Statement II is incorrect
d. Statement II is correct but Statement I is incorrect

8 **Statement I** Boron is always covalent and does not form B^{3+} ions.

Statement II $B(g) \longrightarrow B^{3+}(g) + 3e^-$, ionisation reaction is exothermic.

One or More than One Options Correct Type

Direction (Q. Nos. 9-21) *This section contains 13 multiple choice questions. Each question has four choices (a), (b), (c) and (d), out of which ONE or MORE THAN ONE are correct.*

9 Select the correct statement(s) about boron nitride.

a. It is formed when $Na_2B_4O_7$ reacts with NH_4Cl
b. It is also formed when B_2H_6 reacts with NH_3 at low temperature
c. BN and C_2 are isoelectronic and isostructural in polymeric forms
d. BN is non-conducting while C_2 is conducting

10 B_2O_3 (boric anhydride) is a weak basic oxide. This property is shown by its reaction(s) with

a. H_3PO_4
b. CH_3COOH
c. C_6H_5OH
d. HNO_3

11 A coloured bead is formed with borax in borax-bead test (in oxidising flame or reducing flame) by

a. MnO
b. MgO
c. CoO
d. Cr_2O_3

12 Select the correct statement(s) about the structure of aluminate ion based on pH and concentration.

a. It exists as tetrahedral $[Al(OH)_4]^-$ in dilute solution above pH 13
b. It polymerises using OH bridges such that each Al is octahedrally coordinated between pH 8 and 12
c. It exists as a dimer $[(OH)_3Al—O—Al(OH)_3]^{2-}$ in concentrated solution above 1.5 M and above pH 13
d. It exists as $[Al(H_2O)_6]^{3+}$ at a pH about 2

13 Select the correct statement(s) about alum.

$$M_2SO_4 \cdot M_2'(SO_4)_3 \cdot 24H_2O$$

a. $[M(H_2O)_6]^+$, $[M'(H_2O)_6]^{3+}$ and SO_4^{2-} are in the ratio of 1 : 1 : 2
b. M^+ is any alkali metal ion
c. $(M')^{3+}$ can be Al^{3+}, Fe^{3+}, Cr^{3+}, etc
d. They are used as mordant in dyes

Working Space

14 Select the correct statement(s).

a. Aluminium monochlorides are formed by the reactions of the trihalides with aluminium at 1300 K followed by rapid cooling
b. Aluminium monochloride (red) is formed when aluminium reacts with HCl at 1200 K
c. Aluminium monochloride disproportionates at room temperature
d. Ga_2Cl_4 exists as $Ga^+[GaCl_4]^-$ and is diamagnetic and has crystalline structure

15 Select the correct statement(s) about borazine.

a. It is isoelectronic with benzene
b. It has planar cyclic structure with all B—N distances equal
c. There is complete delocalisation of π electrons
d. Borazine is more reactive than benzene

16 In Hall-Heroult process,

a. Al is formed at anode and O_2 collects at cathode
b. Al is formed at cathode and O_2 collects at anode
c. CO_2 or O_2 can be collected at anode
d. CO_2 or O_2 can be collected at cathode

17 Select the correct statement(s).

a. B_2H_6 has B—H bonds of equal length
b. AlH_3 is a colourless solid and is polymeric containing Al—H—Al bridges
c. $LiAlH_4$ is an ionic compound
d. AlX_3 (X = Cl, Br, I) have low melting point, are covalent and have the halogen-bridged dimeric structure

18 B_2H_6 is shown as

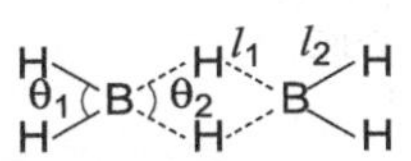

Select the correct statement(s) about the above structure.

a. Bond angle $\theta_1 > \theta_2$
b. Bond angle $\theta_1 < \theta_2$
c. Bond length $l_1 = l_2$
d. Bond length $l_1 > l_2$

19 In the reaction $2X^- + B_2H_6 \longrightarrow [BH_2(X)_2]^+[BH_4]^-$, the amine(s) X is/are **[IIT JEE 2009]**

a. NH_3
b. CH_3NH_2
c. $(CH_3)_2NH$
d. $(CH_3)_3N$

20 The correct statement(s) for orthoboric acid is/are **[JEE Advanced 2014]**

a. It behaves as a weak acid in water due to self ionisation
b. Acidity of its aqueous solution increases upon the addition of ethylene glycol
c. It has three dimensional structure due to hydrogen bonding
d. It is a weak electrolyte in water

21 Pure crystalline boron is obtained by

a. thermal decomposition of B_2H_6
b. pyrolysis of BI_3
c. reduction of BCl_3 with H_2
d. reduction of B_2O_3 with Mg or Na

Matching List Type

Working Space

Direction *(Q. Nos. 22 and 23) Choices for the correct combination of elements from Column I and Column II are given as options (a), (b), (c) and (d), out of which one is correct.*

22 Match the property given in Column I with the related element(s) given in Column II and select the answer from the codes given below.

	Column I		Column II
i.	Unusual low melting point	p.	B
ii.	Extremely toxic	q.	Al
iii.	Forms acidic oxide	r.	Ga
iv.	The most stable in +1 state	s.	In
v.	Semiconductor	t.	Tl
vi.	Forms molecular fluoride		

Codes

	i	ii	iii	iv	v	vi
a.	r	t	p	t	p	p
b.	r	t	p	q	p	t
c.	r	t	p	q	r	s
d.	p	r	s	q	t	p

23 Match the names of the minerals of aluminium in Column I with their formulae in Column II and select the answer from the codes given below.

	Column I		Column II
i.	Bauxite	p.	Al_2O_3
ii.	Kaolin	q.	$KAlSi_3O_8$
iii.	Cryolite	r.	Na_3AlF_6
iv.	Feldspar	s.	$Al_2O_3 \cdot 2SiO_2 \cdot 2H_2O$
v.	Orthoclase	t.	$K_2O \cdot Al_2O_3 \cdot 6SiO_2$
vi.	Corundum	u.	$Al_2O_3 \cdot 2H_2O$

Codes

	i	ii	iii	iv	v	vi
a.	p	q	r	t	s	u
b.	u	s	r	t	q	p
c.	u	s	r	t	p	q
d.	r	s	p	q	t	u

Comprehension Type

Direction *(Q. Nos. 24-30) This section contains 3 paragraphs, each describing theory, experiments, data, etc. Seven questions related to the paragraphs have been given. Each question has only one correct answer among the four given options (a), (b), (c) and (d).*

Passage I

Borax swells on heating and then sets to a glassy transparent bead.

24 Transparent bead (glassy mass) consists of

a. $NaBO_2$
b. B_2O_3
c. Both (a) and (b)
d. None of these

25 Fusion with CuO forms hot green coloured beads in oxidising flame of

a. Cu_2BO_3
b. $CuBO_2$
c. $Cu(BO_2)_2$
d. $Cu_2B_4O_7$

26 Green coloured bead changes to ...... coloured bead in reducing flame due to the formation of ......

a. red, $Cu(BO_2)_2$
b. red, Cu
c. blue, $Cu_2B_4O_7$
d. black, CuO

Working Space

Passage II

Diborane combines with LiH as well as Na.

27 Compound formed with LiH can

a. reduce —CHO, —COOH, —COCl to 1° alcohol

b. reduce $(\gt C{=}C\lt)$ to alkane

c. reduce $(—C{\equiv}C—)$ to alkane

d. All of the above

28 Compound(s) formed with Na is/are

a. $NaBH_4$
b. $Na[B_3H_8]$
c. Both (a) and (b)
d. None of these

Passage III

Compound *X* on reduction with $LiAlH_4$, gives a hydride *Y* containing 21.72% hydrogen along with other products. The compound *Y* reacts with air explosively resulting in boron trioxide.

29 Compound *X* is

a. BCl_3
b. B_2Cl_6
c. $BHCl_2$
d. BH_3

30 Compound *Y* is

a. BH_3
b. BH_4^-
c. B_2H_6
d. BH_2^-

One Integer Value Correct Type

Direction *(Q. Nos. 31-35) This section contains 5 questions. When worked out will result in one integer from 0 to 9 (both inclusive).*

31 When borax is dissolved in water, there is the formation of two units each of $B(OH)_3$ and $[B(OH)_4]^-$ per mole of borax dissolved. How many moles of HCl are required to neutralise total moles of $B(OH)_3$ and $[B(OH)_4]^-$ from one mole of borax?

32 $AlCl_3$ is used as a Lewis acid in Friedel-Crafts reaction

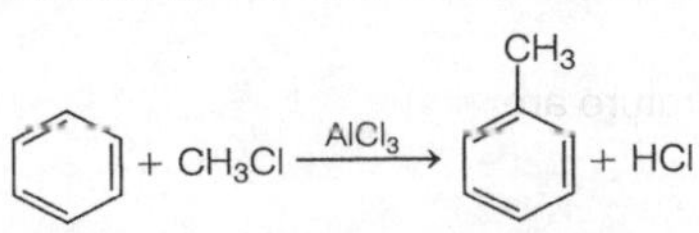

How many bonds does Al form with Cl in the intermediate?

33 Total number of N and B atoms in borazine is ......

34 In borazine, each N-atom is bonded to *x* B-atoms. What is the value of *x*?

35 Borax is $Na_2B_4O_7 \cdot 10H_2O$. How many (OH) units does it contain?

SET - 1

Revisal Problems for JEE Main

One Option Correct Type

Direction (Q. Nos. 1-25) This section contains 25 multiple choice questions. Each question has four choices (a), (b), (c) and (d), out of which ONLY ONE is correct.

Working Space

1 Borax is actually a salt of

a. orthoboric acid
b. metaboric acid
c. pyroboric acid
d. None of these

2 The structure of diborane (B_2H_6) contains **[AIEEE 2005]**

a. four ($2c$ - $2e^-$) bonds and four ($3c$ - $2e^-$) bonds
b. two ($2c$ - $2e^-$) bonds and two ($3c$ - $3e^-$) bonds
c. two ($2c$ - $2e^-$) bonds and four ($3c$ - $2e^-$) bonds
d. four ($2c$ - $2e^-$) bonds and two ($3c$ - $2e^-$) bonds

3 The type of hybridisation of boron in diborane is **[AIEEE 2007]**

a. sp
b. sp^2
c. sp^3
d. sp^3d^2

4 Heating of an aqueous solution of aluminium chloride to dryness will give **[AIEEE 2005]**

a. $Al(OH)Cl_2$
b. Al_2O_3
c. Al_2Cl_6
d. $AlCl_3$

5 Alum is widely used to purify water since **[AIEEE 2002]**

a. it forms complex with clay particles
b. it coagulates the mud particles
c. it exchanges Ca^{2+} and Mg^{2+} ions present in hard water
d. its sulphate ion is water purifier

6 Aluminium is made passive by the formation of oxide layer. The layer is so useful that in industry, it is purposely increased by an electrolytic process called anodising. This is done by

a. conc. HNO_3
b. electrolysing dil. H_2SO_4 with Al as anode
c. electrolysing H_2O with Al as cathode
d. electrolysing H_2O with Al as anode

7 Physical states of BX_3 (X = F, Cl, Br, I) at room temperature are

	BF_3	BCl_3	BBr_3	BI_3
a.	gas	gas	volatile liquid	solid
b.	gas	gas	gas	liquid
c.	gas	volatile liquid	solid	solid
d.	solid	solid	solid	solid

8 Borates can be tested by converting them into

a. BF_3 which gives green-coloured flame
b. $(C_2H_5O)_3B$ which gives green-edged flame
c. Both (a) and (b) are correct
d. None of the above is correct

9 Select the correct statement(s) about the adduct $B_2H_6 \cdot 2NH_3$.

a. It is an ionic compound
b. It forms borazine on heating
c. Both (a) and (b) are correct
d. None of the above is correct

Working Space

10 Pure boron is converted to H_3BO_3 by

a. hot concentrated H_2SO_4 and HNO_3
b. hot Na_2O_2 solution
c. Both (a) and (b)
d. pure boron is unreactive

11 The reaction $2M + 6NH_3 \longrightarrow 2M(NH_2)_3 + 3H_2$, is not shown by

a. B
b. Al
c. Ga
d. Tl

12

$$Mg + B \longrightarrow X$$
$$X + H_2SO_4 \longrightarrow Y$$

X and *Y* respectively are

a. $Mg(BO_2)_2$ and BH_3
b. Mg_3B_2 and B_2H_6
c. Mg_3B_2 and H_3BO_3
d. Mg_3B_2 and HBO_2

13 Angle Cl—*M*—Cl (*M* = B, P, As, Bi) will be in the order

a. $BCl_3 > PCl_3 = AsCl_3 = BiCl_3$
b. $BCl_3 > PCl_3 > AsCl_3 > BiCl_3$
c. $BCl_3 < PCl_3 = AsCl_3 = BiCl_3$
d. $BCl_3 < PCl_3 < AsCl_3 < BiCl_3$

14 H_3BO_3 is

a. monobasic acid and weak Lewis acid
b. monobasic and weak Bronsted acid
c. monobasic and strong Lewis acid
d. tribasic and weak Bronsted acid

15 Borax is used as a primary standard in the titrations against acids.

$$Na_2B_4O_7 \cdot 10H_2O + 2HCl \longrightarrow 2NaCl + H_3BO_3 + 5H_2O$$

Indicator to be used in this titration is

a. phenolphthalein
b. methyl orange
c. methylene blue
d. borax itself

16 H_3BO_3 and HBO_2 do not differ in

a. oxidation number
b. basicity
c. melting point
d. structure

17 Light-emitting diodes (LED), laser diodes and memory chips of computers are made of

a. gallium
b. aluminium
c. arsenic
d. gallium-arsenide

18 Boron carbide is used

a. in nuclear reactor to absorb neutrons
b. as an abrasive
c. Both (a) and (b)
d. None of the above

19 Soda free glass fibres are made of

a. $H_3BO_3 \cdot B_2O_3$ and $Ca_3(BO_3)_2$
b. H_3BO_3 and B_2O_3
c. H_3BO_3
d. B_2O_3

20 Na_3AlF_6 is added to Al_2O_3 to

a. improve the electrical conductivity of the cell
b. increase the rate of production
c. increase the melting point
d. decrease the electrical conductivity

21 Reactivity of borazine (I) is greater than that of benzene (II), since

a. I is polar compound
b. I is non-polar compound
c. I is electron-deficient
d. I has localised electrons in it

Working Space

22 Anion containing both three and four coordinated boron is

a. $[B(OH)_4]^-$ b. $[B_3O_3(OH)_4]^-$
c. $[B_4O_5(OH)_4]^{2-}$ d. $(BO_3)BO_4^{4-}$

23 On passing CO_2 into aqueous solution containing AlO_2^-,

a. $Al_2(CO_3)_2$ is formed
b. $Al(OH)_3$ is precipitated
c. $[Al(OH)_4]^-$ is formed
d. colloidal $Al(OH)_3$ is formed

24 Borax $(Na_2B_4O_7)$ is formed when

a. orthoboric acid is neutralised by sodium carbonate
b. colemanite is boiled with concentrated solution of sodium carbonate
c. sulphur dioxide is passed into sodium metaborate solution
d. All of the above cases

25 CO_2 is passed into mother liquor containing sodium metaborate when

a. B_2O_3 and Na_2CO_3 are formed
b. $Na_2B_4O_7$ and Na_2CO_3 are formed
c. $NaBO_2$ and Na_2CO_3 are formed
d. Na_3BO_3 and Na_2CO_3 are formed

Statement Type

Direction (Q. Nos. 26-33) ***This section is based on Statement I and Statement II. Select the correct answer from the codes given below.***

a. Both Statement I and Statement II are correct and Statement II is the correct explanation of Statement I
b. Both Statement I and Statement II are correct but Statement II is not the correct explanation of Statement I
c. Statement I is correct but Statement II is incorrect
d. Statement II is correct but Statement I is incorrect

26 **Statement I** $Al(OH)_3$ is amphoteric in nature.

Statement II (Al—O) and (O—H) bonds can be broken with equal ease in $Al(OH)_3$.

27 **Statement I** In Al_2Cl_6, all (Al—Cl) bonds are equivalent.

Statement II $AlCl_3$ is a Lewis acid.

28 **Statement I** In water, orthoboric acid behaves as a weak monobasic acid.

Statement II In water, orthoboric acid acts as a proton donor. **[IIT JEE 2007]**

29 **Statement I** NaOH and HNO_3 can be stored in a vessel made of aluminium.

Statement II Aluminium is made passive by HNO_3.

30 **Statement I** All (B—F) bonds in BF_3 are equivalent.

Statement II Dative bond between B and F (with double bond character) is delocalised.

31 **Statement I** Thermite reaction can be stopped by water.

Statement II By high temperature of thermite reaction (about 1000°C), H_2O is decomposed by Al giving H_2.

32 **Statement I** $Al(OH)_3$ is precipitated from its salts by passing CO_2.

Statement II $Al(OH)_3$ is amphiprotic.

33 **Statement I** B_2Cl_6 and Al_2Cl_6 are stable.

Statement II $AlCl_3$ is a Lewis acid.

Working Space

Matching List Type

***Direction** (Q. Nos. 34 and 35) Choices for the correct combination of elements from Column I and Column II are given as options (a), (b), (c) and (d), out of which one is correct.*

34 Match the salts of boron given in Column I with their formulae given in Column II and select the answer from the codes given below.

	Column I		Column II
i.	Prismatic borax obtained by crystallising the salt at ordinary temperature.	p.	$Na_2B_4O_7$
ii.	Octahedral borax obtained by crystallising the salt above 370 K.	q.	$Na_2B_4O_7 \cdot 8H_2O$
iii.	Borax glass	r.	$Na_2B_4O_7 \cdot 5H_2O$
iv.	Jeweller's borax	s.	$Na_2[B_4O_5(OH)_4] \cdot 8H_2O$

Codes

	i	ii	iii	iv
a.	p	q	r	r
b.	r	p	q	q
c.	s	r	p	r
d.	s	p	q	r

35 H_3BO_3 changes to different forms under different conditions. Match the conditions undertaken in Column I with the products formed in Column II and select the answer from the codes given below.

	Column I		Column II
i.	$H_3BO_3 \xrightarrow{100°C}$	p.	$[B_3O_3(OH)_4]^-$
ii.	$H_3BO_3 \xrightarrow{140°C}$	q.	HBO_2
iii.	$H_3BO_3 \xrightarrow{\text{Red hot}}$	r.	$H_2B_4O_7$
iv.	$H_3BO_3(aq) \xrightarrow{\text{Concentrated}}$	s.	B_2O_3

Codes

	i	ii	iii	iv
a.	s	p	q	r
b.	p	s	q	r
c.	q	r	s	p
d.	r	q	p	s

SET - 2

Revisal Problems
for JEE Advanced

One Option Correct Type

Direction (Q. Nos. 1-10) *This section contains 10 multiple choice questions. Each question has four choices (a), (b), (c) and (d), out of which ONLY ONE is correct.*

Working Space

1 An alloy of boron and aluminium is allowed to react with NH_3. Products formed are

a. BN, AlN and H_2
b. $B(NH_2)_3$, $Al(NH_2)_3$ and H_2
c. BN, $Al(NH_2)_3$ and H_2
d. $B(NH_2)_3$, AlN and H_2

2 The barriers to rotation around the (B—N) bond are found to decrease as

a. $H_2B—NR_2 > HB—(NR_2)_2 > B(NR_2)_3$
b. $B(NR_2)_3 > HB—(NR_2)_2 > H_2B—NR_2$
c. $B(NR_2)_3 = HB—(NR_2)_2 = H_2B—NR_2$
d. $B(NR_2)_3 > H_2B—NR_2 > B(NR_2)_3$

3 Borax is used as a buffer, since

a. its aqueous solution contains equal moles of weak acid and its salt
b. its aqueous solution contains equal moles of strong acid and its salt
c. it is easily available
d. borax is used as buffer-this statement is incorrect

4 $E^\circ_{Al^{3+}/Al} = -1.66$ V; $E^\circ_{SO_4^{2-}/SO_2} = 1.17$ V; $E^\circ_{H^+/H_2/Pt} = 0.00$ V

On reaction of Al with H_2SO_4 (conc.),

a. SO_2 gas is formed in preference to H_2 gas
b. H_2 gas is formed in preference to SO_2 gas
c. Both gases are formed at the same time
d. Al does not react with H_2SO_4

5 In H_2O, Ga^+ changes to

a. Ga^{3+}
b. Ga
c. Ga and Ga^{3+}
d. Ga^{2+} and Ga^{3+}

6 For group 13 elements, E° values are ($M^{3+} + 3e^- \longrightarrow M$)

$$E^\circ_{Al^{3+}/Al} = -1.66\text{ V}, E^\circ_{In^{3+}/In} = -0.34\text{ V}$$
$$E^\circ_{Ga^{3+}/Ga} = -0.56\text{ V}, E^\circ_{Tl^{3+}/Tl} = +1.26\text{ V}$$

Thus, most stable $M^{3+}(aq)$ ion is

a. Al^{3+}
b. In^{3+}
c. Ga^{3+}
d. Tl^{3+}

7 Boron also forms halides B_2X_4. Select the correct statement(s) about B_2Cl_4.

a. There is free rotation about (B—B) bond
b. It adopts a non-eclipsed conformation in gaseous and liquid state
c. It adopts a planar structure in the solid state because of crystal force and ease of packing
d. All of the above are correct statements

8 Consider the following statements for diborane.

I. Boron is approximately sp^3-hybridised.
II. B—H—B angle is 180°.
III. There are two terminal (B—H) bonds for each B-atom.
IV. There are only 12 bonding electrons available.

Select the correct statements.

a. I, II and III
b. II, III and IV
c. I, III and IV
d. I, II and IV

Working Space

9 In the following reactions,

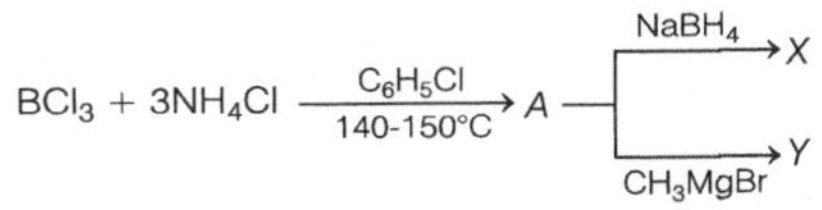

products X and Y respectively are

a. $B_3N_3H_6$ and $Me_3B_3N_3H_3$
b. $Me_3B_3N_3H_3$ and $B_3N_3H_6$
c. $B_3N_3H_6$ and $B_3N_3H_6$
d. $Me_3B_3N_3H_3$ and $Me_3B_3N_3H_3$

10 In the context of the Hall-Heroult process for the extraction of Al, which of the following statements is false? **[JEE Main 2015]**

a. CO and CO_2 are produced in this process
b. Al_2O_3 is mixed with CaF_2 which lowers the melting point of the mixture and increases the conductivity
c. Al^{3+} is reduced at the cathode to form Al
d. Na_3AlF_6 serves as the electrolyte

Statement Type

Direction (Q. Nos. 11 and 12) *This section is based on Statement I and Statement II. Select the correct answer from the codes given below.*

a. Both Statement I and Statement II are correct and Statement II is the correct explanation of Statement I
b. Both Statement I and Statement II are correct but Statement II is not the correct explanation of Statement I
c. Statement I is correct but Statement II is incorrect
d. Statement II is correct but Statement I is incorrect

11 **Statement I** The polarity of (B—X) bond is in order

$$B—F < B—Cl < B—Br$$

Statement II Lewis acidity shows the order

$$BF_3 > BCl_3 > BBr_3$$

12 **Statement I** $Al(OH)_3$ finds use in 'fire extinguisher'.

Statement II The hydroxide has the virtue of stabilising the foam, which covers the fire.

One or More than One Options Correct Type

Direction (Q. Nos. 13-21) *This section contains 9 multiple choice questions. Each question has four choices (a), (b), (c) and (d), out of which ONE or MORE THAN ONE are correct.*

13 $[M(H_2O)_6]^{3+}$ (where, $M =$ Al, Ga) can exist as

a. $[M(H_2O)_5(OH^-)]^{2+}$
b. $[M(H_2O)_4(OH^-)_2]^+$
c. $[M(H_2O)_3(OH^-)_3]$
d. $[M(H_2O)_2(OH^-)_4]^-$

14 The reaction, $2Al(s) + \frac{3}{2}O_2 \longrightarrow Al_2O_3(s)$

a. is exothermic
b. is called thermite reaction and is used in welding
c. can be used to reduce oxides into metals
d. is endothermic

15 Boron resembles aluminium in many properties. Select the property(ies) which is/are different.

	Boron	Aluminium
a.	B_2O_3 is an acidic oxide	Al_2O_3 is an amphoteric oxide
b.	H_3BO_3 is a weak monobasic acid	$Al(OH)_3$ is amphoteric
c.	Halides are easily hydrolysed	Halides are not easily hydrolysed
d.	Boranes are gaseous compounds	$(AlH_3)_n$ is a polymeric solid

Working Space

16 Hydroboronation oxidation of $CH_3CH{=}CH_2$

a. converts it to 1° alcohol
b. converts it to 2° alcohol
c. uses BH_3 followed by the oxidation with H_2O_2 /OH^-
d. uses BF_3 followed by the oxidation with H_2O_2 /OH^-

17 Select the correct statement(s).

a. Kernite is the ore of aluminium
b. ^{10}B and ^{11}B are the isotopes of boron
c. Bauxite and cryolite are the minerals of aluminium
d. Aluminium is the most abundant metal and the third most abundant element in the earth's crust

18 BF_3 can accept lone pair to complete octet of B from which of the following?

a. $AlCl_3$
b. NH_3
c. Et_2O
d. Me_3N

19 BF_3 and BCl_3 differ in

a. products of hydrolysis
b. Lewis acid behaviour
c. melting and boiling points
d. structure

20 Select the species from which B_2H_6 (diborane) can be prepared.

a. Mg_3B_2
b. BF_3
c. $NaBH_4$
d. B_2O_3

21 For thallium metal, $E^\circ_{Tl^{3+}/Tl} = 1.26$ V and $E^\circ_{Tl^{+}/Tl} = -0.34$ V,

Thus,

a. Tl is oxidised to Tl^+
b. Tl is oxidised to Tl^{3+}
c. Tl^+ disproportionates to Tl^{3+} and Tl
d. Tl^{3+} is reduced to Tl^+

Matching List Type

Direction (Q. No. 22) ***Choices for the correct combination of elements from Column I and Column II are given as options (a), (b), (c) and (d), out of which one is correct.***

22 $AlCl_3$ dimerises to Al_2Cl_6 to complete its octet (Al) as shown below

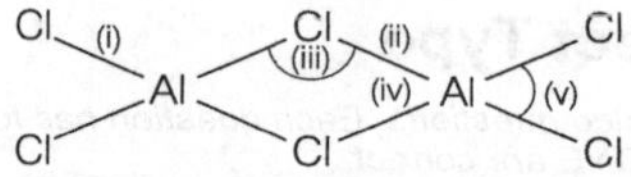

Compare bond lengths (i) and (ii), (given in pm) and bond angles (iii), (iv) and (v) (given in degree) and select the correct answer from codes given below.

	Column I		Column II
i.	Bond length	p.	206
ii.	Bond length	q.	101
iii.	Bond angle	r.	221
iv.	Bond angle	s.	79
v.	Bond angle	t.	118

Codes

	i	ii	iii	iv	v
a.	p	q	r	s	t
b.	p	r	q	s	t
c.	t	q	r	s	p
d.	s	t	q	r	p

Working Space

Comprehension Type

Direction (Q. Nos. 23-28) This section contains 3 paragraphs each describing theory, experiments, data. etc. Six questions related to the paragraphs have been given. Each questions has only one correct answer among the four given options (a), (b), (c) and (d).

Passage I

AlF_3 is insoluble in anhydrous BF_3 but soluble in KF.

23 What happens when AlF_3 is mixed with KF?

a. K_3AlF_6 is formed
b. $KAlF_4$ is formed
c. $AlKF_4$ is formed
d. $Al(KF_2)_3$ is formed

24 When BF_3 is added to a solution of AlF_3 in KF,

a. $AlKF_4$ is formed
b. $BF_3 \cdot AlF_3$ is formed
c. AlF_3 is precipitated
d. $Al(BF_6)$ is formed

Passage II

In the following reaction,

$$\text{Metal}(A) + N_2 \xrightarrow{\Delta} (B) \xrightarrow{H_2O} (C) + (D)$$

C is amphoteric and gas *D* gives white fumes in HCl. *B* contains 34.15% nitrogen and is a binary salt of molar mass 41 g mol^{-1}.

25 *A* and *B* respectively are

a. B and BN
b. Al and AlN
c. Mg and Mg_3N_2
d. Li and Li_3N

26 *C* and *D* respectively are

a. $Al(OH)_3$ and NH_3
b. $B(OH)_3$ and NH_3
c. $Mg(OH)_2$ and NH_3
d. $[Al(H_2O)_6]^{3+}$ and NH_3

Passage III

Reaction of ammonia with diborane gives initially $B_2H_6 \cdot 2NH_3$.

27 $B_2H_6 \cdot 2NH_3$ changes to

a. $BH_3 \cdot NH_3$ by symmetrical cleavage
b. $BH_3 \cdot NH_3$ by unsymmetrical cleavage
c. $[BH_2(NH_3)_2]^+ [BH_4]^-$ by unsymmetrical cleavage
d. $[BH_2(NH_3)_2]^+ [BH_4]^-$ by symmetrical cleavage

28 Heating of $B_2H_6 \cdot 2NH_3$ at high temperature gives

a. BN (Boron nitride)
b. $B_3N_3H_6$ (Borazine-inorganic benzene)
c. BH_3 and NH_3
d. $[BH_2(NH_3)_2]^+ [BH_4]^-$

One Integer Value Correct Type

Direction (Q. Nos. 29-34) This section contains 6 questions. When worked out will result in an integer from 0 to 9 (both inclusive).

29 Number of H-bonds formed by one unit of orthoboric acid is ......

30 Me_3N(3° amine) forms adducts with B_2H_6. Adducts splits into *x* fragments by homolytic cleavage. What is the value of *x*?

31 RNH_2 causes unsymmetrical cleavage of B_2H_6

$$B_2H_6 + nRNH_2 \longrightarrow [BH_2(RNH_2)_n][BH_4]$$

What is the value of *n*?

32 The value of *x* in the molecular formula $Be_xAl_2Si_6O_{18}$ is ...... **[IIT JEE 2010]**

33 Based on structure, there are total of *x* tetrahedral and triangular units in one unit of borax. What is the value of *x*?

34 Maximum number of O-atoms attached to one boron atom in one unit of borax is *y*. What is the value of *y*?

JEE Main & AIEEE Archive

(Compilation of Last 13 Years' Questions)

1 Boron can't form which one of the following anions? **(AIEEE 2011)**

a. BF_6^{3-} b. BH_4^-

c. $[B(OH)_4]^-$ d. BO_2^-

2 The bond dissociation energy of B—F in BF_3 is 646 kJ mol^{-1} whereas that of C—F in CF_4 is 515 kJ mol^{-1}. The correct reason for higher B—F bond dissociation energy as compared to that of C—F is **(AIEEE 2009)**

a. Smaller size of B-atom as compared to that of C-atom

b. Stronger σ-bond between B and F in BF_3 as compared to that between C and F in CF_4

c. Significant $p\pi$-$p\pi$ interaction between B and F in BF_3 whereas there is no possibility of such interaction between C and F in CF_4

d. Lower degree of $p\pi$-$p\pi$ interaction between B and F in BF_3 than that between C and F in CF_4

3 Which one of the following is the correct statement? **(AIEEE 2007)**

a. Boric acid is a protonic acid

b. Beryllium exhibits coordination number of six

c. Chlorides of both beryllium and aluminium have bridged chloride structures in solid phase

d. $B_2H_6 \cdot 2NH_3$ is known as 'inorganic benzene'

4 The type of hybridisation of boron in diborane is **(AIEEE 2007)**

a. sp-hybridisation

b. sp^2-hybridisation

c. sp^3-hybridisation

d. sp^3d^2-hybridisation

5 The structure of diborane (B_2H_6) contains **(AIEEE 2005)**

a. four $2c$ - $2e$ bonds and four $3c$ - $2e$ bonds

b. two $2c$ - $2e$ bonds and two $3c$ - $3e$ bonds

c. two $2c$ - $2e$ bonds and four $3c$ - $2e$ bonds

d. four $2c$ - $2e$ bonds and two $3c$ - $2e$ bonds

6 Heating of an aqueous solution of aluminium chloride to dryness will give **(AIEEE 2005)**

a. $Al(OH)Cl_2$ b. Al_2O_3

c. Al_2Cl_6 d. $AlCl_3$

7 Aluminium chloride exists as dimer, Al_2Cl_6 in solid state as well as in solution of non-polar solvents such as benzene. When dissolved in water, it gives **(AIEEE 2004)**

a. $Al^{3+} + 3Cl^-$ b. $[Al(H_2O)_6]^{3+} + 3Cl^-$

c. $[Al(OH)_6]^{3-} + 3HCl$ d. $Al_2O_3 + 6HCl$

JEE Advanced & IIT JEE Archive

(Compilation of Last 10 Years' Questions)

1 The correct statement(s) for orthoboric acid is/are

(2014 Adv., One or More than One Options Correct Type)

a. It behaves as a weak acid in water due to self ionisation
b. Acidity of its aqueous solution increases upon addition of ethylene glycol
c. It has a three dimensional structure due to hydrogen bonding
d. It is a weak electrolyte in water

Direction *(Q. Nos. 2 and 3) This section is based on Statement I and Statement II. Select the correct answer from the codes given below.*

a. Both Statement I and Statement II are correct and Statement II is the correct explanation of Statement I
b. Both Statement I and Statement II are correct but Statement II is not the correct explanation of Statement I
c. Statement I is correct but Statement II is incorrect
d. Statement II is correct but Statement I is incorrect

2 **Statement I** Boron always forms covalent bond.
Statement II The small size of B^{3+} favours the formation of covalent bond.

(2007, Statement Type)

3 **Statement I** In water, orthoboric acid behaves as a weak monobasic acid.
Statement II In water, orthoboric acid acts as a proton donor. **(2007, Statement Type)**

4 $B(OH)_3 + NaOH \rightleftharpoons NaBO_2 + Na[B(OH)_4] + H_2O$

How can this reaction is made to proceed in forward direction?

(2006, Only One Option Correct Type)

a. Addition of *cis*-1, 2-diol
b. Addition of borax
c. Addition of *trans*-1, 2-diol
d. Addition of Na_2HPO_4

6. Group 14 Elements (Carbon Family)

DPP-1 Periodicity of Group 14

Only One Option Correct Type

Working Space

Direction (Q. Nos. 1-10) *This section contains 10 multiple choice questions. Each question has four choices (a), (b), (c) and (d), out of which ONLY ONE option is correct.*

1 Stability of hydrides of carbon family is in order

a. $CH_4 > SiH_4 > GeH_4 > SnH_4 > PbH_4$
b. $CH_4 > SiH_4 < GeH_4 > SnH_4 < PbH_4$
c. $CH_4 > SnH_4 > GeH_4 > SiH_4 > PbH_4$
d. None of these

2 Catenation property in Group 14 is in order

a. $C < Si < Ge < Sn < Pb$
b. $Pb < Sn < Ge < Si < C$
c. $C = Si < Ge < Sn < Pb$
d. $C = Si > Ge > Sn > Pb$

3 The stability of dihalides of Si, Ge, Sn and Pb increases steadily in the sequence **[AIEEE 2007]**

a. $GeX_2 < SiX_2 < SnX_2 < PbX_2$
b. $SiX_2 < GeX_2 < PbX_2 < SnX_2$
c. $SiX_2 < GeX_2 < SnX_2 < PbX_2$
d. $PbX_2 < SnX_2 < GeX_2 < SiX_2$

4 In which of the following arrangements, the sequence is not strictly according to the property written against it?

a. $CO_2 < SiO_2 < SnO_2 < PbO_2$ (Increasing oxidising power)
b. $CH_4 < SiH_4 < SnH_4 < PbH_4$ (Increasing reducing power of hydrides)
c. C > Si > Ge > Pb > Sn (Decreasing $(IE)_1$)
d. C > Si > Ge > Sn > Pb (Decreasing $(IE)_1$)

5 Which of the following exists as a covalent crystal in the solid state? **[JEE Main 2013]**

a. Iodine
b. Silicon
c. Sulphur
d. Phosphorus

6 Carbon shows allotropic forms due to

a. $p\pi$-$p\pi$ bond formation
b. catenation
c. Both (a) and (b)
d. None of these

7 Which of the following is/are possible?

a. $[SiCl_6]^{2-}$
b. $[SiF_6]^{2-}$
c. PbI_4
d. All of these

8 A metal M forms chloride in $+2$ and $+4$ oxidation states. Which of the following statements about these chlorides is correct? **[AIEEE 2006]**

a. MCl_2 is more volatile than MCl_4
b. MCl_2 is more soluble in anhydrous ethanol than MCl_4
c. MCl_2 is more ionic than MCl_4
d. MCl_2 is more easily hydrolysed than MCl_4

9 The soldiers of Napoleon army while at Alps during freezing winter suffered a serious problems as regards to the tin buttons of their uniforms. White metallic tin buttons got converted to grey powder. This transformation is related to **[AIEEE 2004]**

a. a change in the crystalline structure of tin
b. an interaction with nitrogen of the air at very low temperature
c. a change in the partial pressure of oxygen in the air
d. an interaction with water vapour contained in the humid air

Working Space

10 Metastannic acid is formed when tin reacts with

a. very dilute HNO_3
b. dilute HNO_3
c. conc. HNO_3
d. All of these

One or More than One Options Correct Type

Direction (Q. Nos. 11-15) This section contains 5 multiple choice questions. Each question has four choices (a), (b), (c) and (d), out of which ONE or MORE THAN ONE are correct.

11 $[M(OH)_6]^{2-}$ is formed when hot or cold NaOH reacts with M, where M is

a. Sn
b. Pb
c. Si
d. Ge

12 Select the correct statement(s) about fullerene.

a. It is a carbon allotrope consisting of spherical C_{60} molecules with the shape of a soccer ball
b. It has 12 pentagonal and 20 hexagonal faces with each C-atom sp^2-hybridised
c. It is a constituent of soot
d. It was first prepared (in 1985) by laser vaporisation of graphite followed by sweeping of carbon atoms by a blast of high pressure helium gas and rapid cooling in a vacuum chamber

13 Select the correct statement(s).

a. All Group 14 elements are unaffected by water and steam
b. Sn reacts with steam forming SnO_2 and H_2
c. Pb is unaffected by H_2O probably because of protective oxide film
d. C, Si and Ge are unaffected by dilute acid

14 Element(s) which can form multiple bonds with oxygen, nitrogen and sulphur is/are

a. carbon
b. silicon
c. lead
d. phosphorus

15 Select the correct statement(s).

a. Sn^{2+} salts are better reducing agents
b. Sn^{4+} salts are better oxidising agents
c. Pb^{4+} salts are better oxidising agents
d. Pb^{2+} salts are better reducing agents

Statement Type

Direction (Q. Nos. 16-19) This section is based on Statement I and Statement II. Select the correct answer from the codes given below.

a. Both Statement I and Statement II are correct and Statement II is the correct explanation of Statement I
b. Both Statement I and Statement II are correct but Statement II is not the correct explanation of Statement I
c. Statement I is correct but Statement II is incorrect
d. Statement II is correct but Statement I is incorrect

16 **Statement I** PbO_2 is an acidic oxide.

Statement II PbO_2 is precipitated as $PbSO_4$ and O_2 is formed on reaction with H_2SO_4.

17 **Statement I** IE values for the formation of Pb^+, Pb^{2+}, Pb^{3+} and Pb^{4+} are higher than that of corresponding IE values for the formation of Sn^+, Sn^{2+}, Sn^{3+} and Sn^{4+}.

Statement II There is poor shielding effect of 4*f*- and 5*d*-orbital electrons.

Working Space

18 **Statement I** Silicon exists only in the diamond structure.

Statement II It has tendency to form $p\pi$-$p\pi$ bonds to itself.

19 **Statement I** Graphite is a better electrical conductor than diamond.

Statement II Conductivity of graphite depends on direction.

Comprehension Type

Direction (Q. Nos. 20 and 21) This section contains a paragraph, which describing theory, experiments, data, etc. Two questions related to the paragraph have been given. Each question has only one correct answer among the four given options (a), (b), (c) and (d).

Passage

Covalent radii of Group 14 (carbon family) are given below

Element	C	Si	Ge	Sn	Pb
Radius (pm)	77	117	122	140	146

20 When we go from C to Si, there is large increase in size. It is due to the

a. introduction of 3rd shell
b. screening effect
c. lanthanide contraction
d. metalloidic nature of Si

21 Going from Si to Ge, there is very marginal increase in radius. This is due to which of the following facts?

a. The effective nuclear charge resulting from addition of eighteen protons to the nucleus balances the effect of the additional shell
b. Outermost shell is pulled to the greater extent due to higher charge
c. Both (a) and (b)
d. None of the above

Matching List Type

Direction (Q. No. 22) Choices for the correct combination of elements from Column I and Column II are given as options (a), (b), (c) and (d), out of which one is correct.

22 Match the oxides in Column I with their corresponding property(ies) in Column II and mark the correct option from the codes given below.

	Column I		Column II
i.	PbO	p.	Oxidation state +2
ii.	SnO_2	q.	Oxidation state + 4
iii.	PbO_2	r.	Amphoteric
iv.	CO_2	s.	Basic
v.	GeO	t.	Acidic

Codes

	i	ii	iii	iv	v
a.	p,r	q,r	q,r	q,t	p,t
b.	p,r	p,r	q,t	q,t	q,r
c.	q,r	p	q	q,t	p,r
d.	p,t	q,s	q,s	q,t	p,r

One Integer Value Correct Type

Direction (Q. No. 23) This section contains a question. When worked out will result in an integer from 0 to 9 (both inclusive).

23 How many of the following elements are more stable in lower oxidation states (assuming to show two types of oxidation states)?

B, Al, Ga, Sn, Ge, Pb, C, Tl, Si, As, Sb, Bi

DPP-2 Compounds of Group 14

Only One Option Correct Type

Working Space

Direction *(Q. Nos. 1-7) This section contains 7 multiple choice questions. Each question has four choices (a), (b), (c) and (d), out of which ONLY ONE option is correct.*

1 Which of the following has the highest calorific value?

a. Water gas
b. Producer gas
c. Carbonium oxide
d. Coal gas

2 CO is a Lewis base thus,

a. if forms dative bond between acceptor and itself
b. its bond order is three
c. Both (a) and (b)
d. None of the above

3 B_2H_6 adds to CO to form BH_3CO. In this formation,

a. BH_3 is a Lewis base, CO is a Lewis acid
b. BH_3 is a Lewis acid, CO is a Lewis base
c. BH_3 is a protonic acid and CO is a base
d. BH_3 and CO both are electron-deficient

4 Water glass is

a. glass made up of water
b. sodium silicate
c. water gas
d. crystal carbonate

5 Ultrapure Si for semiconductor is prepared as

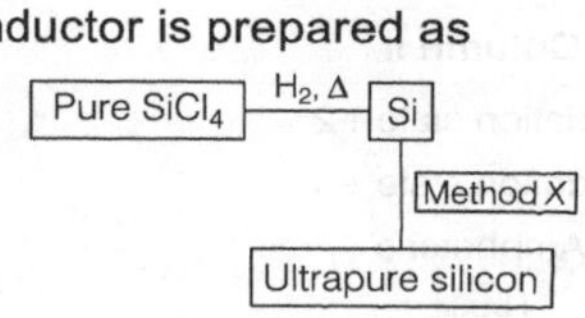

Method X is called

a. oxidation
b. smelting
c. roasting
d. zone refining

6 Tin reacts with dilute HNO_3 forming

a. NO and $Sn(NO_3)_2$
b. NO and $Sn(NO_3)_4$
c. NO_2 and $Sn(NO_3)_2$
d. NH_4NO_3 and $Sn(NO_3)_2$

7 Estimation of CO can be done by

a. I_2
b. I_2O_5
c. O_2
d. All of these

Working Space

One or More than One Options Correct Type

Direction (Q. Nos. 8-13) This section contains 6 multiple choice questions. Each question has four choices (a), (b), (c) and (d), out of which ONE or MORE THAN ONE are correct.

8 Silicones and silicates have which of the following facts?

a. O^- of silicates has been replaced by methyl groups
b. Silicones and silicates both are isoelectronic in anions
c. Silicones and silicates both have same molar mass in anions
d. Silicones and silicates both have different number of electrons

9 Out of the following oxides, oxidising agents are

a. litharge
b. massicot
c. red lead
d. lead dioxide

10 The great thermal and chemical stability of silicones is attributed to the high strength of

a. (Si—C) bond
b. (Si—O) bond
c. (Si—Si) bond
d. (C—C) bond

11 Select the correct statement(s) about zeolites.

a. They are formed by the replacement of some of the silicon atoms of the SiO_2 lattice by aluminium
b. They have a more closed structures than feldspar
c. They can absorb CO_2, NH_3 and EtOH
d. They can separate straight chain hydrocarbon from a mixture of straight chain and branched chain hydrocarbons

12 Conc. H_2SO_4 reacts with carbon forming

a. CO_2
b. SO_2
c. H_2CO_3
d. H_2SO_3

13 Silly putty

a. is a silicone polymer
b. has a composition intermediate between silicone oils and silicone rubbers
c. is an asbestos
d. is not related to any chemical

Statement Type

Direction (Q. Nos. 14 and 15) This section is based on Statement I and Statement II. Select the correct answer from the codes given below.

a. Both Statement I and Statement II are correct and Statement II is the correct explanation of Statement I
b. Both Statement I and Statement II are correct but Statement II is not the correct explanation of Statement I
c. Statement I is correct but Statement II is incorrect
d. Statement II is correct but Statement I is incorrect

14 **Statement I** Lamellar compounds are conducting in nature.

Statement II The high electrical conductivity may arise from transfer of electrons from the metal to the π-orbitals of the carbon plane, where they are free to move under an electric field.

15 **Statement I** Carbon does not form CF_6^{2-} but silicon can form SiF_6^{2-}.

Statement II Glass reacts with HF to form Na_2SiF_6.

Working Space

Comprehension Type

***Direction** (Q. Nos. 16-19) This section contains 2 paragraphs, each describing theory, experiments, data, etc. Four questions related to the paragraphs have been given. Each question has only one correct answer among the four given options (a), (b), (c) and (d).*

Passage I

The silicate anion in the mineral kinoite is a chain of three SiO_4 tetrahedra that share corners with adjacent tetrahedra in a linear (chain) structure. The mineral also contains Ca^{2+}, Cu^{2+} and H_2O in 1 : 1 : 1 ratio.

16 Anion is

a. $Si_{13}O_{12}^{24-}$ b. $Si_3O_{10}^{8-}$

c. $Si_3O_9^{6-}$ d. $Si_3O_9^{8-}$

17 Mineral is

a. $Ca_2Cu_2Si_3O_{10} \cdot 2H_2O$ b. $CaCuSi_{13}O_{18} \cdot H_2O$

c. $CaCuSi_{13}O_{12} \cdot 2H_2O$ d. None of these

Passage II

CO forms volatile metal carbonyl which can be decomposed to pure metal.

$$M + CO \xrightarrow{T_1} M(CO)_4 \xrightarrow{T_2} M + 4CO\uparrow$$

18 Name the metal M and the process.

a. Ni, Mond b. Fe, Mond

c. Ni, Peterson d. Fe, Peterson

19 Temperatures, T_1 and T_2 (in kelvin scale) respectively are

a. 301, 401 b. 301, 503

c. 28, 230 d. 28, 503

One Integer Value Correct Type

***Direction** (Q. Nos. 20-30) This section contains 11 questions. When worked out will result in an integer from 0 to 9 (both inclusive).*

20 10 moles of CO react quantitatively with I_2O_5 forming CO_2 and I_2. How many moles of $Na_2S_2O_3$ are required to reduce I_2?

21 How many of the following types of silicate minerals contain (SiO_4^{4-}) anion?

1. Orthosilicate
2. Pyrosilicate
3. Pyroxene
4. Amphibole
5. Mica
6. Zeolite
7. Single chain silicate
8. Double chain silicate
9. Sheet like silicate

22 What is the negative charge on the silicate (anion) represented as?

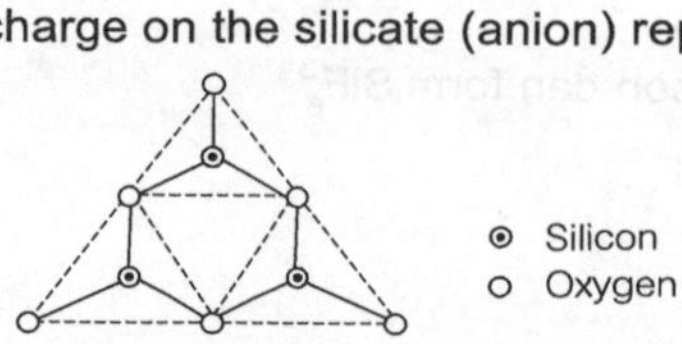

Working Space

23 The starting materials for the manufacture of silicones are alkyl or aryl substituted silicon chlorides.

$$R_nSiCl_{4-n}$$

What is the maximum value of n?

24 How many of the following fuel gases have CO and H_2 both as combustible gases?

LPG, water gas, producer gas, coal gas, CNG.

25 How many of the following have equal number of σ and π bonds?

$$C_2^{2-}, C_3^{4-}, CN_2^{2-}, CO_2, C_2H_4, CN^-, C_3O_2$$

26 How many elements in Group 14 are used as semiconductor?

27 When malonic acid is dehydrated using P_4O_{10} at 150° C, species has total number of atoms..........

28 Out of elements C, Si, Ge, Sn, Pb, S, Cl, only +4 oxidation state is shown by how many of the elements?

29 How many of the following are amphotoric oxides?

$$CO, CO_2, SiO_2, GeO, GeO_2, SnO, SnO_2, PbO, PbO_2, Pb_3O_4$$

30 Lead content (in percentage) in lead pencil is......

SET - 1

Revisal Problems for JEE Main

Only One Option Correct Type

Working Space

Direction (Q. Nos. 1-33) *This section contains 33 multiple choice questions. Each question has four choices (a), (b), (c) and (d), out of which ONLY ONE is correct.*

1 $SnCl_2$ dissolves in

a. KCl b. H_2O
c. $FeCl_3$ d. $HgCl_2$

2 Mellitic acid $C_{12}H_6O_{12}$ (benzene hexacarboxylic acid) is formed when hot conc. HNO_3 reacts with

a. diamond b. graphite
c. Both (a) and (b) d. None of these

3 Main content of gases responsible for green house effect is

a. CO_2 b. CH_4
c. N_2 d. CFC

4 $SnCl_2$ is a reducing agent. Which of the following is reduced to a metallic state by $SnCl_2$?

a. $FeCl_3$ b. $CuCl_2$
c. $HgCl_2$ d. $FeCl_2$

5 Following are the some facts about SiC.

I. It has diamond like structure in which C and Si-atoms each are tetrahedrally surrounded by four of the other kinds of atoms.
II. It is called carborundum.
III. It is used as an abrasive and in cutting tools.
IV. C and Si-atoms are sp^2 hybridised.

Mark the correct statements.

a. Both I and III b. I, II, III and IV
c. Both II and III d. I, II and III

6 CO finds use as given below but not as a/an

a. reducing agent b. fuel
c. ligand d. oxidising agent

7 Select the pair of methanides out of the following.

a. CaC_2, Mg_2C_3 b. Al_4C_3, Be_2C
c. Al_4C_3, Mg_2C_3 d. Be_2C, Mg_2C_3

8 In the structure of CO_2,

a. C-atom is *sp* hybridised
b. two electrons of C-atom are involved in $p\pi$-$p\pi$ bonding with O-atoms
c. (C—O) bond lengths are equal and dipole moment is zero
d. All are the correct statements

9 Which of the following helps to maintain the pH of blood between 7.26 to 7.42?

a. $H_3BO_3 + H_2BO_3^-$ b. $H_2CO_3 + HCO_3^-$
c. $H_3PO_4 + H_2PO_4^-$ d. $HCO_3^- + CO_3^{2-}$

10 More than one —OH group is not possible on the same carbon. Select the species that can have two —OH groups at the same carbon.

a. Carbonic acid b. Chloral hydrate
c. Both (a) and (b) d. None of these

Working Space

11 CO is toxic in nature and can be fatal in mines. Hence, rescue teams in mines are provided with a breathing apparatus which may contain

a. MnO_2/CuO
b. K_2O_2
c. Both (a) and (b)
d. None of these

12 Bond order of C_2^{2-} (acetylide) ion is

a. 2
b. 3
c. 1
d. 2.5

13 In silicon dioxide, **[AIEEE 2005]**

a. there are double bonds between silicon and oxygen atoms
b. silicon atom is bonded to two oxygen atoms
c. each silicon atom is surrounded by two oxygen atoms and each oxygen atom is bonded to two silicon atoms
d. each silicon atom is surrounded by four oxygen atoms and each oxygen atom is bonded to two silicon atoms

14 Which is the respiration process?

a. $6CO_2 + 6H_2O \longrightarrow 6O_2 + C_6H_{12}O_6$
b. $C_6H_{12}O_6 + 6O_2 \longrightarrow 6CO_2 + 6H_2O$
c. Both (a) and (b)
d. None of the above

15 Solid CO_2 is called

a. dry ice
b. cardice
c. Both (a) and (b)
d. None of these

16 Which of the following structures is assigned to silicates in which three oxygen atoms of $[SiO_4]^{4-}$ are shared?

a. Pyrosilicate
b. Sheet silicate
c. Linear chain silicate
d. Three-dimensional silicate

17 Which of the following is/are correct contents in gases given below?

	Gas	Content
a.	Water gas	$CO + H_2$
b.	Producer gas	$CO + N_2$
c.	Coal gas	$CO + H_2 + CH_4 + CO_2$
d.	All are correct contents	

18 CO is toxic because

a. it is a Lewis base
b. it forms complex with haemoglobin in the blood
c. it burns with blue flame
d. it is a Lewis acid

19 Consider the following statements,

I. Silicon becomes *p*-type semiconductor when doped with Ga.
II. Silicon becomes *n*-type semiconductor when doped with P.
III. Hyperfine silicon is used in the production of transistors.

Mark the correct statements.

a. Both I and II
b. Both II and III
c. Both I and III
d. I, II and III

Working Space

20 If ⊙ represents silicon and ○ represents oxygen then this silicate is

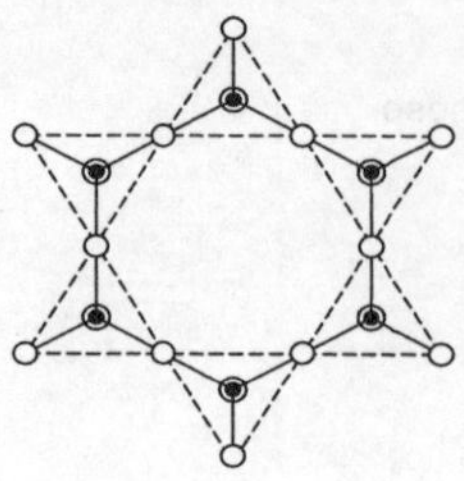

a. $Si_6O_{24}^{24-}$
b. $Si_6O_{18}^{24-}$
c. $Si_6O_{18}^{12-}$
d. $Si_6O_{24}^{12-}$

21 When a mixture of CO and CO_2 is passed into slaked lime,

a. both are absorbed in it
b. only CO is absorbed in it
c. only CO_2 is absorbed in it
d. None is absorbed in it

22 Magnesium silicide reacts with ammonium bromide to form

a. magnesium nitride
b. silane
c. silane and ammonia
d. ammonia

23 $2CH_3Cl + Si \xrightarrow[570\text{ K}]{\text{Cu powder}} \xrightarrow{H_2O} \xrightarrow{\Delta} A$

A is called

a. silanol
b. silicone
c. disilanol
d. trisilanol

24 An amorphous form of silica is used in filtration plants. This form is called as

a. piezoelectric
b. kieselguhr
c. silane
d. silicones

25 Consider some properties of silicones.

I. Hydrophobic in nature.
II. High thermal stability.
III. High dielectric strength.
IV. Resistance to oxidation and chemicals.
V. Hydrophilic in nature.

Select the correct property(ies).

a. II, III and V
b. III, IV and V
c. I, II, III and IV
d. Only V

26 ZSM-5 (a type of zeolites) is used

a. to convert alcohols directly into gasoline
b. in ion exchangers in softening of hard water
c. Both (a) and (b)
d. None of the above

27 Aqueous solution of SnO liberates CO_2 with $NaHCO_3$. This shows that SnO is

a. an acidic oxide
b. a basic oxide
c. an amphoteric oxide
d. a Lewis acid

28 The glass having higher refractive index is prepared using

a. NiO
b. CoO
c. PbO
d. CaO

29 Stable oxides of carbon are

a. CO, CO_2 and C_3O_2
b. CO, C_3O_2 and C_2O
c. CO and CO_2
d. C_2O and C_3O_2

30 Thermodynamically, most stable form of carbon is

a. graphite
b. diamond
c. lonsdaleite
d. fullerene

Working Space

31 When a mixture of carbon monoxide and chlorine is exposed to sunlight, the product formed is

a. thionyl chloride
b. phosgene
c. phosphine
d. carbon tetrachloride

32 Which of the following cuts ultraviolet rays?

a. Soda glass
b. Crooks glass
c. Pyrex
d. None of these

33 White lead is

a. $PbSO_4 \cdot PbO$
b. $PbCO_3 \cdot PbO$
c. $PbCO_3$
d. $Pb(OH)_2 \cdot PbCO_3$

Statement Type

***Direction** (Q. Nos. 34 and 35) This section is based on Statement I and Statement II. Select the correct answer from the codes given below.*

a. Both Statement I and Statement II are correct and Statement II is the correct explanation of Statement I
b. Both Statement I and Statement II are correct but Statement II is not the correct explanation of Statement I
c. Statement I is correct but Statement II is incorrect
d. Statement II is correct but Statement I is incorrect

34 **Statement I** Graphite is a good conductor of electricity.

Statement II In graphite, all the four valence electrons are used in bonding.

35 **Statement I** Lead often appears more noble (unreactive) than expected.

Statement II Lead does not dissolve in dilute H_2SO_4.

SET - 2

Revisal Problems for JEE Advanced

Only One Option Correct Type

Direction (Q. Nos. 1-13) *This section contains 13 multiple choice questions. Each question has four choices (a), (b), (c) and (d), out of which ONLY ONE is correct.*

Working Space

1 In the presence of $NaBH_4/THF$, CO reacts with B_2H_6 forming

a. BH_3CO
b. CH_3OH
c. $(CH_3)_3B_3O_3$
d. None of these is correct

2 Select the correct statement(s).

a. Oxides of carbon family (MO_2) are all network solids with octahedral coordination
b. Silicon dioxide (SiO_2, silica) is a network solid with tetrahedral coordination and is a giant molecule
c. GeO_2, SnO_2 and PbO_2 all are acidic oxides
d. All of the above are correct statements

3 Select the correct statement(s).

a. Hydrides of boron and silicon are volatile and catch fire on exposure to air
b. Oxides of B and Si (B_2O_3 and SiO_2) are acidic in nature
c. Borates and silicates have tetrahedral BO_4 and SiO_4 structural units
d. All of the above are correct statements

4 Holme's signals for guiding of ships are given by using

a. $CaC_2 + CaCN_2$
b. $CaC_2 + Ca_3P_2$
c. $CaC_2 + CaCO_3$
d. $Ca_3P_2 + CaCN_2$

5 The gas evolved by heating of potassium ferrocyanide crystals with conc. H_2SO_4, is

a. CO
b. CO_2
c. SO_2
d. SO_3

6 When zeolite is treated with hard water, there is exchange reaction between calcium/magnesium ion and

a. aluminium
b. sodium ion
c. water of hydration
d. sulphate ion

7 Carbogen is a/an

a. mixture of CO and CO_2
b. mixture O_2 and CO_2
c. pure form of carbon
d. unsaturated organic compound

8 Carbon monoxide (CO) is a poisonous gas, the antidote used for this poisoning is

a. pure oxygen
b. carbonic acid
c. carborundum
d. carbogen

9 Red lead used as a primer for iron to prevent it from rusting is

a. Pb_3O_4
b. PbO
c. PbO_2
d. Pb_2O

10 Structures of different allotropes of carbon have been compared. Which represents the incorrect comparison?

a. Polyne (*sp*- hybridised carbon)
b. Buckyball (*sp* - hybridised carbon)
c. Graphite (sp^2 - hybridised carbon)
d. Diamond (sp^3 - hybridised carbon)

Working Space

11 Some of the facts about zeolites are

I. They are aluminosilicates formed by replacing some of Si-atoms by Al.

II. They can be represented as

$$M_{x/n}[(AlO_2)_x \cdot (SiO_2)_y] \cdot mH_2O$$

where, $M = Na^+/K^+/Ca^{2+}$

III. They are widely used as catalyst in petrochemical industries.

Mark the correct facts.

a. Both I and II
b. Both I and III
c. Both II and III
d. I, II and III

12 The chain length of the silicone (polymer) can be controlled by adding

a. CH_3SiCl_3
b. $(CH_3)_2SiCl_2$
c. $(CH_3)_3SiCl$
d. None of these

13 In photosynthesis process of glucose, O_2 formed is from

a. CO_2
b. H_2O
c. Both (a) and (b)
d. None of these

One or More than One Options Correct Type

Direction *(Q. Nos. 14-23) This section contains 10 multiple choice questions. Each question has four choices (a), (b), (c) and (d), out of which ONE or MORE THAN ONE are correct.*

14 Mark the correct statement(s) about graphite and diamond.

a. Graphite uses sp^2 hybrid orbitals and diamond uses sp^3 hybrid orbitals in bonding
b. The graphite hexagonal layers can slide over one another and thus is lubricant white diamond is hard
c. Both are insulators
d. Graphite is thermodynamically more stable

15 Select the pair of elements which are metalloids and are used as semiconductors?

a. Carbon and silicon
b. Silicon and germanium
c. Boron and silicon
d. Silicon and lead

16 Which is/are insoluble lead salt(s) (in water) at room temperature?

a. PbS
b. $Na_2[Pb(OH)_6]$
c. $PbSO_4$
d. $PbCl_2$

17 Due to greenhouse effect of CO_2,

a. pH of water increases
b. pH of water decreases
c. fertility of soil increases
d. fertility of soil decreases

18 CO is formed in which of the following reactions?

a. $HCOOH \xrightarrow[\text{Conc. } H_2SO_4]{373\text{ K}}$

b. $\begin{matrix} COOH \\ | \\ COOH \end{matrix} \xrightarrow{\Delta}$

c. $CH_4 + O_2 \xrightarrow{\Delta}$

d. $\text{Graphite} + O_2 \longrightarrow$

19 Select the correct statement(s) about carbon suboxide.

a. It is anhydride of succinic acid
b. It is anhydride of malonic acid
c. C-atom is *sp* hybridised
d. C content is 52.94%

20 Carboxylic acid salt(s) is/are formed in which of the following reactions?

a. $CO + NaOH \longrightarrow$
b. $CO + CH_3ONa \longrightarrow$
c. $CO + B_2H_6 \longrightarrow$
d. $CO + Na \xrightarrow{NH_3}$

Working Space

21 Which is/are hydrolysed (with H_2O)?

a. CCl_4
b. $SiCl_4$
c. $SnCl_2$
d. $PbCl_2$

22 Mark the correct statement(s).

a. Cyanamide ion (CN_2^{2-}) is isoelectronic with CO_2 and has the same linear structure
b. BeC_2 and Mg_2C_3 form alkyne on reaction with H_2O
c. Be_2C and Al_4C_3 are called methanides
d. CaC_2 has NaCl type structure

23 Which of the following statements is/are correct about zeolites?

a. They are used as cations exchangers
b. They have open structures which enables them to take up small molecules
c. They are aluminosilicates having three-dimensional network
d. Some of the SiO_4^{4-} units are replaced by AlO_4^{5-} and AlO_6^{9-} ions

Statement Type

Direction *(Q. Nos. 24 and 25)* ***This section is based on Statement I and Statement II. Select the correct answer from the codes given below.***

a. Both Statement I and Statement II are correct and Statement II is the correct explanation of Statement I
b. Both Statement I and Statement II are correct but Statement II is not the correct explanation of Statement I
c. Statement I is correct but Statement II is incorrect
d. Statement II is correct but Statement I is incorrect

24 **Statement I** Silica in its normal form is almost non-reactive.

Statement II Silica and even glass is attached by HF and NaOH.

25 **Statement I** $N(CH_3)_3$ has pyramidal structure while $N(SiH_3)_3$ has planar structure.

Statement II Both have lone pair on nitrogen atom.

Matching List Type

Direction *(Q. Nos. 26 and 27)* ***Choices for the correct combination of elements from Column I and Column II are given as options (a), (b), (c) and (d), out of which one is correct.***

26 Match the silicate minerals in Column I with their formulae in Column II and mark the correct option from the codes given below.

	Column I		Column II
i.	Zircon olivine	p.	$LiAl(SiO_3)_2$
ii.	Beryl	q.	$CaMg(SiO_3)_2$
iii.	Spodumene	r.	$[Mg_3Si_4O_{10}(OH)_2]_n$
iv.	Asbestos	s.	$Be_3Al_2Si_6O_{18}$
v.	Talc	t.	$[Ca_2Mg_5(Si_4O_{11})_2(OH)_2]_n$
vi.	Diopside	u.	$ZrSiO_4 \cdot Mg_2SiO_4$

Codes

	i	ii	iii	iv	v	vi
a.	u	s	p	t	r	q
b.	u	s	p	r	t	q
c.	u	s	q	p	r	t
d.	u	s	r	p	q	t

Working Space

27 Match the species in Column I with their corresponding property(ies) in Column II and mark the correct option from the codes given below.

	Column I		Column II
i.	CO_2	p.	Dry ice
ii.	$Al_2(SO_4)_3$	q.	Fire extinguisher
iii.	KCN	r.	Acidic
iv.	CaC_2	s.	Ligand
v.	Mg_2C_3	t.	*sp* hybridisation
		u.	Green house effect

Codes

	i	ii	iii	iv	v
a.	p,q,r,t,u	q,r	s,t	t	t
b.	p,q,t	q,r	t	t	t
c.	p,q,r	q,r	s	t	s
d.	p,q,r	r	u	t	s

Comprehension Type

Direction (Q. Nos. 28-35) *This section contains 3 paragraphs, each describing theory, experiments, data, etc. Eight questions related to the paragraphs have been given. Each question has only one correct answer among the four given options (a), (b), (c) and (d).*

Passage I

Pb_3O_4 (red lead) on heating forms PbO and O_2.

28 Pb_3O_4 is 1 : 2 mixture of

a. PbO, PbO_2
b. PbO_2, PbO
c. Pb, PbO_2
d. PbO_2, Pb

29 Pb_3O_4 is insoluble in H_2SO_4 forming

a. $PbSO_4$, O_2
b. $Pb(SO_4)_2$, O_2
c. $PbSO_4$, $Pb(SO_4)_2$
d. None of these is correct

30 Pb_3O_4 reacts with HNO_3 forming

a. $Pb(NO_3)_2$
b. PbO_2
c. Both (a) and (b)
d. None of these

Passage II

$$A + H_2O \longrightarrow CH{\equiv}CH + B$$

On passing CO_2 into *B*, milkiness appears.

$$A + N_2 \longrightarrow C + D$$
$$C + H_2O + CO_2 \longrightarrow NH_2CONH_2 + CaCO_3$$

31 *A* and *B* respectively are

a. CaC_2 and $Ca(OH)_2$
b. Al_4C_3 and $Al(OH)_3$
c. Be_2C and $Be(OH)_2$
d. None of these

32 *C* and *D* respectively are

a. $CaCN_2$ and $Ca(CN)_2$
b. $Ca(CN)_2$ and $CaCN_2$
c. $CaCN_2$ and C
d. C and $CaCN_2$

Passage III

Consider the following process of synthesis of chlorosilanes.

$$RMgCl + SiCl_4 \longrightarrow A + MgCl_2$$
$$A + RMgCl \longrightarrow B + MgCl_2$$
$$B + RMgCl \longrightarrow C + MgCl_2$$

A, *B* and *C* are different types of chlorosilanes.

Working Space

33 $A + H_2O \xrightarrow{\Delta} A'$

A' is a

a. dimer
b. long chain polymer
c. cross-linked polymer
d. None of these

34 $B + H_2O \longrightarrow \xrightarrow{\Delta} B'$

a. B' is a long chain polymer
b. B' are used in making cement work and water-repellent fibres
c. If R is phenyl (in toluene solvent) then silicone resin is formed
d. All of the above

35 Dimer (silicones) is formed from hydrolysis followed by heating

a. A
b. B
c. C
d. All of these

One Integer Value Correct Type

Direction (Q. Nos. 36-40) ***This section contains 5 questions. When worked out will result in an integer from 0 to 9 (both inclusive).***

36 There are x hexagonal faces and y pentagonal faces in C_{60} (fullerene). What is the value of $(x - y)$?

37 How many of the following are volatile liquids?

$AlCl_3, CCl_4, SiCl_4, GeCl_4, SnCl_4, PbCl_4, Hg_2Cl_2, HgCl_2, PbCl_2, CaCl_2, NaCl$

38 10 L of a mixture of CO and CO_2 gas is passed over heated coke when volume increases to 12 L. What is the volume of CO in the mixture (in litre)?

39 How many of the following are ionic carbides?

$SiC, B_4C_3, Be_2C, Mg_2C_3, CaC_2, Al_4C_3, WC, Cr_4C_3$

40 How many water molecules are formed per molecule of glucose synthesised in a photosynthesis process?

JEE Main & AIEEE Archive

(Compilation of Last 13 Years' Questions)

1 Among the following substituted silanes, the one which will give rise to cross-linked silicone polymer on hydrolysis is **(AIEEE 2007)**

a. R_4Si b. $RSiCl_3$
c. R_2SiCl_2 d. R_3SiCl

2 The stability of dihalides of Si, Ge, Sn and Pb increases steadily in the sequence **(AIEEE 2007)**

a. $GeX_2 < SiX_2 < SnX_2 < PbX_2$ b. $SiX_2 < GeX_2 < PbX_2 < SnX_2$
c. $SiX_2 < GeX_2 < SnX_2 < PbX_2$ d. $PbX_2 < SnX_2 < GeX_2 < SiX_2$

3 A metal, M forms chlorides in its +2 and +4 oxidation states. Which of the following statements about these chlorides is correct? **(AIEEE 2006)**

a. MCl_2 is more volatile than MCl_4
b. MCl_2 is more soluble in anhydrous ethanol than MCl_4
c. MCl_2 is more ionic than MCl_4
d. MCl_2 is more easily hydrolysed than MCl_4

4 Which of the following oxides is amphoteric in character? **(AIEEE 2005)**

a. SnO_2 b. SiO_2
c. CO_2 d. CaO

5 In silicon dioxide, **(AIEEE 2005)**

a. there are double bonds between silicon and oxygen atoms
b. silicon atom is bonded to two oxygen atoms
c. each silicon atom is surrounded by two oxygen atoms and each oxygen atom is bonded to two silicon atoms
d. each silicon atom is surrounded by four oxygen atoms and each oxygen atom is bonded to two silicon atoms

6 The soldiers of Napolean army while at Alps during freezing winter suffered a serious problem as regards to the tin buttons of their uniforms. White metallic tin buttons got converted to grey powder. This transformation is related to **(AIEEE 2004)**

a. a change in the crystalline structure of tin
b. an interaction with nitrogen of the air at very low temperature
c. a change in the partial pressure of oxygen in the air
d. an interaction with water vapour contained in the humid air

7 Glass is a **(AIEEE 2003)**

a. microcrystalline solid b. supercooled liquid
c. gel d. polymeric mixture

8 Graphite is a soft solid lubricant extremely difficult to melt. The reason for this anomalous behaviour is that graphite **(AIEEE 2003)**

a. is a non-crystalline substance
b. is an allotropic form of diamond
c. has molecules of variable molecular masses like polymers
d. has carbon atoms arranged in large plates of rings of strongly bound carbon atoms with weak interplate bonds

JEE Advanced & IIT JEE Archive

(Compilation of Last 10 Years' Questions)

1 With respect to graphite and diamond, which of the statement(s) given below is (are) correct? **(2012, One or More than One Options Correct Type)**

a. Graphite is harder than diamond

b. Graphite has higher electrical conductivity than diamond

c. Graphite has higher thermal conductivity than diamond

d. Graphite has higher C—C bond order than diamond

2 The value of n in the molecular formula $Be_nAl_2Si_6O_{18}$ is **(2010, Integer Type)**

3 **Statement I** Pb^{4+} compounds are stronger oxidising agents than Sn^{2+} compounds.

Statement II The higher oxidation states for the group 14 elements are more stable for the heavier members of the group due to inert pair effect. **(2008, Statement Type)**

a. Both Statement I and Statement II are correct and Statement II is the correct explanation of Statement I

b. Both Statement I and Statement II are correct but Statement II is not the correct explanation of Statement I

c. Statement I is correct but Statement II is incorrect

d. Statement II is correct but Statement I is incorrect

4 Match the Column I with Column II and mark the correct option from the codes given below. **(2006, Matching Type)**

	Column I		Column II
i.	$Bi^{3+} \rightarrow (BiO)^+$	p.	Heat
ii.	$[AlO_2]^- \rightarrow Al(OH)_3$	q.	Hydrolysis
iii.	$SiO_4^{4-} \rightarrow Si_2O_7^{6-}$	r.	Acidification
iv.	$(B_4O_7^{2-}) \rightarrow [B(OH)_3]$	s.	Dilution by water

Codes

	i	ii	iii	iv
a.	r,s	p	q	r
b.	p	p	q	r,s
c.	q	r	p	q,r
d.	p	p	q	r,s

5 Name the structure of silicates in which three oxygen atoms of $[SiO_4]^{4-}$ are shared. **(2005,Only One Option Correct Type)**

a. Pyrosilicates

b. Sheet silicates

c. Linear chain silicates

d. Three dimensional silicates

Answers with Explanations

Answers

1. Periodic Table

DPP-1 Classification of Elements

1. *d* 2. *c* 3. *b* 4. *b* 5. *d* 6. *d* 7. *c* 8. *a* 9. *d* 10. *a*
11. *b* 12. *a* 13. *b* 14. *b* 15. *c* 16. *a* 17. *c* 18. *b* 19. *b* 20. *d*
21. *c* 22. *b* 23. *a* 24. *c* 25. *a*

DPP-2 Periodic Trends-I (Radii, Screening Effect, Hydration Energy, etc)

1. *c* 2. *a* 3. *b* 4. *c* 5. *d* 6. *c* 7. *d* 8. *c* 9. *a* 10. *d*
11. *a* 12. *d* 13. *c* 14. *c* 15. *d* 16. *c* 17. *c* 18. *b* 19. *c* 20. *d*
21. *a* 22. *a* 23. *a* 24. *a* 25. *c* 26. *b* 27. *b* 28. *b* 29. *a* 30. *a*

DPP-3 Periodic Trends-II (IE, Electron Affinity and Electronegativity)

1. *a* 2. *b* 3. *c* 4. *b* 5. *a* 6. *b* 7. *c* 8. *a* 9. *c* 10. *c*
11. *b* 12. *c* 13. *d* 14. *c* 15. *c* 16. *b* 17. *c* 18. *d* 19. *c* 20. *b*
21. *d* 22. *a* 23. *b* 24. *c* 25. *c* 26. *b* 27. *c* 28. *b* 29. *a* 30. *d*
31. *a*

Revisal Problems (JEE Main)

1. *d* 2. *d* 3. *a* 4. *d* 5. *d* 6. *d* 7. *d* 8. *a* 9. *a* 10. *c*
11. *c* 12. *a* 13. *b* 14. *d* 15. *d* 16. *c* 17. *b* 18. *a* 19. *b* 20. *d*
21. *a* 22. *d* 23. *d* 24. *a* 25. *d* 26. *d* 27. *d* 28. *a* 29. *d* 30. *a*
31. *d* 32. *c* 33. *b* 34. *a* 35. *b* 36. *a* 37. *c* 38. *d* 39. *b* 40. *b*
41. *b* 42. *a* 43. *b* 44. *a* 45. *c* 46. *d* 47. *b* 48. *a*

JEE Main & AIEEE Archive

1. *c* 2. *b* 3. *c* 4. *a* 5. *c* 6. *d* 7. *b* 8. *c* 9. *a* 10. *b*
11. *d* 12. *d* 13. *c* 14. *c* 15. *c* 16. *c* 17. *c* 18. *d* 19. *c* 20. *d*
21. *b* 22. *b*

2. Hydrogen

DPP-1 Hydrogen and Its Isotopes

1. *b* 2. *a* 3. *c* 4. *b* 5. *b* 6. *b* 7. *d* 8. *d* 9. *b* 10. *b*
11. *c* 12. *d* 13. *d* 14. *b* 15. *b* 16. *a* 17. *a* 18. *b* 19. *d* 20. *c*
21. *a* 22. *d* 23. *a* 24. *b* 25. *c* 26. *a* 27. *d* 28. *d* 29. *d* 30. *d*
31. *d* 32. *a* 33. *b* 34. *a* 35. *a*

DPP-2 Hydrides

1. *a* 2. *d* 3. *b* 4. *b* 5. *b* 6. *b* 7. *b* 8. *b* 9. *b* 10. *d*
11. *c* 12. *d* 13. *d* 14. *a* 15. *c* 16. *d* 17. *b* 18. *d* 19. *c* 20. *b*
21. *a* 22. *c* 23. *b* 24. *d* 25. *a*

DPP-3 Soft, Hard and Heavy Water

1. *a* 2. *c* 3. *c* 4. *b* 5. *c* 6. *d* 7. *d* 8. *b* 9. *a* 10. *c*
11. *d* 12. *a* 13. *b* 14. *d* 15. *c* 16. *a* 17. *d* 18. *a* 19. *c* 20. *c*
21. *d* 22. *b* 23. *a* 24. *d* 25. *a* 26. *b* 27. *b* 28. *d* 29. *b* 30. *b*

DPP-4 Hydrogen Peroxide

1. (i) $\frac{68x}{22400}$ g mL^{-1} (ii) $\frac{68x}{22.4}$ g L^{-1} (iii) $\frac{68x}{224}$ g/100mL (iv) $\frac{x}{5.6}$ g equiv·L^{-1} (v) $\frac{x}{11.2}$ mol mL^{-1}

2. *d* 3. *b* 4. *a* 5. *d* 6. *b* 7. *c* 8. *a* 9. *a* 10. *d* 11. *c*
12. *c* 13. *b* 14. *d* 15. *d* 16. *c* 17. *d* 18. *a* 19. *a* 20. *c* 21. *d*
22. *d* 23. *a* 24. *b* 25. *a* 26. *d* 27. *c* 28. *b* 29. *b* 30. *a*

Revisal Problems (JEE Main)

1. *d* 2. *d* 3. *c* 4. *b* 5. *c* 6. *b* 7. *c* 8. *b* 9. *d* 10. *b*
11. *a* 12. *b* 13. *a* 14. *c* 15. *a* 16. *c* 17. *d* 18. *c* 19. *d* 20. *c*
21. *c* 22. *d* 23. *b* 24. *d* 25. *c* 26. *b* 27. *b* 28. *b* 29. *c* 30. *b*
31. *b* 32. *b* 33. *b* 34. *a* 35. *d* 36. *a* 37. *a* 38. *a* 39. *b* 40. *b*

JEE Main & AIEEE Archive

1. *d* 2. *a* 3. *d* 4. *a* 5. *d* 6. *d* 7. *d* 8. *a* 9. *b,c,d*

3. Alkali Metals (Group 1)

DPP-1 Periodicity in Group 1

1. *b* 2. *d* 3. *a* 4. *a* 5. *a* 6. *b* 7. *b* 8. *d* 9. *a* 10. *a*
11. *a* 12. *b* 13. *a* 14. *b* 15. *b* 16. *b* 17. *b* 18. *b* 19. *c* 20. *a*
21. *a,b,c* 22. *b,c,d* 23. *a,b,c,d* 24. *c,d* 25. *a,b,c* 26. *a* 27. *a* 28. *c* 29. *b* 30. *b*
31. 6

DPP-2 Compounds of Alkali Metals

1. *c* 2. *a* 3. *b* 4. *b* 5. *b* 6. *b* 7. *c* 8. *c* 9. *c* 10. *c*
11. *d* 12. *c* 13. *d* 14. *d* 15. *b* 16. *a* 17. *a* 18. *a* 19. *d* 20. *a,b*
21. *a,b,d* 22. *a,b* 23. *a,b,c* 24. *a,c,d* 25. *a* 26. *a* 27. *b* 28. *a* 29. *b* 30. *a*
31. *b* 32. *a* 33. *c* 34. *a* 35. *a* 36. *a* 37. 3 38. 6

Revisal Problems (JEE Main)

1. *b* 2. *b* 3. *b* 4. *d* 5. *b* 6. *c* 7. *b* 8. *c* 9. *d* 10. *b*
11. *c* 12. *d* 13. *d* 14. *d* 15. *d* 16. *c* 17. *c* 18. *b* 19. *c* 20. *c*
21. *d* 22. *d* 23. *c* 24. *c* 25. *c* 26. *a* 27. *d* 28. *a* 29. *b* 30. *a*
31. *c* 32. *d* 33. *c* 34. *b* 35. *c*

Revisal Problems (JEE Advanced)

1. *c* 2. *d* 3. *a* 4. *c* 5. *d* 6. *c* 7. *b* 8. *d* 9. *b* 10. *a*
11. *a,c* 12. *b,c* 13. *a,b* 14. *a,b,c,d* 15. *a,c* 16. *a,b,c,d* 17. *a,b* 18. *a,b,c* 19. *a,b,d*
20. *a,b,d* 21. *a,b* 22. *a* 23. *a* 24. *d* 25. *c* 26. *b* 27. *a* 28. *d* 29. *a*
30. *b* 31. *a* 32. *b* 33. 4 34. 9 35. 2

JEE Archive

1. *b* 2. *c* 3. *a* 4. *a* 5. *a,b,c* 6. *a,b* 7. *b*

4. Alkaline Earth Metals (Group 2)

DPP-1 Periodicity in Group 2

1. *b* 2. *a* 3. *b* 4. *a* 5. *b* 6. *b* 7. *a* 8. *a* 9. *b* 10. *c*
11. *b* 12. *d* 13. *a* 14. *b* 15. *b* 16. *d* 17. *a,b,c,d* 18. *b,c,d* 19. *a* 20. *a*
21. *c*

DPP-2 Compounds of Group 2

1. *a* 2. *d* 3. *c* 4. *b* 5. *c* 6. *c* 7. *a* 8. *d* 9. *c* 10. *b*
11. *a* 12. *b* 13. *a* 14. *d* 15. *a* 16. *a* 17. *a* 18. *b* 19. *b* 20. *d*
21. *b,d* 22. *b,c,d* 23. *a,b,c* 24. *a,b,c,d* 25. *a,b,c* 26. *d* 27. *b* 28. *c* 29. *d*
30. *c* 31. *a* 32. 7 33. 6 34. 2 35. 7 36. 6

Revisal Problems of (JEE Main)

1. *b* 2. *b* 3. *d* 4. *d* 5. *b* 6. *a* 7. *a* 8. *b* 9. *c* 10. *d*
11. *a* 12. *a* 13. *b* 14. *b* 15. *d* 16. *a* 17. *c* 18. *b* 19. *d* 20. *b*
21. *a* 22. *d* 23. *b* 24. *c* 25. *a*

Revisal Problems (JEE Advanced)

1. *c* 2. *c* 3. *a* 4. *a* 5. *a* 6. *b* 7. *d* 8. *d* 9. *b* 10. *b*
11. *a,b,c,d* 12. *a,b,c,d* 13. *a,b,c,d* 14. *a,c,d* 15. *b,c,d* 16. *a* 17. *a,c,d* 18. *a,c,d* 19. *d* 20. *b*
21. *c* 22. *b* 23. *a* 24. *b* 25. *c* 26. *a* 27. *c* 28. 4 29. 5 30. 4
31. 4 32. 5 33. 5 34. 4

JEE Main & AIEEE Archive

1. *c* 2. *c* 3. *a* 4. *b* 5. *b* 6. *b* 7. *a,b* 8. *b* 9. *b*

5. Group 13 Elements

DPP-1 Periodicity in Group 13

1. *b* 2. *a* 3. *a* 4. *c* 5. *a* 6. *a* 7. *a* 8. *c* 9. *a* 10. *a*
11. *c* 12. *a* 13. *b* 14. *b,c* 15. *a,b* 16. *a,b,c* 17. *b* 18. *c* 19. *a* 20. 4

DPP-2 Compounds of Group 13

1. *b* 2. *c* 3. *a* 4. *c* 5. *c* 6. *c* 7. *a* 8. *c* 9. *a,c,d*
10. *a,b,d* 11. *a,c,d* 12. *a,b,c,d* 13. *a,c,d* 14. *a,b,c,d* 15. *a,b,d* 16. *b,c* 17. *b,c,d* 18. *a,d*
19. *a,b,c* 20. *b,d* 21. *a,b,c* 22. *a* 23. *b* 24. *c* 25. *c* 26. *b* 27. *a* 28. *c*
29. *a* 30. *a* or *c* 31. 2 32. 4 33. 6 34. 2 35. 4

Revisal Problems (JEE Main)

1. *c* 2. *d* 3. *c* 4. *b* 5. *b* 6. *b* 7. *a* 8. *c* 9. *c* 10. *c*
11. *a* 12. *b* 13. *b* 14. *a* 15. *b* 16. *a* 17. *d* 18. *c* 19. *a* 20. *a*
21. *d* 22. *b* 23. *b* 24. *d* 25. *b* 26. *a* 27. *d* 28. *c* 29. *d* 30. *a*
31. *d* 32. *c* 33. *d* 34. *c* 35. *c*

Revisal Problems (JEE Advanced)

1. *c* 2. *a* 3. *a* 4. *a* 5. *c* 6. *a* 7. *d* 8. *c* 9. *a* 10. *b*
11. *c* 12. *b* 13. *a,b,c,d* 14. *a,b,c* 15. *a,b,d* 16. *a,c* 17. *b,c,d* 18. *b,c,d* 19. *a,b,c*
20. *a,b,c,d* 21. *b,d* 22. *b* 23. *a* 24. *c* 25. *b* 26. *a* 27. *c* 28. *b* 29. 6
30. 2 31. 2 32. 3 33. 4 34. 4

JEE Main & AIEEE Archive

1. *a* 2. *c* 3. *c* 4. *c* 5. *d* 6. *b* 7. *b*

JEE Advanced & IIT JEE Archive

1. *b,d* 2. *a* 3. *c* 4. *a*

6. Group 14 Elements (Carbon Family)

DPP-1 Periodicity of Group 14

1. *a* 2. *b* 3. *c* 4. *d* 5. *b* 6. *c* 7. *b* 8. *c* 9. *a* 10. *c*
11. *a,b* 12. *a,b,c,d* 13. *b,c,d* 14. *a,b* 15. *a,c* 16. *b* 17. *a* 18. *a* 19. *b* 20. *a*
21. *a* 22. *a* 23. 3

DPP-2 Compounds of Group 14

1. *d* 2. *c* 3. *b* 4. *b* 5. *d* 6. *d* 7. *b* 8. *a,b* 9. *c,d* 10. *a,b*
11. *a,c,d* 12. *a,b,c,d* 13. *a,b* 14. *a* 15. *b* 16. *b* 17. *a* 18. *a* 19. *b* 20. 4
21. 4 22. 6 23. 3 24. 2 25. 4 26. 2 27. 5 28. 2 29. 5 30. 0

Revisal Problems (JEE Main)

1. *a* 2. *b* 3. *a* 4. *c* 5. *d* 6. *d* 7. *b* 8. *d* 9. *b* 10. *c*
11. *c* 12. *b* 13. *d* 14. *b* 15. *c* 16. *b* 17. *d* 18. *b* 19. *d* 20. *c*
21. *c* 22. *c* 23. *b* 24. *b* 25. *c* 26. *c* 27. *c* 28. *c* 29. *c* 30. *a*
31. *b* 32. *b* 33. *d* 34. *c* 35. *b*

Revisal Problems(JEE Advanced)

1. *c* 2. *b* 3. *d* 4. *b* 5. *a* 6. *b* 7. *b* 8. *d* 9. *a* 10. *b*
11. *d* 12. *c* 13. *b* 14. *a,b,d* 15. *b,c* 16. *a,c,d,* 17. *b,d* 18. *a,b* 19. *b,c,d* 20. *a,b*
21. *b,c,d* 22. *a,b,c,d* 23. *a,b,c* 24. *b* 25. *b* 26. *a* 27. *a* 28. *b* 29. *a* 30. *c*
31. *a* 32. *c* 33. *c* 34. *d* 35. *c* 36. 8 37. 5 38. 8 39. 4 40. 6

JEE Main & AIEEE Archive

1. *b* 2. *c* 3. *c* 4. *a* 5. *d* 6. *a* 7. *b* 8. *d*

JEE Advanced & IIT JEE Archive

1. *b,d* 2. 3 3. *a* 4. *c* 5. *b*

1. Periodic Table

DPP-1 Classification of Elements

1. (d)

I. H resembles alkali metals as it forms monovalent cation as Na^+. H resembles halogen family as it forms H^-. Thus, its position is not defined. Thus, correct.

II. Elements (58-71) called lanthanides and (90-103) called actinides do not find proper place in the table.
Thus, correct.

III. Members of zero group (noble gases) were not discovered at the time of Mendeleev hence, their position was not given. Thus, correct.

IV. Isotopes have same atomic number but different atomic mass. Since, classification was based on atomic mass, isotopes would be placed at different positions. But they were placed at the same position. Thus, correct.

2. (c) Magic numbers are 2, 8, 8, 18, 18 and 32.
Periodic properties are repeated after 2, 8, 8, 18, 18 and 32 elements.
After first period, properties repeated after 2 elements.
Second period properties repeated after 8 elements.
in all = 10 elements
Similarly, in third period, properties repeated after 8 elements, in all = 18 elements
In fourth period, properties repeated after 18 elements, in all = 36 elements

3. (b) Lanthanides have three shells (4th, 5th, and 6th) incomplete.
and last filling electron goes into 4f-subshell.
Elements are Ce(58) - Lu (71), total 14 elements

$$Ce(58) - [Xe]_{54}\ 4f^1\ 5d^1\ 6s^2$$
$$Gd(64) - [Xe]_{54}\ 4f^7\ 5d^1\ 6s^2$$
$$Lu(71) - [Xe]_{54}\ 4f^{14}\ 5d^1\ 6s^2$$

4. (b)

Z	EC	Block	Group
113	$[Rn]_{86}\ 5f^{14}\ 6d^{10}\ 7s^2 7p^1$	p	13
115	$[Rn]\ 5f^{14}\ 6d^{10}\ 7s^2 7p^3$	p	15

5. (d) Elements (58-71) are called **lanthanides** and elements (90-103) are called **actinides**.

6. (d)

		New block	Original	Block
I (Ca)	$1s^2 2s^2 2p^6 3s^2 3p^6 3d^2$	d	s	Changes
II (Zn)	$1s^2 2s^2 2p^6 3s^2 3p^6 3d^{10} 4s^2$	s	d	Changes
III (Fe)	$1s^2 2s^2 2p^6 3s^2 3p^6 3d^8$	d	d	No
IV (K)	$1s^2 2s^2 2p^6 3s^2 3p^6 3d^1$	d	s	Changes

Thus, I, II and IV have different blocks if Aufbau rule is not followed.

7. (c) New element discovered in 2003 was named as **Darmstadtium** (Ds) on August 16, 2003, by IUPAC.

8. (a) $\underset{115}{X} \longrightarrow \underset{113}{Y} + {}^4_2He$ (α-decay)

9. (d) Valence shell electronic configuration is $3d^5\ 4s^1$. Thus, differentiating electron goes into $(n-1)d$. It is chromium ($Z = 24$).
It is d-block element.
Total valence electrons = 6
Group = 6

10. (a) Group is based on number of electrons in neutral atom (i.e. number of protons)

$$Sc^{3+}\ [Ne]\ 3s^2 3p^6$$
$$Sc: [Ne]\ 3s^2 3p^6\ 4s^2 3d^1$$

Valence electrons = 3
Group = 3 (d-block element)

11. (b) Group 16 elements have valence shell configuration $ns^2\ np^4$.
Thus, they are p-block elements called chalcogens.

12. (a) If shells n and $(n-1)$ are vacant and last electron is being filled into $(n-1)d$ suborbit then, elements are called d-block or transition elements.

Sc (21)	$[Ar]\ 4s^2\ 3d^1$
Ti (22)	$[Ar]\ 4s^2\ 3d^2$
V (23)	$[Ar]\ 4s^2\ 3d^3$
Cr (24)	$[Ar]\ 4s^1\ 3d^5$
Mn (25)	$[Ar]\ 4s^2\ 3d^5$
⋮	⋮
Cu (29)	$[Ar]\ 4s^1\ 3d^{10}$
Zn (30)	$[Ar]\ 4s^2\ 3d^{10}$

13. (b) Element has $5s^2 5p^3$ valence configuration.
Thus, it belongs to N-family group VA (15). It is antimony (Sb).

14. (b) Element lying below the element with atomic number 25 has atomic number = 43 (Tc).
Its electronic configuration is $[Kr]_{36}\ 5s^2 4d^5$.

15. (c) Group 16 means valence shell has $ns^2 np^4$ configuration with $n = 5$.

$$1s^2\ 2s^2\ 2p^6\ 3s^2 3p^6\ 4s^2 3d^{10}\ 4p^6\ 5s^2\ 4d^{10} 5p^4$$

It is Te, total electrons = 52
Thus, atomic number = 52

16. (a) Element below Hg (80) has atomic number = 112
Thus, it is Uub.

17. (c)

I. It is in first excited state. It represents element of atomic number 20 (Ca).
Its ground state EC is [Ar] $4s^2$. Thus, s-block element.

II. It is in first excited state. It represents element of atomic number 10 (Ne).
Its ground state EC is $1s^2 2s^2 2p^6$. Thus, p-block element.

III. It represents ground state of element having atomic number 29 (Cu). Due to more stability of paired electrons, 4s electron goes to 3d. Thus, d-block element.

IV. It represents ground state of element with atomic number 71 (Lu) (lanthanide). Thus, f-block element.

18. (b) $\sqrt{v} = aZ - ab$
This represents a straight line, $y = mx - c$
When $\sqrt{v} = 0, aZ = ab$

$\therefore\quad Z = b = OA = 1$

When $Z = 0$, $\sqrt{v} = -ab = -a$ (since, $b = 1$)
$= \tan 45°$ (slope) $= 1$

Thus, $a = 1$

$\therefore\quad \sqrt{v} = Z - 1$

$\sqrt{225} = Z - 1$

$\therefore\quad Z = 16$

EC of atomic number 16 is = $1s^2 2s^2 2p^6 3s^2 3p^4$

Thus, it is *p*-block element thus, option (a) is correct.

Group = 16, valence electrons = 6

Period = 3 Thus, option (d) is correct.

$(P + 1) = 4 > 6$ (valence electrons)

Thus, it is non-metal.

Thus, option (c) is correct.

Thus, option (b) is incorrect.

19. (b) Modern periodic table has total 18 vertical columns.

$$\underbrace{1-2}_{s}$$

can take 2 electrons

Thus, (c) is correct.

$$\underbrace{3-4-5-6-7-8-9-10-11-12}_{d}$$

can take 10 electrons

Thus, (b) is incorrect.

$$\underbrace{13-14-15-16-17-18}_{p}$$

6 electrons

Thus, (a) is correct.

Block is that in which differentiating electron enters based on azimuthal quantum number. Thus, (d) is correct.

20. (d)

(a) Elements in various subshells are filled based on ascending energy rules and this decides block of the elements. Thus, correct.

(b) Based on number of valence electrons, valency of the element can be predicted

ns^1 1

ns^2 – 2

ns^2np^1 – 3

ns^2np^2 – 4

ns^2np^3 – 5 etc. Thus, correct

(c) Correct

(d) It does not predict the nature of bonds between two elements. Thus, incorrect.

21. (c) Series (in *d*-block and *f*-block) is that in which differentiating electron is being filled.

		Last electron in	Series
(i)	Sc-Zn	3*d*	3*d*
(ii)	Y-Cd	4*d*	4*d*
(iii)	La-Hg	5*d*	5*d*
(iv)	Ac-Ds	6*d*	6*d*
(v)	Ce-Lu	4*f*	4*f*
(vi)	Th-Lr	5*f*	5*f*

Hence, (i)→ (s), (ii) → (r), (iii) → (q), (iv) → (p), (v) → (u), (vi) → (t)

22. (b)

	Z	EC	Block
(i)	43	[Kr] $5s^2 4d^5$	*d*
(ii)	52	[Kr] $5s^2 4d^{10} 5p^4$	*p*
(iii)	64	[Xe] $4f^7 5d^1 6s^2$	*f*
(iv)	88	[Rn] $7s^2$	*s*

Hence, (i) → (r), (ii) → (q), (iii) → (s), (iv) → (p)

23. (a)

	Atomic no.	EC	Group
(i)	11	[Ne] $3s^1$	1
(ii)	15	[Ne] $3s^2 3p^3$	15
(iii)	18	[Ne] $3s^2 3p^6$	18
(iv)	28	[Ar] $4s^2 3d^8$	10

Hence, (i) → (p), (ii) → (r), (iii) → (s), (iv) → (q)

24. (c)

(i) *s*-block – representative elements

(ii) *p*-block – representative elements

(iii) *d*-block – transition metals

(iv) *f*-block – Lanthanides (4*f*-series)

Actinides (5*f*-series)

Hence, (i) → (p), (ii) → (p), (iii) → (q), (iv) → (r,s)

25. (a)

(i) 105 – Unp

(ii) 107 – Uns

(iii) 110 – Uun

(iv) 112 – Uub

(v) 115 – Uup

(vi) 125 – Ubp

IUPAC coding

0	nil	n
1	un	u
2	bi	b
3	tri	t
4	quad	q
5	pent	p
6	hex	h
7	sept	s
8	oct	o
9	enn	e

Hence, (i)→ (s), (ii) → (t), (iii) → (u), (iv) → (p), (v) → (q), (vi) → (r)

DPP-2 **Periodic Trends-I**
(Radii, Screening Effect, Hydration Energy, etc)

1. (c)

$$\frac{AB}{2} = r_{\text{covalent}}$$
$$\frac{CD}{2} = r_{\text{crystal}}$$
$$\frac{EF}{2} = r_{\text{van der Waals'}}$$

2. (a)

(a) Be (4)
(having four electrons)

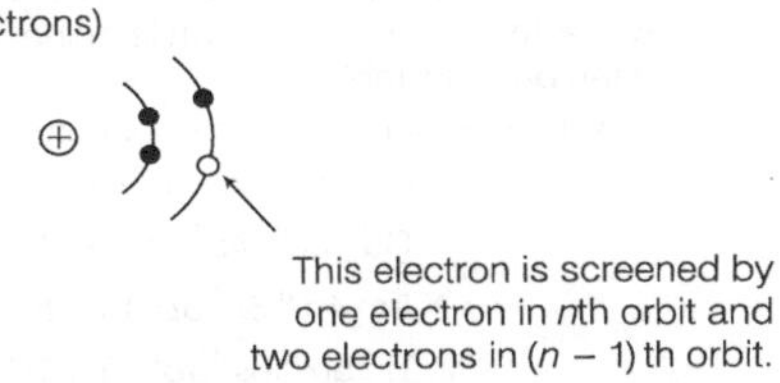

$$Z_{\text{eff}} = 4 - (1 \times 0.35 + 2 \times 0.85)$$
$$= 4 - (0.35 + 1.70)$$
$$= 1.95$$

(b) Be^+ (4)
(having three electrons)

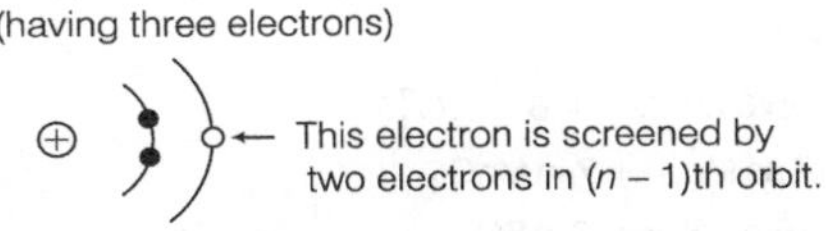

Thus, $Z_{\text{eff}} = 4 - (2 \times 0.85)$
$= 4 - 1.70$
$= 2.30$

(c) Be^{3+} (4)
(having only one electron)

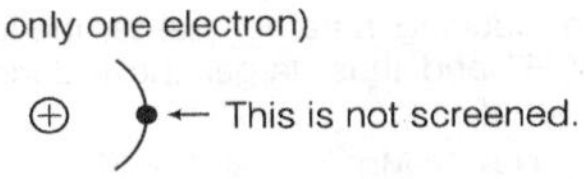

$$Z_{\text{eff}} = 4 - 0$$
$$= 4$$

3. (b) Ionic or atomic radius increases while moving down the group and decreases while moving along a period left to right.

r_n (Increases ↓, Decreases →)

Li^+ $Z = 3$	Be^{2+} $Z = 4$
Na^+ $Z = 11$	Mg^{2+} $Z = 12$

$Li^+ > Be^{2+}$
$Be^{2+} < Mg^{2+}$
$Li^+ < Na^+$

Thus, $Na^+ > Li^+ > Mg^{2+} > Be^{2+}$

4. (c) Ionic radius, $r_n = \frac{n^2 a_0}{Z}$

If species are isoelectronic, n is same

$$r_n \propto \frac{1}{Z}$$

Species	Electrons	Z
Cl^-	18	17
Ca^{2+}	18	20 (least size)
K^+	18	19
S^{2-}	18	16 (maximum size)

$$Ca^{2+} < K^+ < Cl^- < S^{2-}$$

5. (d) $r_n = \frac{n^2 a_0}{Z}$

It is applicable for cases having one electron in which screening effect is zero (due to no screening) and $Z = Z_{\text{eff}} = 1 - 0 = 1$
He^+, Li^{2+} and Be^{3+} are isoelectronic of hydrogen, hence

$$r_n(He^+) = \frac{a_0}{2}$$
$$r_n(Li^{2+}) = \frac{a_0}{3}$$
$$r_n(Be^{3+}) = \frac{a_0}{4}$$

6. (c) $Li(g) \longrightarrow Li^+(g) + e^-$

In Li, valence electron in second orbit is screened by two electrons in first orbit. Thus, effective nuclear charge.

$$Z_{\text{eff}} = Z - \sigma$$

where, Z = atomic number,
and σ = Slater's screening constant
= 0.35 for electrons in nth orbit
= 0.85 for electrons in $(n - 1)$ th orbit
= 1.00 for electrons in $(n - 2)$, $(n - 3)$

For 1s electron, $\sigma = 0.30$
Thus, $Z_{\text{eff}} = 3 - 2 \times 0.85$
$= 3 - 1.70$
$= 1.3$

When $Li(g)$ takes up electron,
Z_{eff} = 3 – (screening by one electron in nth orbit + screening by two electrons in $(n - 1)$ th orbit
$= 3 - (1 \times 0.35 + 2 \times 0.85)$
$= 0.95$

7. (d) $r_n = \frac{n^2 a_0}{Z_{\text{eff}}}$

Valence electron in Li, Na and K is screened by electrons in $(n - 1)$, $(n - 2)$...shell. Thus, effective nuclear charge decreases.

Li ($Z = 3$)
Z_{eff} = 3 – screening effect of two electrons in $(n - 1)$ orbit
$= 3 - 2 \times 0.85$
$= 1.3$

Na ($Z = 11$)
Z_{eff} = 11 – screening effect of 8 electrons in $(n - 1)$ and 2 electrons in $(n - 2)$ orbit
$= 11 - (8 \times 0.85 + 2 \times 1.0)$
$= 2.2$

K ($Z = 19$)
Z_{eff} = 19 – screening effect of 8 electrons in $(n - 1)$, 8 electrons in $(n - 2)$ and 2 electrons in $(n - 3)$
$= 19 - (8 \times 0.85 + 8 \times 1.0 + 2 \times 1.0)$
$= 19 - 16.8 = 2.2$

Thus, for Na and K, r_n is dependent on the value of n (orbit) but is independent of Z_{eff} (being 2.2).

8. (c) When an electron is added to the neutral atom, anion is formed.

$$X(g) + e^- \longrightarrow X^-(g)$$

Due to increase in number of electrons, force of attraction of electrons increases. This pulls the outer shell outwards hence, size of anion increases. Also, nuclear charge per electron is thus reduced and the electron cloud is held less tightly by the nucleus resulting in the expansion of the outer shell.

9. (a) $O^{2-} > O^- > O$

Since, $F^- > F$, $O > F$

Thus, $O^- > F^-$

and $O^{2-} > F^-$

Thus, $O^{2-} > O^- > F^- > F$

10. (d) $Ln \longrightarrow Ln^{3+} + 3e^-$

In Ln, screening power of 4*f* electrons is equal to unity, but after Ln^{3+} is formed, this screening power is reduced to 0.85 (since, it is now $(n-1)$, penultimate shell) and effective nuclear charge is increased by 0.15.

$$r_n \propto \frac{1}{Z_{eff}^2}$$

Thus, ionic radius decreases continuously from Ce^{3+} to Lu^{3+}.

11. (a) An *s*-electron penetrates nearer to the nucleus and is therefore more tightly held than a *p*-electron, and a *p*-electron is more tightly held than *d*-electron and so on. Thus, relative extent to which the various orbitals penetrate the electron clouds of other orbitals is

$$s > p > d > f$$

12. (d) Screening effect on the valence electron is experienced by the core electrons. He^+, Li^{2+} and Be^{3+} each has one electron. Thus, there is no core electron hence, no screening effect is observed.

13. (c) As given in graph I, valency increases from 1 to 7 while moving along a group in *s* and *p*-block. It is w.r.t. oxygen. From Graph II, valency increases from 1 to 4 and then decreases to 1. It is w.r.t. hydrogen.

	s-block		*p*-block				
Group	1	2	13	14	15	16	17
Elements	Na	Mg	Al	Si	P	S	Cl
Compounds with oxygen	Na_2O	MgO	Al_2O_3	SiO_2	P_2O_5	SO_3	Cl_2O_7
Valency w.r.t. oxygen	1	2	3	4	5	6	7
Compounds with hydrogen	NaH	MgH_2	AlH_3	SiH_4	PH_3	H_2S	HCl
Valency w.r.t. hydrogen	1	2	3	4	3	2	1

14. (c) As electrons in $(n-1)d$-orbitals are in an energy state comparable to *ns*-electrons, hence variable oxidation states are due to involvement of $(n-1)d$- and *ns*-electrons.

Mn (25): [Ar] $4s^2 3d^5$

Mn^{2+} if 4*s* electrons are involved,

Mn^{7+} if 4*s* and 3*d*-electrons are involved.

15. (d)

	Valence electrons	Valency
$X : ns^2np^1$	3	3
$Y : ns^2p^4$	6	-2

(*Y* is more electronegative than *X*)

Thus, $X^{3+}\ Y^{2-}$

Compound formed is X_2Y_3.

16. (c) Due to inert pair effect, stability of *N*th state decreases and that of $(N-2)$th state increases while going down the group in *p*-block of the periodic table. [*N* is the number of valence electrons], Valency

As : $3d^{10}4s^24p^3$, $N = 5 + 5, + 3$

Sb : $4d^{10}5s^25p^3$, $N = 5 + 5, + 3$

Bi : $5d^{10}6s^26p^3$ $N = 5 + 5, + 3$

Stability of $As^{5+} > Sb^{5+} > Bi^{5+}$

and that of $Bi^{3+} > Sb^{3+} > As^{3+}$

Thus, $BiCl_3$ is most stable.

17. (c) By inert pair effect, stability of *N*th state decreases and that of $(N-2)$th. state increases while going down the group in *p*-block of the periodic table.

(*N* is the number of valence electrons)

$$(n-1)d^{10}ns^2np^1, N = 3$$

Ga : $3d^{10}4s^24p^1$ Ga^+, Ga^{3+}

In : $4d^{10}5s^25p^1$ In^+, In^{3+}

Tl : $5d^{10}6s^26p^1$ Tl^+, Tl^{3+}

$Ga^{3+} > In^{3+} > Tl^{3+}$

$Ga^+ < In^+ < Tl^+$

Since, Tl^+ is most stable, hence $Tl^{3+} < Tl^+$ (stability) and reaction $Tl^{3+} + 2e^- \longrightarrow Tl^+$ is most spontaneous $\Delta G < 0$ (most negative).

18. (b) Group 6 as Cr = + 6 in CrO_4^{2-}

Group 7 as Mn = + 7 in MnO_4^-

Group 16 as S = + 6 in SO_3

Group 18 as Ne = 0

19. (c) Hydrides are compounds with H^- anion.

$$H^- \longrightarrow H + e^-$$

Thus, they are reducing in nature. Larger the ionic nature, larger the source of H^- and thus, larger the reducing nature. Since, polarising power of

$$Na^+ < Mg^{2+} < Al^{3+} < Si^{4+}$$

Thus, ionic nature of

$$NaH > MgH_2 > AlH_3 > SiH_4$$

Thus, reducing nature of

$$SiH_4 < AlH_3 < MgH_2 < NaH$$

20. (d) All of these oxides are amphoteric, soluble in base as well as in acid.

$$BeO + 2NaOH \longrightarrow Na_2BeO_2 + H_2O$$
$$BeO + 2HCl \longrightarrow BeCl_2 + H_2O$$

(Similarly for ZnO)

$$Al_2O_3 + 2NaOH \longrightarrow 2NaAlO_2 + H_2O$$
$$Al_2O_3 + 6HCl \longrightarrow 2AlCl_3 + 3H_2O$$
$$SnO_2 + 2NaOH \longrightarrow Na_2SnO_3 + H_2O$$
$$SnO_2 + 4HCl \longrightarrow SnCl_4 + 2H_2O$$

(Similarly for PbO_2)

21. (a) $Na^+H^- + H_2O \longrightarrow \underset{\text{Basic}}{NaOH} + H_2O$

22. (a) $NaH \longrightarrow Na^+ + H^-$

$$H^- \longrightarrow H + e^-$$
$$Na^+ + e^- \longrightarrow Na$$

23. (a) In a period from left to right, basic nature of oxides decreases. In a group top to bottom, basic nature of oxides increases.

	→ Acidic		
Basic ↓	Na_2O	MgO	Al_2O_3 (amphoteric)
	K_2O		

$$Na_2O < K_2O \text{ (basic)}$$
$$Na_2O > MgO > Al_2O_3 \text{ (basic)}$$

Basic oxides are soluble in acid.
Acidic oxides are soluble in base.
Amphoteric oxides are soluble in acid as well as in base.

$$\underset{\text{Basic}}{Na_2O} + 2HCl \longrightarrow 2NaCl + H_2O$$
$$\underset{\text{Basic}}{MgO} + 2HCl \longrightarrow MgCl_2 + H_2O$$
$$\underset{\text{Basic}}{Al_2O_3} + 6HCl \longrightarrow 2AlCl_3 + 3H_2O$$
$$\underset{\text{Acidic}}{Al_2O_3} + 4NaOH \longrightarrow 2NaAlO_2 + 2H_2O$$

Correct order is $Al_2O_3 < MgO < Na_2O < K_2O$

24. (a) Ions belong to alkaline earth metals

	Size	Charge	Hydration	Heat of hydration
Be^{2+}			Maximum ↑	Maximum ↑
Mg^{2+}				
Ca^{2+}				
Sr^{2+}				
Ba^{2+}	↓ Maximum	↓		
	Increases	Same	Decreases	

Smaller size, larger charge then larger hydration and hydration energy.

Thus, $Ca^{2+} > Sr^{2+} > Ba^{2+}$

25. (c) In aqueous medium, ion migrates towards oppositely charged electrode. Cations migrate to cathode and anions migrate to anode under the application of electric field. Smaller the size of cation (Li^+ in (a) and Al^{3+} in (b)), larger the charge, larger the hydration, then, larger the size of the hydrated ion, heavier the hydrated ion and smaller the ionic mobility.

$$Li^+ < Na^+ < K^+$$
$$\xrightarrow{\text{Size of the ion}}$$

Charge	Same
Hydration	$K^+ < Na^+ < Li^+$
Size of hydrated ion	$K^+ < Na^+ < Li^+$
Ionic mobility	$Li^+ < Na^+ < K^+$
Similarly,	$Al^{3+} < Mg^{2+} < Na^+$

26. (b) Due to inert pair effect, Ga, In and Tl show two types of oxidation state + 3 and +1.

Due to this, $Ga^{3+} > In^{3+} > Tl^{3+}$

and $Ga^+ < In^+ < Tl^+$

Also $Ga^+ < Ga^{3+}$

$$Ga^+ \longrightarrow Ga^{3+} + 2e^-$$

Thus, spontaneous and $\Delta G° < 0$.

Thus, Statement I and Statement II are correct but Statement II is not the correct explanation of Statement I.

27. (b)

$$Na \longrightarrow Na^+ + e^-$$

⊕))) (2, 8, 1) ⊕)) (2, 8)

$$Mg \longrightarrow Mg^+ \longrightarrow Mg^{2+}$$

⊕))) (2, 8, 2) ⊕))) (2, 8, 1) ⊕)) (2, 8)

When a positive ion is formed, there may be removal of outermost shell (as in Na to Na^+ and Mg^+ to Mg^{2+}). Also the number of positive charges on the nucleus exceeds the number of orbital electrons and the effective nuclear charge is increased. This results in the increase in inward attraction of electrons and thus, reducing size of cation.

Thus, $Na^+ < Na$ (Size)

$$Mg^{2+} < Mg^+ < Mg$$

Thus, Statement I and Statement II are correct but Statement II is not the correct explanation of Statement I.

28. (b) Fluorine (9) $1s^2 2s^2 2p^5$

It is one short of electron than octet. Hence, it can form F^- ion by taking one electron. It can also form (F—F) covalent bond, and covalent radius is considered.

Neon (10) : $1s^2 2s^2 2p^6$

It has its octet complete. Thus, it is monoatomic gas.It is noble gas and van der Waals' radius is considered.

van der Waals' > covalent radius hence, Statement I and Statement II are correct but Statement II is not the correct explanation of Statement I.

29. (a) The lanthanide contraction cancels almost exactly the normal size increase on descending a group of transition elements. Thus, covalent and ionic radii of Nb and Ta and also that of Zr and Hf are same. Thus, Statement I and Statement II are correct and Statement II is the correct explanation of Statement I.

30. (a)

(i) Na_2O dissolves in acid forming salt thus, it is a basic oxide.

$$\underset{\text{Basic}}{Na_2O} + \underset{\text{Acid}}{2HCl} \longrightarrow 2NaCl + H_2O$$

Thus, **(i) → (s)**

(ii) Al_2O_3 dissolves in acid as well as in base thus, it is an amphoteric oxide.

$$\underset{\text{Basic}}{Al_2O_3} + \underset{\text{Acid}}{6HCl} \longrightarrow 2AlCl_3 + 3H_2O$$
$$\underset{\text{Acidic}}{Al_2O_3} + \underset{\text{Base}}{2NaOH} \longrightarrow 2NaAlO_2 + H_2O$$

Thus, **(ii) → (q)**

(iii) Cl_2O_7 dissolves in base thus, it is an acidic oxide.

$$\underset{\text{Acidic}}{Cl_2O_7} + \underset{\text{Base}}{2NaOH} \longrightarrow 2NaClO_4 + H_2O$$

Thus, **(iii) → (r)**

(iv) CO is a neutral oxide. It is neither soluble in acid nor in base.

Thus, **(iv) → (p)**

DPP-3 **Periodic Trends-II**
(IE, Electron Affinity and Electronegativity)

1. (a) Force of attraction between electron and nucleus is

$$F = \frac{k Z_1 Z_2 e^2}{r_n^2}$$

and $\quad$ IE = –Energy of electron

$$= -\left(-\frac{E_H Z^2}{n^2}\right)$$

$$(IE)_z = \frac{E_H Z^2}{n^2}$$

where, E_H (= – 13.6 eV) is the energy of electron of H-atom.

In alkali metals, effective nuclear charge almost remains constant.

Alkali metals	Z	Z_{eff}
Li	3	1.3
Na	11	2.2
K	19	2.2
Rb	37	2.2
Cs	55	2.2
Fr	81	2.2

Thus, IE is dependent mainly on r_n (radius) and is independent of the values of effective nuclear charge.

2. (b)

I. $Na(g) \longrightarrow Na^+(g) + e^-$; IE = 5.1 eV

II. $Na^+(g) + e^- \longrightarrow Na$

$$\Delta H_{eg} = x$$

Process II is just reverse of I and represents in which electron is gained. Energy change is called electron gain enthalpy.

Thus, $\quad x = -5.1$ eV

3. (c) Mg with paired electrons in $3s^2$ has greater IP as compared to Na as well as Al.

IP increases while moving along a period from left to right.

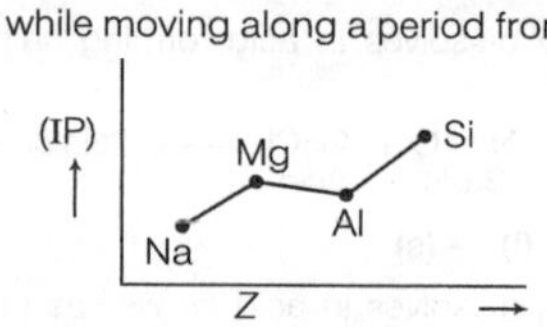

4. (b) IP increases while moving along a period and decreases while moving along a group.

(b) [Ne] $3s^2 3p^3$ with three unpaired half-filled electrons has maximum IP .

(d) [Ar] $3d^{10}\ 4s^2 4p^3$ is in same group as (b) but lies below (b). Thus, IP of (d) is smaller than that of (b).

(a) [Ne] $3s^2$ | 1 | | |

(b) [Ne] $3s^2$ | 1 | 1 | 1 |

(c) [Ne] $3s^2$ | 1 | 1 | |

5. (a) By Mulliken- Jaffe electronegativity rule,

$$\text{Electronegativity} = \left(\frac{\text{Ionisation potential} + \text{Electron affinity}}{2}\right)$$

$$E_1 = \left(\frac{E_2 + E_3}{2}\right)$$

$\therefore\ E_3$ (electron affinity) = $(2E_1 - E_2)$

6. (b) $(IE)_1 < (IE)_2 << (IE)_3 < (IE)_4$

$(IE)_3$ is very high thus, M^{2+} is stable with inert gas configuration.

Thus, element belongs to alkaline earth metals.

7. (c) $M(g) \longrightarrow M^+(g) + e^-$; $(IE)_1 = 100$ eV

$M^+(g) \longrightarrow M^{2+}(g) + e^-$; $(IE)_2 = x$ eV

$M(g) \longrightarrow M^{2+}(g) + 2e^-$

$(IE)_1 + (IE)_2 = 100 + x = 250$ eV

$x = (IE)_2 = 150$ eV

Also $(IE)_2$ of $M(g)$ is also the $(IE)_1$ of $M^+(g)$.

Thus, (a) is correct,

(b) is correct; (c) is incorrect; (d) is correct.

8. (a) As we move from left to right in $3d$ and $4d$-series, successive addition of d-electrons screens the outer electrons (ns) from the inward pull of the nucleus. As a result of this, size does not vary too much. Hence, IE values which depend on the effective nuclear charge and radius, do not vary too much.

Elements	Sc	Ti	V	Cr	Mn	Fe	Co	Ni	Cu	Zn
Radius (pm)	164	147	135	130	135	126	125	125	128	137
IE (kJ mol^{-1})	631	656	650	652	717	762	758	736	775	906

9. (c) IE of $5d$-elements are higher than those of the $3d$ and $4d$-elements. This is due to

I. Greater effective nuclear charge thus, (a) is incorrect.

II. Size of $5d > 3d$ and $4d$ thus, (b) is incorrect.

III. Weak shielding of the nucleus by $4f$-electrons thus, (c) is correct.

10. (c) Zn, Cd, Hg — $ns^2\ (n-1)d^{10}$

Ga, In, Tl — $ns^2\ (n-1)d^{10}\ np^1$

In the formation of $Ga^+/In^+/Tl^+$, unpaired electron from np orbital is removed while in the formation of $Zn^+/Cd^+/Hg^+$, paired electron from ns orbital is removed.

Since, energy of $np > ns$ and unpaired electron is removed more easily than paired electron, hence following the transition elements, $(IE)_1$ drops abruptly in Ga, In and Tl.

11. (b)

Li	Be	B	C	N
Na	Mg		Si	
K	Ca		Ge	

IP decreases along a group moving top to bottom and increases along a period from left to right.

But IP of Be > B

Thus, (a) Li > Na > K. Thus, given order is incorrect.

(b) Be > Mg > Ca. Thus, given order is correct.

(c) B < C < N. Thus, given order is incorrect.

(d) Ge < Si < C. Thus, given order is incorrect.

12. (c) When an electron is added to a negative ion, a strong repulsion is felt and energy of the system increases.

$$O^- + e^- \longrightarrow O^{2-}$$

$$EA = 702 \text{ kJ mol}^{-1}$$

Thus, process is unfavourable in the gas phase because the resulting increase in electron-electron repulsion overweighs the stability gained by achieving the noble gas configuration. However, O^{2-} (oxide ion) is common in ionic compounds and O^{2-} in solids is stabilised by neighbouring ions.

Na^+ stabilises O^{2-} in Na_2O.

Ca^{2+} stabilises O^{2-} in CaO.

Thus, (a) and (b) are correct.

$$O + e^- \longrightarrow O^-$$
$$\Delta H_{eg} = -142 \text{ kJ mol}^{-1}$$
$$S + e^- \longrightarrow S^-$$
$$\Delta H_{eg} = -200 \text{ kJ mol}^{-1}$$

Thus, EA of S > O.

Thus, (c) is incorrect.

13. (d)

	Electron affinity (kJ mol^{-1})
$F(g) + e^- \longrightarrow F^-(g)$,	−328
$Cl(g) + e^- \longrightarrow Cl^-(g)$,	−348
$S(g) + e^- \longrightarrow S^-(g)$,	−200
$P(g) + e^- \longrightarrow P^-(g)$,	−74

Thus, most negative electron affinity is for Cl and least negative electron affinity is for P.

14. (c) IP increases from left to right, but IP also depends whether electrons are paired or unpaired.

Paired electron requires higher IP than unpaired electron.

Half-filled and completely filled orbital are stable and have higher IP.

$$C(6) \xrightarrow{(IP)_1} C^+ \xrightarrow{(IP)_2} C^{2+}$$
[1| |] $2s^1$

$$N(7) \xrightarrow{(IP)_1} N^+ \xrightarrow{(IP)_2} N^{2+}$$
[1|1|]

$$O(8) \xrightarrow{(IP)_1} O^+ \xrightarrow{(IP)_2} O^{2+}$$
[1|1|1]

$$F(9) \xrightarrow{(IP)_1} F^+ \xrightarrow{(IP)_2} F^{2+}$$
[⥮|1|1]

O^+ is most stable that has maximum $(IP)_2$. In other cases, $(IP)_2$ depends on atomic number Z, greater the value of Z, larger the $(IP)_2$.

Thus, $\underset{8}{O} > \underset{9}{F} > \underset{7}{N} > \underset{6}{C}$

15. (c) Variation of IP along a period with atomic number (Z) is shown below

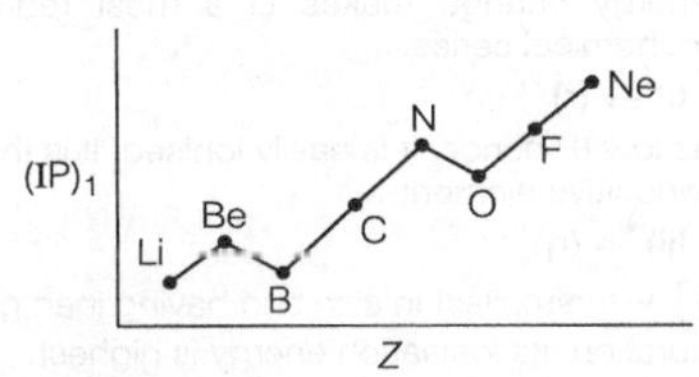

N has completely half-filled orbitals and is most stable in given elements

N [1|1|1] Completely half-filled

O_8 [⥮|1|1]

(IP) of O < (IP) of N

If O changes to O^+, all electrons are half-filled and becomes more stable.

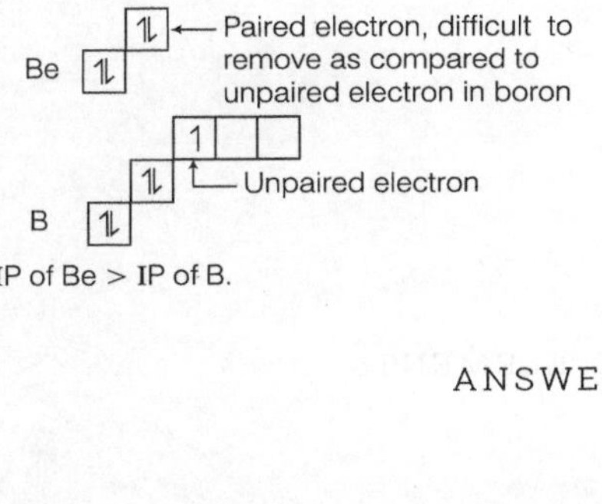

Thus, IP of Be > IP of B.

16. (b)

(a) Al (13) : [Ne] $3s^2$ [1| |] Unpaired electron in Al is removed easily.

Mg (12) : [Ne] [⥮] $3s^2$ Paired electron in Mg is difficult to remove easily.

Thus, $(IP)_1$ of Al < $(IP)_1$ of Mg. Thus, correct.

(b) Mg (12) : [Ne] [⥮] $3s^2$ $\xrightarrow{(IP)_1}$ [Ne] [1] $3s^1$ $\xrightarrow{(IP)_2}$ [Ne] $\xrightarrow{(IP)_3}$

Na (11) : [Ne] [1] $3s^1$ $\xrightarrow{(IP)_1}$ [Ne] $1s^2 2s^2p^6$ $\xrightarrow{(IP)_2}$ $1s^2 2s^2 p^5$ $\xrightarrow{(IP)_3}$

Unpaired electron is removed by $(IP)_2$ of Mg.

While paired electron from stable configuration of Ne is to be removed by $(IP)_2$ of Na.

Thus, $(IP)_2$ of Mg << $(IP)_2$ of Na

Thus, Statement (b) is incorrect.

(c) As in (b), $(IP)_1$ of Na < $(IP)_1$ of Mg. Thus, correct.

(d) As in (b), $(IP)_3$ of Mg > $(IP)_3$ of Na as paired electron from stable configuration is to be removed from Mg by $(IP)_3$.

Thus, correct.

17. (c) IE increases along a period (L to R) and decreases along a group top to bottom.

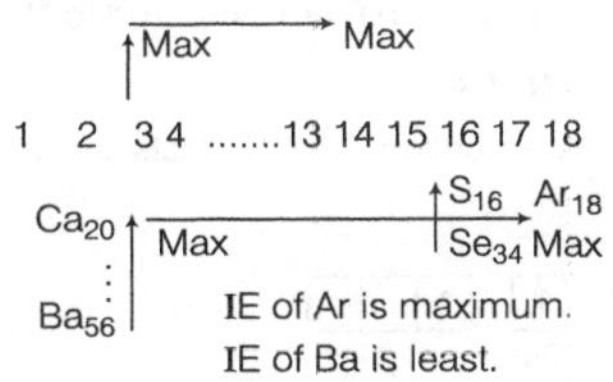

Thus, Ba < Ca < Se < S < Ar

18. (d)

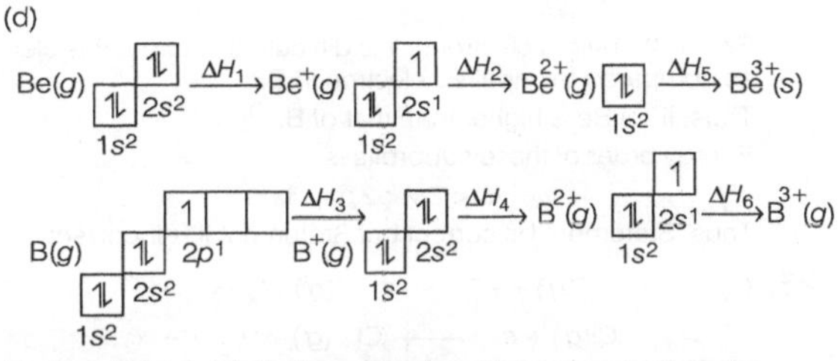

Removal of paired electrons requires higher energy than the unpaired electrons. Also inert gas has maximum IE.

Thus,
$$\Delta H_1 > \Delta H_3$$
$$\Delta H_4 > \Delta H_2$$
$$\Delta H_5 >> \Delta H_6$$

(Be^{2+} has [He] configuration)

19. (c) EC of an alkali metal is

[Inert gas] ns^1

By $(IE)_1$, alkali metal changes to monovalent cation M^+ with stable inert gas configuration from which removal of electron requires very high energy. Thus, $(IE)_2 >>> (IE)_1$. There is high jump from $(IE)_1$ to $(IE)_2$ in element R.

Thus, element R is alkali metal.

20. (b) $M(g) \xrightarrow{(IE)_1} M^+(g) + e^-$

$$M^+(g) \xrightarrow{(IE)_2} M^{2+}(g) + e^-$$

When an electron is removed from gaseous atom by $(IE)_1$, $M^+(g)$ formed has more force of attraction for outermost electron hence, removal of electron from $M^+(g)$ requires higher energy than $(IE)_1$.

In general, $(IE)_1 < (IE)_2 < (IE)_3 \ldots$

21. (d)

(a) Be (4) [⥮] ← Paired electrons in Be requires more energy than unpaired electron in B

[⥮] $2s^2$

$1s^2$

B(5) [1][ ][ ] $2p^1$ Thus, (a) is correct

[⥮] $2s^2$

[⥮] $1s^2$

(b) N(7) [1][1][1] Stable being half-filled $2p^3$

O(8) [⥮][1][1] Paired electrons in O(8) is removed easily as O^+ attains stable configuration.

Thus, (b) is correct.

(c) As we go down the group top to bottom, atomic size increases, effective nuclear charge remains almost constant. Thus, ionisation energy decreases. Thus, (c) is correct.

(d) He $\longrightarrow$ $He^+ + e^-$, $(IE)_1$

$He^+ \longrightarrow He^{2+} + e^-$, $(IE)_2$

$(IE)_2 > (IE)_1$,

Thus, (d) is incorrect.

22. (a) Effective nuclear charge of F is higher than that of N, thus added electron is more in F as compared to N.

23. (b) EN of Y > EN of X

Thus, X exists as cation and Y exists as anion, forming ionic compound X^+Y^-.

24. (c)

B_5 [⥮] $1s^2$ [⥮] $2s^2$ [1][ ][ ] $2p^1$

Be_4 [⥮] $1s^2$ [⥮] $2s^2$

Be has $2s$ paired electrons. It is difficult to take out this electron as compared to unpaired electron in B.

Thus, IE of Be is higher than that of B.

Energy order of these suborbits is

$$1s < 2s < 2p < 3s \ldots$$

Thus, Statement I is correct but Statement II is incorrect.

25. (c)

$$F(g) + e^- \longrightarrow F^-(g), H_{eg} = -x_1$$

$$Cl(g) + e^- \longrightarrow Cl^-(g), \Delta H_{eg} = -x_2$$

Electron affinity is the energy released ($\Delta H < 0$) when an electron is added to the gaseous atom to form gaseous anion.

Electron affinity depends on

(i) Force of attraction, $F = \dfrac{Z_1 Z_2 e^2}{r^2}$

(ii) Atomic number (Z).

Smaller the size of the atom, greater the force of attraction and thus, larger the energy released (more negative).

Electron affinity of F should be higher than that of Cl.

But ΔH_{eg} (F) = -328 kJ mol^{-1}

ΔH_{eg} (Cl) = -349 kJ mol^{-1}

It is due to the greater effectiveness of $2p$ electrons in small F-atom to repel the additional electron entering the atom than $3p$-electrons in the larger Cl-atom.

Thus, Statement I is correct and Statement II is incorrect.

26. (b) Oxidation state of Mn in $MnO_4^- = +7$

Oxidation state of I in $IF_7 = +7$

Thus, maximum oxidation state is in its oxide and fluoride. Oxygen and fluorine are of smaller size and are of higher electronegativity. Thus, they can oxidise the element to its highest oxidation state. As we move along a period from left to right, size decreases, and tendency to take electron (electronegativity) increases. Thus, Statement I and Statement II are correct but Statement II is not the correct explanation of Statement I.

27. (c)

Li → Be (EN) increases

Na Mg ↓ (EN) decreases

While moving from Li to Be, electronegativity (EN) increases and while moving from Be to Mg, electronegativity decreases such that Li and Mg have same values of EN. Thus, they resemble each other in many properties.

Li_2CO_3 and $MgCO_3$ are decomposed on heating.

$$Li_2CO_3 \xrightarrow{\Delta} Li_2O + CO_2$$

$$MgCO_3 \xrightarrow{\Delta} MgO + CO_2$$

Li_2CO_3 differ from Na_2CO_3 which is stable to heat.

Thus, Statement I is correct and Statement II is incorrect.

28. (b) Be → B (EN) increases

Mg Al ↓ (EN) decreases

While moving from Be to B, electronegativity (EN) increases and while moving from B to Al, EN decreases such that Be and Al show similarity in many properties. It is called diagonal relationship.

$$Be_2C + 4H_2O \longrightarrow 2Be(OH)_2 + CH_4$$

$$Al_4C_3 + 12H_2O \longrightarrow 4Al(OH)_3 + 3CH_4$$

Both are decomposed by H_2O forming CH_4. Thus, Statement I and Statement II are correct but Statement II is not the correct explanation of Statement I.

29. (a) (i) Li $\longrightarrow$ $Li^+ + e^-$

Li is strongest reducing agent as Li^+ is maximum hydrated.

$$Li \longrightarrow Li^+ + e^-$$

$$Li^+ + aq \longrightarrow Li^+(aq)$$

Net energy change makes Li a most reducing agent in electrochemical series.

Thus, **(i) → (r)**

(ii) Ca has low IP hence, it is easily ionised. It is thus electropositive element.

Thus, **(ii) → (q)**

(iii) He [⥮] with smallest in size and having inert gas configuration, its ionisation energy is highest.

Thus, **(iii) → (s)**

(iv) Cl has highest electron gain enthalpy among halogen family.

Thus, **(iv) → (p)**

30. (d)

(i) → (p,q,r,s)

(ii) → (p,q,r,s)

(iii) → (p,q,r,s)

(iv) → (p,q)

Revisal Problems (JEE Main)

1. (d)

(a) Successive addition of d-electron screens the outer electrons ($4s$) from the inward pull of the nucleus. As a result of this, the size of the atom does not alter much. Thus, Fe, Co, Ni are approximately equal size.

(b) Zr and Hf and (c) also Nb, Ta have equal size due to lanthanide contraction.

2. (d) Size decreases

Size decreases →

O^{2-}	F^-
S^{2-}	Cl^-
Se^{2-}	Br^-
Te^{2-}	I^-

↓ Size increases

$$O^{2-} > F^-$$
$$I^- > Br^- > F^-$$
$$Se^{2-} > O^{2-}$$
$$I^- > Se^{2-}$$

Thus, order is $I^- > Se^{2-} > Br^- > O^{2-} > F^-$

3. (a) There are three types of radii; crystal, van der Waals' and covalent. Their order being

Covalent < Crystal < van der Waals'

90 pm   100 pm   112 pm

4. (d)

(a) Correct (Repulsive forces just balances attractive forces)

(b) Correct (Repulsive forces overcomes attractive forces)

(c) Correct (normal gradation)

(d) Ti < Zr, thus correct (normal gradation)

Zr = Hf, thus incorrect (due to lanthanide contraction)

Thus, (d) is incorrect.

5. (d) Consider the case of alkali metals

$Li_3 = 1s^2 2s^1$ (No d-orbital)

(Size) ↓ (EN) ↑ (IP) ↑

$Na_{11} = 1s^2 2s^2 2p^6 3s^1$

$K_{19} = [Ar] 4s^1$ (Vacant d-orbitals present)

$Rb_{37} = [Ar] 4s^2 3d^{10} 4p^6 5s^1$

Li lacks d-orbital.

Li is smallest in size.

Li has maximum (EN) and (IP).

It differs from other alkali metals.

6. (d)

(a) Properties of elements are periodic function of atomic numbers. Thus, correct.

(b) Number of non-metals in periodic table is lesser than that of metals. Thus, correct.

(c) IE of

$$Li < Be > B < C < N > O < F < Ne$$

(Be > B: Irregular; N > O: Irregular)

Thus, correct

(d) $V(23) = [Ar]\, 3d^3 4s^2$

$Cr(24) = [Ar]\, 3d^5 4s^1$

$Ni(28) = [Ar]\, 3d^8 4s^2$

$Cu(29) = [Ar]\, 3d^{10} 4s^1$

Thus, d-subshell is not filled monotonically.

Thus, Statement (d) is incorrect.

7. (d) Group 14 contains C, Si, Ge, Sn and Pb out of Ge, Sn and Pb are strongly electropositive elements.

8. (a) $4f$-series is called lanthanide series and differentiating electron goes into $4f$.

$$Ce(58) = [Xe]_{54}\, 4f^1\, 5d^1\, 6s^2$$
$$Pr(59) = [Xe]_{54}\, 4f^3\, 5d^0\, 6s^2$$
$$\vdots$$
$$Gd(64) = [Xe]_{54}\, 4f^7\, 5d^1\, 6s^2$$

due to more stability of all unpaired electrons of $4f$ orbitals.

$$Yb(70) = [Xe]_{54}\, 4f^{14}\, 5d^0\, 6s^2$$
$$Lu(71) = [Xe]_{54}\, 4f^{14}\, 5d^1\, 6s^2$$

due to more stability of all paired electrons in $4f$ orbitals.

9. (a) $Cu(29) = 1s^2 2s^2 2p^6 3s^2 3p^6 3d^{10} 4s^1$

If Aufbau rule is not followed, last filling electron (called differentiating electron) goes into $4s$. Thus, s-block element.

10. (c) $5s^2\, 5p^4$

Valence electrons = 6

Thus, group = 10 + 6 = 16

$n = 5$ Thus, period = 5

11. (c) p-orbital is being filled, thus it is a p-block element.

$$[Ar]\, 3d^{10}\, 4s^2\, 4p^3$$

Group = 15 (V A)

Valence electrons (x) = 5

Period (P) = 4

If (P + 1) = x then metalloid.

(P + 1) > x then metal.

(P + 1) < x then non-metal.

12. (a)

Element	Na	Mg	Al	Si	P	S	Cl
Compound with oxygen	Na_2O	MgO	Al_2O_3	SiO_2	P_2O_5	SO_3	Cl_2O_7
Valency w.r.t. O-atom (taking it –2)	1	2	3	4	5	6	7

Thus, maximum valency w.r.t. O-atom is shown by halogen family.

13. (b) Best oxidising agent is that one which is reduced most easily or reduced state is more stable than oxidised state.

$$Ge^{4+} + 2e^- \longrightarrow Ge^{2+}$$
$$Pb^{4+} + 2e^- \longrightarrow Pb^{2+}$$
$$Sn^{4+} + 2e^- \longrightarrow Sn^{2+}$$

Due to inert-pair effect, stability of

$$Ge^{2+} < Sn^{2+} < Pb^{2+}$$

and $$Ge^{4+} > Sn^{4+} > Pb^{4+}$$

Thus, Pb^{2+} is most stable in +2 state and Ge^{4+} is most stable in +4 state.

Thus, oxidising nature

$$Ge^{4+} < Sn^{4+} < Pb^{4+}$$

and reducing nature

$$Ge^{2+} > Sn^{2+} > Pb^{2+}$$

14. (d) As number of electrons in d increases, screening effect increases.

$$s < p < d < f$$

Thus, d-electrons in Cu^{2+} shield the nucleus poorly, increasing force of attraction. Also, $Cu^{2+} < Ca^{2+}$ (size)

Hence, Cu^{2+} has stronger polarising power than Ca^{2+}.

15. (d) F_2 has the most negative $\Delta G°$ value, hence it is the strongest oxidising agent.

$$\frac{1}{2}F_2(g) \longrightarrow F(g),\ x_1$$

$$F(g) + e^- \longrightarrow F^-(g),\ x_2$$

$$F^-(g) + aq \longrightarrow F^-(aq),\ x_3$$

$$\Delta G° = x_1 + x_2 + x_3$$

where, x_1 = free energy of dissociation

x_2 = electron affinity

x_3 = free energy of hydration

	$\Delta G°$ (kJ mol^{-1})
F_2	−730
Cl_2	−592
Br_2	−54.85
I_2	−486.5

$F^-(g)$ with smallest size has maximum free energy of hydration thus, maximum oxidising power.

16. (c) For Ni : $(IE)_1 + (IE)_2 + (IE)_3 + (IE)_4 = 11.29$ kJ mol^{-1}

and for Pt : $(IE)_1 + (IE)_2 + (IE)_3 + (IE)_4 = 9.36$ kJ mol^{-1}

Thus, $(IE)_1 + (IE)_2$ for Ni is smaller than that for Pt, thus nickel (II) compounds tend to be thermodynamically more stable than platinum (II).

$(IE)_{total}$ for Ni (IV) is higher than that for Pt (IV), thus Pt(IV) compounds are thermodynamically stable than Ni (IV) compounds.

17. (b) MgO is most stable.

Stability is based on the $\Delta H_f°$ value per oxygen atom. Most negative value of $\Delta H_f°$ per oxygen atom indicates most stable oxide.

$\Delta H_f°$ per oxygen atom

Na_2O = − 416 kJ mol^{-1}

MgO = − 602 kJ mol^{-1} (most negative)

$$Al_2O_3 = -\frac{1676}{3} = -\ 559 \text{ kJ mol}^{-1}$$

$$SiO_2 = -\frac{911}{2} = -\ 456 \text{ kJ mol}^{-1}$$

$$P_4O_{10} = \frac{-\ 2984}{10} = -\ 298.4 \text{ kJ mol}^{-1}$$

$$SO_3 = -\frac{395}{3} = -\ 132 \text{ kJ mol}^{-1}$$

18. (a) Oxide of *A* is ionic thus, ion of *A* is largest in size and smallest atomic number. Oxide of *C* is giant molecule (covalent) thus it is covalent, thus smallest in size and largest atomic number.

Oxide of *B* is amphoteric thus having atomic number in between *A* and *C*.

$$A < B < C$$

19. (b) K^+ (first), Na^+, Li^+

Smaller the size of the cation, larger the charge on the cation, then larger the hydration and hydration energy and larger the size of the hydrated ion. Ion becomes heavier and ionic velocity is retarded when electric field is applied.

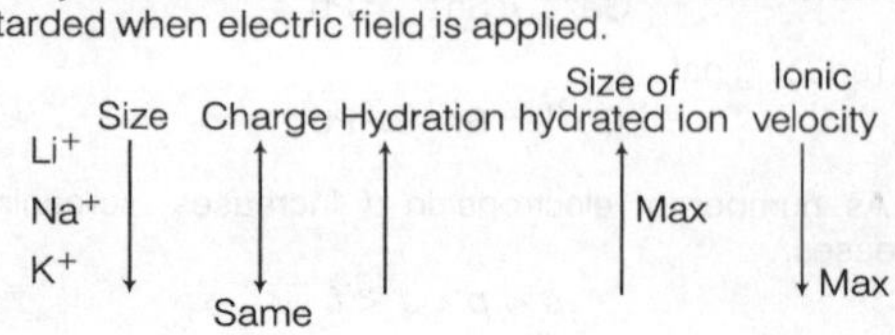

20. (d) Li_2O

Na_2O MgO Al_2O_3 SiO_2 P_2O_5 SO_3 Cl_2O_7

K_2O

Basic → Acidic

(a) Na_2O, MgO (Basic)

Al_2O_3, (Amphoteric)

$SiO_2 < P_2O_5 < SO_3 < Cl_2O_7$ (Acidic nature)

Thus, (a) is correct.

(b) $Li_2O < Na_2O < K_2O$ (Basic nature)

Thus, (b) is correct.

(c) H_2O is oxide of hydrogen. It is H^+ ion donor as well as H^+ ion acceptor, thus it is amphiprotic.

Thus, (c) is correct.

$$H_2O + H_2O \longrightarrow H_3O^+ + OH^-$$

(H^+ transferred; $-H^+$)

21. (a) When values of bond energies are in kcal mol^{-1} then by Pauling's electronegativity scale

$$\chi_F - \chi_H = 0.182\sqrt{\Delta_{H-F}}$$

where, Δ_{H-F} (stabilisation energy)

$$= (BE)_{HF} - \sqrt{(BE)_{H-H}\ (BE)_{F-F}}$$

$$= 134.6 - \sqrt{104.2 \times 36.6}$$

$$= 72.84$$

$$\therefore \quad \chi_F = \chi_H + 0.182\sqrt{72.84}$$

$$= 2.05 + 1.55 = 3.60$$

22. (d) When EN difference of *X* and H is 1.7 then H*X* is 50% ionic and 50% covalent.

23. (d) For *X*, $(IE)_2$ is very high, thus X^+ attains inert gas configuration. It is Na.

$$Na(11) \longrightarrow Na^+ + e^-$$

[Ne] $3s^1$ [Ne]

Stable, $(IE)_2$ is very high

For *Y*, $(IE)_3$ is very high thus, Y^{2+} attains inert gas configuration. It is Mg.

$$Mg(12) \longrightarrow Mg^{2+} + 2e^-$$

[Ne] $3s^2$ [Ne]

Stable, $(IE)_3$ is very high.

For *Z*, $(IE)_1$ is very high. It is [Ar].

24. (a) 376×10^3 J of energy absorbed cause ionisation of N_0 atoms to Cs^+.

1 J of energy absorbed causes ionisation of

$$= \frac{N_0}{376 \times 10^3} \text{ atoms to } Cs^+$$

$$= \frac{6.02 \times 10^{23}}{376 \times 10^3}$$

$$= 1.60 \times 10^{18} \text{ atoms to } Cs^+$$

25. (d) Let ionisation energy per atom = IE and electron affinity = EA

$$X(g) \longrightarrow X^+(g) + e^-,\ \Delta H = IE$$

$$\text{For } \frac{N_0}{2} \text{ atoms, } \Delta H = \Delta H_1° = \frac{N_0}{2}(IE)$$

$$\therefore \quad (IE) = \frac{2\Delta H_1°}{N_0} \text{ per atom}$$

$$X(g) \longrightarrow X^+(g) + e^-, \Delta H = \frac{N_0}{2}\text{(IE)}$$

$$X(g) + e^- \longrightarrow X^-(g), \Delta H = -\frac{N_0}{2}\text{(EA)}$$

$$\frac{N_0\text{(IE)}}{2} - \frac{N_0}{2}\text{(EA)} = \Delta H_2^\circ$$

$$\Delta H_1^\circ - \frac{N_0}{2}\text{(EA)} = \Delta H_2^\circ$$

$$-\frac{N_0}{2}\text{(EA)} = \Delta H^\circ_2 - \Delta H_1^\circ$$

$$\therefore \quad -\text{(EA)} = 2\frac{(\Delta H_2^\circ - \Delta H_1^\circ)}{N_0}$$

26. (d) By Fajan's rule, smaller the size of cation, larger the size of anion, larger the charge on the ion. d, f-electrons present then, larger the polarising power and larger the distortion hence, larger the covalent nature of compounds. Si^{4+} with largest charge and smallest size has maximum polarising power.

27. (d)

Period	Suborbit	Orbitals	Electrons (elements)	Total
2	$2s$	1	$1 \times 3 = 3$	
	$2p$	3	$3 \times 3 = 9$	12
4	$4s$	1	$1 \times 3 = 3$	
	$3d$	5	$5 \times 3 = 15$	
	$4p$	3	$3 \times 3 = 9$	27

28. (a) The properties of elements are periodic functions of atomic numbers. If elements are arranged in increasing atomic numbers, there is repetition of properties after 2, 8, 8, 18, 18 and 32 elements. Based on modern periodic law, these numbers are called **magic numbers**.

29. (d)

Period	Suborbit	Orbitals	Elements	Total
3	$3s$	1	3	
	$3p$	3	9	27
	$3d$	5	15	
4	$4s$	1	3	
	$4p$	3	9	48
	$4d$	5	15	
	$4f$	7	21	
5	$5s$	1	3	75
	$5p$	3	9	
	$5d$	5	15	
	$5f$	7	21	
	$5g$	9	27	

30. (a) Electrons in various subshells have been filled based on Aufbau rule (ascending energy rule). Suborbits in increasing order are placed as

$$1s < 2s < 2p < 3s < 3p < 4s < 3d < 4p < 5s <$$
$$4d < 5p < 6s < 4f < 5d < 6p < 7s < 5f \ldots$$

Thus, $3d$-suborbit which is filled after $4s$ is a part of 4th period. $4d$-suborbit is filled after $5s$ and is a part of 5th period and so on. Maximum capacity of $s = 2$; $p = 6$; $d = 10$, $f = 14$.

Period	Suborbit	Elements	Total elements	Comment
1	$1s$	2	2	
2	$2s$	2	8	
	$2p$	6		
3	$3s$	2	8	
	$3p$	6		
4	$4s$	2		
	$3d$	10	18	$3d$ is in 4th period
	$4p$	6		
5	$5s$	2		
	$4d$	10	18	$4d$ is in 5th period
	$5p$	6		
6	$6s$	2		
	$4f$	14	32	$4f$ and $5d$ are in 6th period
	$5d$	10		
	$6p$	6		

31. (d) M^{2+} (divalent cation) has 24 electrons.
Thus, 26 protons.

$$Z = 26$$
$$\text{EC} = [\text{Ar}]\, 4s^2 3d^6$$

Thus, group = 8 (VIII B)

32. (c) $(n-1)\, s^2p^6d^x ns^y$

$$n = 4$$
$$x = 6$$
$$y = 2$$
$$M : 3s^2p^6d^6 4s^2$$
$$M^{2+} : 3s^2p^6d^6$$

Protons = Z (atomic number) = 26

33. (b) Number of elements in a group = $2n^2$. There is no separate block for lanthanides and actinides, thus

n (period)	Elements
1	2
2	8
3	18
4	32
5	50
6	02
	112

In all six periods, two elements being in sixth period.

34. (a) Mass number of B is an average of A and C.

$$m_B = 23 = \frac{m_A + m_C}{2}$$
$$= \frac{m_A + 39}{2}$$
$$m_A = 46 - 39 = 7$$

35. (b)

Species	Electrons	Protons	Atomic number	(EC)	Group
Na^+	10 (Ne)	11	11	$[Ne]\ 3s^1$	1
Mg^{2+}	10 (Ne)	12	12	$[Ne]\ 3s^2$	2
F^-	10 (Ne)	9	9	$1s^2 2s^2 2p^5$	17
O^{2-}	10 (Ne)	8	8	$1s^2 2s^2 2p^4$	16

Isoelectronic species have same EC but placed in different groups. Thus, Statement I and Statement II are correct but Statement II is not the correct explanation of Statement I.

36. (a) Al (13) = $1s^2 2s^2 2p^6 3s^2 3p^1$

Ga (31) = $[Ar]\ 3d^{10} 4s^2 4p^1$

In Ga (31), one extra shell is added and one expects increase in radius of Ga. But additional 10 electrons in 3*d* orbitals offer poor screening effect for the outer electrons and, thus force of attraction is increased and, thus size of Ga is less than that of Al. Thus, Statements I and Statement II are correct and Statement II is the correct explanation of Statement I.

37. (c) By inert-pair effect, stability of $Pb^{4+} < Pb^{2+}$.

Thus, $$Pb^{4+} + 2e^- \longrightarrow Pb^{2+}$$

Thus, Pb^{4+} can be reduced to Pb^{2+} (being more stable). Hence, Pb^{4+} (as in PbO_2) is a good oxidising agent and itself reduced to Pb^{2+} (as in PbO).

Thus, Statement I is correct and Statement II is incorrect.

38. (d) P : $[Ne]\ 3s^2 3p^3$

Ground state [⇅] [1|1|1]
$3s^2$ $3p^3$

Three unpaired electrons hence, PCl_3 is formed.

Excited state [1] [1|1|1] [1| | | |]
$3s^1$ $3p^3$ $3d^1$

Five unpaired electrons hence, PCl_5 is formed.

N : $[He]\ 2s^2 2p^3$ [⇅] [1|1|1]
2s 2p

Three unpaired electrons hence, NCl_3 is formed.

Since, there is no (2*d*) orbital in N (7), hence 2*s* electron cannot be excited. Thus, NCl_5 is not formed.

Thus, Statement I is incorrect and Statement II is correct.

39. (b) Isoelectronic species (containing equal number of electrons) will have different size due to different values of *Z*.

Species	Z	Electrons	e/Z
N^{3-}	7	10	1.43
O^{2-}	8	10	1.25
F^-	9	10	1.11
Ne	10	10	
Na^+	11	10	0.91
Mg^{2+}	12	10	0.83
Al^{3+}	13	10	0.77
Si^{4+}	14	10	0.71

Thus, greater the value of (e/Z), greater the size of isoelectronic species.

Thus, Statement I and Statement II are correct but Statement II is not the correct explanation of Statement I.

40. (b) $$O^-(g) + e^- \longrightarrow O^{2-}(g)$$
$$\Delta G^\circ > 0$$
$$\Delta H^\circ = 702\ \text{kJ mol}^{-1}$$

This process is unfavourable in the gas phase because the resulting increase in electron-electron repulsion overweighs the stability gained by achieving the noble gas configuration. O^{2-} of smaller size is common in ionic compounds in solid state and is stabilised by neighbouring cations. Thus, Statement I and Statement II are correct but Statement II is not the correct explanation of Statement I.

41. (b) As from graph, sum of $(IE)_1$ and $(IE)_2$ is linear for 3*d*-series; but for chromium and copper, these values deviate and are higher than expected. Chromium has $[Ar]\ 3d^5 4s^1$ and copper has $[Ar]\ 3d^{10} 4s^1$ configuration. An increasing value of effective nuclear charge affects 3*d*-orbital more than 4*s*-orbital and $(IE)_2$ for an electron from 3*d*-orbital is higher than expected.

Thus, Statement I and Statement II are correct but Statement II is not the correct explanation of Statement I.

42. (a) In the periodic table, electronegativity increases while moving along a period left to right (with some exception).

Li Be B C N O F Ne
F: Maximum, Ne: Zero

Effective nuclear charge on N

$$Z_{eff} = Z - S$$
S = shielding effect

In fluorine, added electron is screened by seven electrons in second shell and two electrons in first shell.

$$F + e^- \longrightarrow F^-$$

⊕ 2 7 — $(n-1)$ n

$$S = 7 \times 0.35 + 2 \times 0.85$$
$$= 2.45 + 1.70$$
$$= 4.15$$
$$\therefore \quad Z_{eff} = 9 - 4.15 = 4.85$$

In nitrogen, added electron is screened by five electrons in second shell and two electrons in 1st shell.

$$N + e^- \longrightarrow N^-$$

⊕ 2 5 $\quad S = 5 \times 0.35 + 2 \times 0.85 = 3.45$

Thus, effective nuclear charge on F > N

Thus, electronegativity of N > F

Thus, Statement I and Statement II are correct and Statement II is the correct explanation of Statement I.

43. (b) Lithium and magnesium resemble in many properties due to same EN values. Lithium is placed diagonally to magnesium. Thus, Statement I and Statement II are correct but Statement II is not the correct explanation of Statement I.

44. (a)

(i) $[Ar]\ 3d^{10} 4s^2 4p^6 5s^1$

It has one unpaired electron, hence paramagnetic and last electron goes into *s*-orbital.

Thus, *s*-block

Period = 5

Thus, (i) → (p)

(ii) $[Ne]\ 3s^2 3p^6$

It is an inert gas (Ar). It is diamagnetic and it is *p*-block element in zero group.

Thus, (ii) → (q,r,t)

(iii) [Ar] $4s^2$ [↑↓ | ↑ | ↑ | ↑ | ↑]
$3d^6$

Last electron goes into $3d$-orbital, thus d-block. It has unpaired electrons, thus paramagnetic.

It lies in 4th period.

Thus, (iii) → (p, s, u)

(iv) [Xe] $4f^{14}5d^{10}6s^2 6p^3$ [↑ | ↑ | ↑]

It has unpaired electrons in p-orbital thus paramagnetic and p-block.

Thus, (iv) → (p, r)

45. (c) Aufbau rule, Hund's rule and Pauli's exclusion rule are not considered.

	EC	Group	Ans.
Na (11)	$1s^2, 2s^2, 2p^6, 3s^1$	1	(i) → (p)
S (16)	$1s^2, 2s^2, 2p^6, 3s^2, 3p^4$	16	(ii) → (s)
Ca (20)	$1s^2, 2s^2, 2p^6, 3s^2, 3p^6, 3d^2$	10	(iii) → (q)
As (33)	$1s^2, 2s^2, 2p^6, 3s^2, 3p^6, 3d^{10}, 4s^2, 4p^3$	15	(iv) → (r)

46. (d)

(i) [↑↓]
ns^2

Paired electron has higher $(IE)_1$ as compared to np^1 (unpaired electron).

(ii) [↑↓] [↑ | |]
ns^2 np^1

(iii) [↑↓] [↑ | ↑ | ↑]
ns^2 np^3

Stable as each p-orbital is half-filled, thus $(IE)_1$ values are higher than that of ns^2 and ns^2np^1.

(iv) [↑↓] [↑↓ | ↑↓ | ↑↓]
ns^2 np^6

Most stable (inert gas) maximum $(IE)_1$.

Thus, (i) →(r), (ii) → (s), (iii) → (q), (iv) → (p)

47. (b)

(i) [Ar] $3d^7 4s^2$

It is d-block element (after $4s$, $3d$ is being filled). It belongs to group 9 (cobalt).

Thus, **(i) → (s)**

(ii) [Ar] $3d^5 4s^1$

It is d-block element (after $4s$, $3d$ is being filled). It belongs to group 6 (chromium).

Thus, **(ii) → (r)**

(iii) [Kr] $4d^2 5s^2$

It is d-block element (after $5s$, $4d$ is being filled). It belongs to group 4 (it is zirconium).

Thus, **(iii) → (q)**

(iv) [Xe] $4f^3 6s^2$

It is f-block element, $4f$-orbital is being filled. Since, it belongs to lanthanide series and placed in group 3.

Thus, **(iv) → (p)**

48. (a) (i) $S + e^- \longrightarrow S^-$

It is exothermic process and enthalpy change is called electron gain enthalpy.

Thus, **(i) → (p)**

(ii) $O^- + e^- \longrightarrow O^{2-}, \Delta H > 0$

[↑↓ | ↑↓ | ↑]   [↑↓ | ↑↓ | ↑↓]

Resulting increase in electron-electron repulsion overweighs the stability gained by achieving noble gas configuration.

Also, O^{2-} has inert gas configuration.

Thus, **(ii) → (q,r)**

(iii) $Mg \longrightarrow Mg^{2+} + 2e^-$

It is ionisation process, $\Delta H > 0$.

Mg^{2+} attains inert gas configuration.

Thus, **(iii) → (q,r)**

(iv) $N \longrightarrow N^+ + e^-$

[↑ | ↑ | ↑]   [↑ | ↑ |]

This is ionisation process.

N loses half-filled configuration.

Thus, **(iv) → (q,t)**

(v) $O \longrightarrow O^+ + e^-$

[↑↓ | ↑ | ↑]   [↑ | ↑ | ↑]

This is ionisation process, $\Delta H > 0$.

Half-filled configuration is attained.

Thus, **(v) → (q,s)**

JEE Main & AIEEE Archive

1. (c) Ionisation energy increases along a period from left to right and decreases down a group. The position of given elements in the periodic table is as

2	16	18
Ca	S	Ar
Ba	Se	

Thus, the order of increasing ΔH_{IE_1} is

$$Ba < Ca < Se < S < Ar$$

2. (b) $Na \longrightarrow Na^+ + e^-$ (First IE)

$Na^+ + e^- \longrightarrow Na$ (Electron gain enthalpy of Na^+)

Because reaction is reverse so

$$\Delta H_{eg} = -5.1 \text{ eV}$$

3. (c) For isoelectronic species, $r_n \propto \frac{1}{Z}$

Here, Z is atomic number.

Species	Z	Electrons
Cl^-	17	18
Ca^{2+}	20	18
K^+	19	18
S^{2-}	16	18

Thus, ionic size is in order

$$Ca^{2+} < K^+ < Cl^- < S^{2-}$$

4. (a) In a period, the nature of oxides varies from basic to acidic while in a group, basic nature of oxides increases.

Thus,

Oxides ← Decreases — In a period, basic nature changes to acidic nature

↓ Basic nature

Increases in a group

Group	1	2	13	
Period 3 →	Na_2O	MgO	Al_2O_3	(Amphoteric oxide)
Period 4 →	K_2O			(Strongly basic)

Hence, basic nature of oxides are as follows

$$Al_2O_3 < MgO < Na_2O < K_2O$$

5. (c) As we go down the group in periodic table, atomic size increases, force of attraction for the added electron decreases. Thus, the added electron becomes loosely bound. Hence, electron gain enthalpy decreases down a group.

$$X(g) + e^- \longrightarrow X^-(g)$$

Actual order is Cl > F > Br > I

The fact that fluorine has a less electron gain enthalpy than chlorine seems to be due to the relatively greater effectiveness of $2p$-electron in the small F-atom to repel the additional electron entering the atom than $3p$-electrons in the larger Cl-atom.

6. (d) $O^{2-}, F^-, Na^+, Mg^{2+}$ and Al^{3+} are isoelectronic species.

For isoelectronic species, $r \propto \frac{1}{Z}$

Thus, larger the atomic number (effective nuclear charge), smaller is the ionic radii.

7. (b) The ionic radius in general increases moving top to bottom and further decreases moving left to right. So, the correct order is

$$\underset{0.98\,\text{Å}}{Na^+} > \underset{0.68\,\text{Å}}{Li^+} > \underset{0.65\,\text{Å}}{Mg^{2+}} > Be^{2+}$$

8. (c) Higher the charge/size ratio, larger is the polarising power.

$\therefore \quad K^+ < Ca^{2+} < Mg^{2+} < Be^{2+}$

9. (a) Isoelectronic means having same number of electrons, $K^+, Cl^-, Ca^{2+}, Sc^{3+}$ (all are having 18 electrons).

10. (b) Li^+ is having largest hydrated ionic size while Rb^+ is having smallest.

Smaller the size, greater is the mobility.

11. (d) Examine the positions in periodic table.

Group	13	14	15	16	17
	B	C	N	O	F
			P	S	

Ionisation enthalpy increases along a period due to increased nuclear charge and decreased atomic radii. While in a group, it decreases from top to bottom. Along with this trend, ionisation enthalpy is highest for the elements possessing half-filled and stable electronic configuration.

Since, phosphorus is having stable half-filled configuration. Hence, order is B < S < P < F.

12. (d) The chemical reactivity of alkali metals is due to their ability to lose electron more readily than other elements. Since, the tendency to lose electron by alkali metals increases down the group, therefore, their chemical reactivity increases down the group.

On the other hand, chemical reactivity of halogens is due to their ability to gain extra electron more readily than other elements. Since, the tendency to gain electron by halogens decreases down the group, hence their chemical reactivity decreases down the group.

13. (c)

(a) Metallic radii increases in a group from top to bottom.
$\therefore$ Li < Na < K < Rb. Thus, true.

(b) Electron gain enthalpy of Cl > F and decreases along the group.
$\therefore \quad I < Br < F < Cl$
Thus, true.

(c) Ionisation enthalpy increases along a period from left to right but due to presence of half filled orbitals in N, ionisation enthalpy of N > O.
$\therefore$ B < C < O < N. Thus, given order is incorrect.

(d) For isoelectronic species, $r_n \propto \frac{1}{Z}$,
$Al^{3+} < Mg^{2+} < Na^+ < F^-$. Thus, true.

14. (c) As we go down in the group, ionic character increases, hence melting point of halides should increase but NaCl has the highest melting point (800°C) due to its high lattice energy.

15. (c) All the ions belong to same period, thus for them cations will be smaller than anions. Now, O^{2-} and F^- are isoelectronic and

$r_n \propto \frac{1}{Z}$

Thus, ionic radius of O^{2-} $(Z = 8) > F^-$ $(Z = 9)$

16. (c) $O^-(g) + e^- \longrightarrow O^{2-}(g);\ \Delta H° = 844$ kJ mol^{-1}

This process is unfavourable in the gas phase because the resulting increase in electron-electron repulsion overweighs the stability gained by achieving the noble gas configuration.

17. (c) F_2 has the most negative $\Delta G°$ value which is dependent on hydration enthalpy.

18. (d) While moving along a group from top to bottom, acidic nature of oxides decreases and along a period from left to right, acidic nature of oxides increases.

	Al	Si	P	S
Z	13	14	15	16
	Al_2O_3	SiO_2	P_2O_3	SO_2
	Amphoteric	Acidic		Max. acidic

Thus, $Al_2O_3 < SiO_2 < P_2O_3 < SO_2$

19. (c) All physical and chemical properties of elements are periodic function of their atomic numbers— Modern Periodic Law.

20. (d) Due to lanthanide contraction, there occurs net decrease in size. Only, 0.85 Å is smaller one.

21. (b) Fe^+ (26) = [Ar] $3d^6, 4s^1$

Mn^+ (25) = [Ar] $3d^5, 4s^1$

V^+ (23) = [Ar] $3d^2, 4s^2$

Cr^+ (24) = [Ar] $3d^5, 4s^0$ [By first IP]

The electronic configuration of Cr^+ is most stable, hence formation of Cr^{2+} by second IP requires maximum enthalpy.

22. (b) r_n (radius) $\propto \frac{1}{Z}$

Thus, $Yb^{3+} < Pm^{3+} < Ce^{3+} < La^{3+}$.

2. Hydrogen

DPP-1 Hydrogen and Its Isotopes

1. (b)
(a) $H(g) \longrightarrow H^+(g) + e^-$

$$IP = \frac{2\pi^2 mZ^2e^4k^2}{n^2h^2}$$

$H(g)$ is smallest in size, hence electron is tightly bound to nucleus. Thus, IP is very high.

$(IP)_H = 13.6\text{ eV atom}^{-1} = 1.312 \times 10^6\text{ J mol}^{-1}$

Thus, (a) is correct.

(b) H_2 is collected at both cathode and anode.

H^+ is discharged at cathode

$$H^+ + e^- \longrightarrow H$$
$$H \longrightarrow \frac{1}{2}H_2$$

H^- is discharged at anode

$$H^- \longrightarrow H + e^-$$
$$H \longrightarrow \frac{1}{2}H_2$$

Thus, (b) is incorrect.

(c) Na^+H^-, oxidation state of H = – 1

H^+Cl^-, oxidation state of H = + 1

Thus, (c) is correct.

(d) EN of H = 2.1, Br = 2.8, F = 4.0, I = 2.5, Cl = 3.0

Thus, EN values are comparable (except F).

Thus, (d) is correct.

2. (a)

	Ionic mass	Electron	Proton	Neutron
H^+	1	0	1	0
D^+	2	0	1	1
T^+	3	0	1	2

Thus, number of electrons is same (zero) in all the species.

3. (c) Average atomic mass, $\overline{A} = \dfrac{A_1x_1 + A_2x_2}{x_1 + x_2}$

$$1.00014 = \frac{1 \times x + 2\,(100 - x)}{x + 100 - x} = \frac{x + 200 - 2x}{100}$$

$$100.014 = 200 - x$$

$$x = 99.986\%\ ^1_1H$$

Thus, $(100 - x) = 0.014\%\ ^2_1H$

4. (b) Reactivity of Cl_2 with H_2 or D_2 depends on the energy of dissociation.

$$BE\,(H_2) = 436.0\text{ kJ mol}^{-1}$$
$$BE\,(D_2) = 438.0\text{ kJ mol}^{-1}$$

Since, BE of H_2 is lower than that of BE of D_2 hence, H_2 reacts with Cl_2 at faster rate than D_2.

5. (b) High purity (about 99.95%) H_2 is formed by electrolysis of warm aqueous $Ba(OH)_2$ solution between nickel electrodes.

6. (b) Mixture of $CO(g)$ and $H_2(g)$ is called synthesis gas or syn gas, as it is used in the synthesis of methanol and other hydrocarbons. The process of producing '*syn* gas' from coal is called 'coal gasification'.

7. (d)

(a) $AgCl + H \longrightarrow Ag + H^+Cl^-$ (Oxidation: H → H^+; Reduction: AgCl → Ag)

H reduces AgCl to metallic Ag and itself oxidised to H^+.

(b) $Ca + 2H \longrightarrow CaH_2(Ca^{2+}H^-)$

H oxidises Ca to Ca^{2+} and itself reduced to H^-.

(c) $H + H \longrightarrow H_2,\ \Delta H = -436\text{ kJ mol}^{-1}$

In the stream of H_2, atomic H combines forming H_2. A large amount of heat is liberated which can be used in welding.

8. (d) At high temperatures in an electric arc or under UV light, H_2 dissociates into atomic hydrogen. Thus, (I) is correct.

Atomic hydrogen is short-lived and combines immediately to form molecular H_2 with liberation of large amount of heat.

$$2H \longrightarrow H_2, \qquad \Delta H = -436\text{ kJ mol}^{-1}$$

This reaction is catalysed by transition metals. Thus, (IV) is correct.

Atomic H also combines with metals forming metallic hydrides.

$$M + H \longrightarrow MH, MH_2 \cdots$$

Thus, (III) is correct.

When atomic H is formed, the atoms set free are carried away by stream of incoming H_2 gas. H atoms combine at once on combining with metal electrode and large amount of heat is liberated raising the temperature to 4000-5000 K. This principle is utilised in making of the **atomic hydrogen welding torch** (used for welding at high temperature but in a reducing atmosphere of H_2).

9. (b)

(a) $Cl_2 + H_2O \longrightarrow 2HCl + O$ (nascent O)

(b) $SO_2 + 2H_2O \longrightarrow H_2SO_4 + 2H$ (nascent H)

10. (b) Nascent hydrogen is more reactive than molecular H_2.

(a) Molecular H_2 does not cause reduction of $FeCl_3$ to $FeCl_2$.

(b) Acidic solution of aqueous $FeCl_3$ reacts with zinc producing nascent hydrogen (H), which being more reactive causes reduction of $FeCl_3$ to $FeCl_2$.

$$FeCl_3 + 3H_2O \longrightarrow Fe(OH)_3 + 3HCl$$
$$Zn + 2HCl \longrightarrow ZnCl_2 + 2H\text{(nascent)}$$
$$FeCl_3 + H \longrightarrow FeCl_2 + HCl$$

11. (c) Hydroformylation reaction is

$$CO + 2H_2 \xrightarrow{\text{Cobalt}} CH_3OH$$

and is used in the industrial preparation of methanol (methyl alcohol).

12. (d)

$$\underset{\text{Propane}}{\overset{1}{C}H_3-\overset{2}{C}H_2-\overset{3}{C}H_3} \xrightarrow[\text{Catalyst}]{875\text{ K}} \underset{\text{Propene}}{CH_3-CH{=}CH_2} + H_2$$

$$\text{Propane} \longrightarrow \underset{\text{Ethene}}{CH_2{=}CH_2} + CH_4$$

$CH_3CH{=}CH_2$ is formed by cleavage of (C—H) bond. $CH_2{=}CH_2$ and CH_4 are formed by cleavage of (C — C) bond. Both are redox reactions.

13. (d)

(a) H_2O dissolves CO_2 but H_2 remains insoluble.

$$H_2O + CO_2 \rightleftharpoons H_2CO_3$$

(b) Lime water $[Ca(OH)_2]$ dissolves CO_2. H_2 remains insoluble.

$$Ca(OH)_2 + CO_2 \longrightarrow CaCO_3 + H_2O$$

(c) $\dfrac{r(H_2)}{r(CO_2)} = \sqrt{\dfrac{44}{2}} = \sqrt{22} = 4.7$

Rate of diffusion of H_2 is 4.7 times more than CO_2, hence first content is rich in H_2.

14. (b) Pure H_2 for the first time was isolated by English chemist Henry Cavendish based on the reaction of zinc (or tin) on dilute H_2SO_4 or HCl.

$$Zn + H_2SO_4 \longrightarrow ZnSO_4 + H_2$$

15. (b)

		Reaction with NaOH	Reaction with HCl
I.	Zn	$\longrightarrow Na_2ZnO_2 + H_2$	$\longrightarrow ZnCl_2 + H_2$
II.	Mg	$\longrightarrow$ No reaction	$\longrightarrow MgCl_2 + H_2$
III.	Al	$\longrightarrow NaAlO_2 + H_2$	$\longrightarrow AlCl_3 + H_2$
IV.	Be	$\longrightarrow Na_2BeO_2 + H_2$	$\longrightarrow BeCl_2 + H_2$

Thus, Zn, Al and Be react with NaOH and HCl both forming H_2.

16. (a)

I. Aqueous solution of $FeCl_3$ is acidic due to hydrolysis. When zinc is added, nascent H is formed which reduces $FeCl_3$ to $FeCl_2$.

$$FeCl_3 + 3H_2O \longrightarrow Fe(OH)_3 + 3HCl$$
$$Zn + 2HCl \longrightarrow ZnCl_2 + 2H$$
$$FeCl_3 + H \longrightarrow \underset{(X)}{FeCl_2} + HCl$$

$FeCl_2$ gives blue colour with $K_3[Fe(CN)_6]$.

$$FeCl_2 + K_3[Fe(CN)_6] \longrightarrow \underset{\text{Turnbull's blue}}{Fe_3[Fe(CN)_6]_2}$$

II. Molecular H_2 is less reactive and does not reduce $FeCl_3$ to $FeCl_2$. *Y* is simply $FeCl_3$.

17. (a)

I. $Zn + HCl(dil.) \longrightarrow ZnCl_2 + 2H$

II. $CH_3OH + Na \longrightarrow CH_3ONa + H$

III. $\underset{\text{(Acidified)}}{2H_2O} \xrightarrow{\text{Current}} \underset{\text{(At cathode)}}{2H_2} + \underset{\text{(At anode)}}{O_2}$

IV. $H_2 \longrightarrow 2H$ (atomic)

18. (b)

(a) $PH_3 + H_2O \rightleftharpoons PH_4^+OH^-$

(b) $B_2H_6 + 6H_2O \longrightarrow 2B(OH)_3 + 6H_2$

(c) $CH_4 + H_2O \longrightarrow$ Insoluble

(d) $H_2S + H_2O \longrightarrow$ Insoluble

19. (d) $C \text{ (red hot)} + H_2O \longrightarrow CO + H_2$

$$CO + H_2O \longrightarrow CO_2 + H_2$$

CO changes to hydrocarbons as

$$CH{\equiv}CH, CH_2{=}CH_2, CH_4, \text{etc.}$$

NaOH absorbs CO and CO_2

$$CO + NaOH \longrightarrow HCOONa$$
$$CO_2 + 2NaOH \longrightarrow Na_2CO_3 + H_2O$$

Cu_2Cl_2/NH_4OH absorbs $CH{\equiv}CH$ (if any)

$$CH{\equiv}CH + Cu_2Cl_2 \longrightarrow C_2Cu_2 + 2HCl$$

Thus, gas *X* is H_2 which remains insoluble.

20. (c)

(a) $NaH(aq) \rightleftharpoons Na^+(aq) + H^-(aq)$

At anode $H^- \longrightarrow H + e^-$
$2H \longrightarrow H_2$

At cathode $Na^+ + e^- \longrightarrow Na$
$2Na + 2H_2O \longrightarrow 2NaOH + H_2$

H_2 gas is formed both at cathode and anode.

(b) $HCOONa(aq) \rightleftharpoons HCOO^- + Na^+(aq)$

At anode $HCOO^- \longrightarrow H + CO_2 + e^-$
$2H \longrightarrow H_2$

At cathode $Na^+ + e^- \longrightarrow Na$
$2Na + 2H_2O \longrightarrow 2NaOH + H_2$

H_2 gas is formed both at anode and cathode.

(c) $NaCl(aq) \rightleftharpoons Na^+(aq) + Cl^-$

At anode $2Cl^- \longrightarrow Cl_2 + 2e^-$

At cathode $Na^+ + e^- \longrightarrow Na$
$2Na + 2H_2O \longrightarrow 2NaOH + H_2$

H_2 is formed only at cathode.

(d) As in (a)

Thus, NaCl is only the exception.

21. (a) Nuclear spin quantum number of

$$H = \frac{1}{2}, \quad D = 1, \quad T = \frac{1}{2}$$

22. (d) H_2 is adsorbed on palladium and other transition metals as they have free valencies for adsorption.

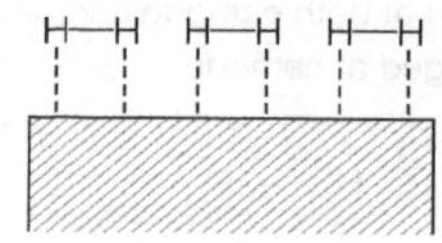

Adsorbed H_2 on metal surface is called occluded H_2.

23. (a) This is a nuclear fusion reaction in which smaller nuclei fuse to form heavier nuclei at high temperature and high pressure.

24. (b)

(a) $Na_2O_2 + 2HCl \longrightarrow 2NaCl + H_2O_2$

(b) $Mg + H_2O(steam) \longrightarrow MgO + H_2$

(c) $BaO_2 + 2HCl \longrightarrow BaCl_2 + H_2O_2$

(d) $H_2S_2O_8 + 2H_2O \longrightarrow H_2O_2 + 2H_2SO_4$

Note $H_2S_2O_8$ *is a peroxy disulphuric acid having two units of sulphuric acid linked by peroxy bond.*

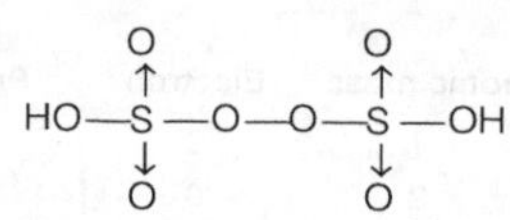

25. (c) $Zn + 2H^+ \longrightarrow Zn^{2+} + H_2$

H^+ is reduced and Zn is oxidised. With HNO_3, NO_3^- is reduced to NO in preference to H^+. Hence, H_2 is not formed.

26. (a) *ortho* and *para* H_2 differ in spins of nuclei (called spin isomers). Thus, they differ in physical properties but have same chemical properties.

27. (d) $CO + \underset{\text{(Steam)}}{H_2O} \longrightarrow CO_2 + H_2$

$$CO_2 + 2KOH \longrightarrow K_2CO_3 + H_2O$$

28. (d) Welding with a flame resulting from the combustion of oxygen and fuel gas (as H_2) is done by lime light. The temperature is high enough the jet of oxygen alone can continue to melt the metals such as platinum.

29. (d) In fuel cell, fuels like CH_4, H_2., etc., are combusted producing electrical energy.

$$2H_2(g) + O_2(g) \longrightarrow 2H_2O(l)$$

30. (d)

(a) Reaction takes place without the evolution of heat hence called cold combustion thus, correct.

(b) In acidic medium,

At anode $H_2(g) \longrightarrow 2H^+ + 2e^-, \quad E° = 0.00\,V$

At cathode $2H^+ + \frac{1}{2}O_2(g) + 2e^- \longrightarrow H_2O(l), \; E° = 1.23\,V$

$$H_2(g) + \frac{1}{2}O_2(g) \longrightarrow H_2O(l), \quad E° = 1.23\,V$$

In basic medium,

At anode $H_2(g) + 2OH^-(aq) \longrightarrow H_2O(l) + 2e^-, E° = 0.83V$

At cathode $H_2O(l) + \frac{1}{2}O_2(g) + 2e^- \longrightarrow 2OH^-(aq),$ $E° = 0.40V$

$$H_2(g) + \frac{1}{2}O_2(g) \longrightarrow H_2O(l), \quad E° = 1.23V$$

Thus, same reactions occur in acidic and alkaline medium and thus, same value of EMF.

Thus, (b) and (c) are correct.

31. (d)

(a) $2H_2 + O_2 \longrightarrow 2H_2O$

Due to combustion, water is formed hence, it is pollution free and thus is a green fuel. Thus, correct.

(b) Energy released (in kJ/g)

$H_2(g) = 143$
$H_2(l) = 142$
LPG = 50
$CH_4 = 53$

Thus, correct.

(c) H_2 can be transported in the form of metallic hydrides which on heating liberate H_2.

$$MgH_2 \xrightarrow{\Delta} Mg + H_2$$

$$TiH_2 \xrightarrow{\Delta} Ti + H_2$$

Thus, correct.

Note *The combination of all these potential uses of H_2 could bring about a fundamental change in our way of life, giving rise to what is called hydrogen economy.*

32. (a) Bond between H-atoms is formed by sharing two electrons between them. Bond pair lies symmetrical and thus, H_2 is non-polar in nature. Thus, intermolecular forces are weak.

Thus, Statement I and Statement II are correct and Statement II is the correct explanation of Statement I.

33. (b) The bonding forces within the H_2 molecule are exceptionally strong and H_2 molecule is thermally stable. Stability is due to when H-atoms approach each other in the formation of H—H bond and energy is released.

$$H + H \longrightarrow H_2, \quad \Delta H = -436 \text{ kJ mol}^{-1}$$

Thus, Statement I and Statement II are correct but Statement II is not the correct explanation of Statement I.

34. (a) Hydrogen is the ninth most abundant element on a mass basis (0.9%). In earth's atmosphere, its abundance is very small being lightest element. The earth's gravitational field is too small to hold the light element. Thus, Statement I and Statement II are correct and Statement II is the correct explanation of Statement I.

35. (a)

(i) ${}^1_1H$ - a stable isotope

Thus, (i) → (s)

(ii) ${}^2_1H$ - 0.014%, stable isotope

Thus, (ii) → (r,s)

(iii) ${}^3_1H$ – radioactive isotope

Thus, (iii) → (p)

(iv) D_2O is used as moderator in nuclear reactor.

Thus, (iv) → (q)

DPP-2 **Hydrides**

1. (a) NaH, CaH_2 and BaH_2 are ionic hydrides.

(a) Group 1 and group 2 metals form ionic hydrides (Be and Mg form non-stoichiometric hydride).

(b) *p*-block elements form covalent hydrides as BH_3, AlH_3, SiH_4.

(c) *d*-block and *f*-block elements form metallic (stoichiometric and non-stoichiometric) hydrides.

(d) HF, HCl, HBr, HI are strong acids but they are source of protons.

2. (d)

(a) $2Li(s) + H_2(g) \xrightarrow{973\text{ K}} 2LiH(s)$

$Ca(s) + H_2(g) \xrightarrow{973\text{ K}} CaH_2(s)$

(b) They are ionic, hence in fused state, they are good conductor.

$$LiH \longrightarrow Li^+ + H^-$$

$$H^+ + e^- \longrightarrow H \longrightarrow H_2$$

$$Li^+ + e^- \longrightarrow Li$$

(c) Being ionic solids, they have high melting points.

3. (b) In this reaction, H^- accepts proton (H^+) from H_2O.

H^+ ion acceptor is a Lowry-Bronsted base and changes to acid.

H_2O donates proton, thus it is a Lowry-Bronsted acid and changes to base.

$$\underset{\text{Lowry-Bronsted base}_1}{H^-(aq)} + \underset{\text{Acid}_2}{H_2O} \longrightarrow \underset{\text{Base}_2}{OH^-(aq)} + \underset{\text{Acid}_1}{H_2(g)}$$

4. (b) AlH_3 is electron-deficient (octet of Al is incomplete). H^- is a lone pair donor and is thus, a Lewis base.

AlH_3 is a Lewis acid.

$$\underset{\text{Lewis base}}{H:^-} \longrightarrow \underset{\text{Lewis acid}}{AlH_3}$$

5. (b) Based on ΔH_f° (standard enthalpy of formation), energetic stability decreases along a group top to bottom and increases along a period left to right.

→ Stability (maximum)

LiH BeH_2 BH_3 CH_4

↑ (maximum)

MgH_2
CaH_2
SrH_2
BaH_2

$BeH_2 > MgH_2 > CaH_2 > SrH_2 > BaH_2$

6. (b)

	Atomic number	(H—*M*) bond length	(H—*M*) bond energy
H—O			
H—S			
H—Se			
H—Te	↓ Increases	↓ Increases	↓ Decreases

Thus, H_2O is not a reducing agent

$$\underset{\text{Minimum}}{H_2O} < H_2S < H_2Se < \underset{\text{Maximum}}{H_2Te}$$

7. (b) Mass of one atom $= \dfrac{A}{N_0} = \dfrac{\text{Atomic mass}}{\text{Avogadro's number}}$

$$\frac{A}{N_0} = 6.66 \times 10^{-23} g$$

$\therefore$ Atomic mass of the atom $= N_0 \times 6.66 \times 10^{-23}$ g mol^{-1}

$= 6.02 \times 10^{23} \times 6.66 \times 10^{-23}$

$= 40.0$ g mol^{-1}

	%	Atomic mass	Moles	Ratio
M	95.24	40.0	2.38	1
H	4.76	1.0	4.76	2

Thus, hydride is MH_2.

8. (b) H^- is a very strong base (H^+ ion acceptor) and accepts protons from a proton donor as water.

$$\underset{\text{Base}_1}{H^-} + \underset{\text{Acid}}{H_2O} \longrightarrow \underset{\text{Base}}{OH^-} + \underset{\text{Acid}_1}{H_2}$$

H^- changes to H_2. Thus, H^- is a conjugate base of acid H_2.

9. (b) $NaH(s) + H_2O(l) \longrightarrow NaOH(aq) + H_2(g)$

$CaH_2(s) + 2H_2O(l) \longrightarrow Ca(OH)_2 + 2H_2(g)$

$H_2(g)$ is combustible in nature and catches fire spontaneously. CO_2 cannot be used as carboxylates are formed.

$$CO_2 + NaOH \longrightarrow \underset{\text{Sodium formate}}{HCOONa}$$

$$2CO_2 + Ca(OH)_2 \longrightarrow \underset{\text{Calcium formate}}{(HCOO)_2Ca}$$

Only sand can be used to put off fire.

10. (d)

(a) $LiH + H_2O \longrightarrow LiOH + H_2$

Hydrides are decomposed by H_2O forming H_2 and solution is alkaline. H_2 can be recovered being insoluble in alkaline solution. Thus, correct.

(b) For organic synthesis, reactants should be moisture free. Traces of moisture can be removed by saline hydrides.

$$CaH_2 + H_2O \longrightarrow Ca(OH)_2 + H_2\uparrow$$

Reactants (volatile) can be separated by distillation. Thus, correct.

(c) Metallic hydrides have H_2 adsorbed on metal surface. Thus, they can be transported easily without any risk. On decomposition, H_2 is formed and metal is set free. Thus, correct.

11. (c) Electrolysis of $NaCl(aq)$

$$NaCl(aq) \rightleftharpoons Na^+(aq) + Cl^-(aq)$$

At anode $2Cl^- \longrightarrow Cl_2 + 2e^-$

At cathode $Na^+ + e^- \longrightarrow Na$

$2Na + 2H_2O \longrightarrow 2NaOH + H_2$

Electrolysis of $NaH(aq)$

$$NaH(aq) \rightleftharpoons Na^+(aq) + H^-(aq)$$

At anode $2H^- \longrightarrow H_2 + 2e^-$

At cathode $Na^+ + e^- \longrightarrow Na$

$2Na + 2H_2O \longrightarrow 2NaOH + H_2$

	At anode	At cathode	Nature of solution
NaCl	Cl_2	H_2	Basic
NaH	H_2	H_2	Basic

Thus, H_2 gas is obtained at anode and cathode only from NaH.

12. (d)

(a) $r_{H^-} = 208\text{pm}\left(r_n \propto \dfrac{1}{Z_{\text{eff}}}\right),$

$r_{F^-} = 133$ pm, $r_{Cl^-} = 181$ pm

$r_{Br^-} = 196$ pm, $r_{I^-} = 220$ pm

$$F^- < Cl^- < Br^- < H^- < I^-$$

Thus, (a) is correct.

(b) Hydrides are reducing agents since, they are oxidise to H_2

$NaH \longrightarrow Na^+ + H^-$

$H^- \longrightarrow H + e^-$, $H \longrightarrow H_2$

$LiAlH_4$ is a reducing agent for organic compounds.

$$CH_3CHO \xrightarrow{LiAlH_4} CH_3CH_2OH$$

Thus, (b) is correct.

(c) Boranes have incomplete octet in B thus, they are electron deficient (Lewis acid or electron pair acceptor).

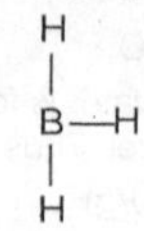

Thus, (c) is correct.

13. (d) Hydrogen reacts with many transition metals, lanthanides and actinides forming metallic hydrides.

(a) H_2 is thought to be accommodated in the interstices in the metal lattice thus, metallic hydrides are also called interstitial hydrides. Thus, correct.

(b) Being deficient in hydrogen, they are non-stoichiometric in nature.

$$TiH_{1.5\text{-}1.8}, \; NiH_{0.6\text{-}0.7}$$

Thus, (b) is correct.

(c) Hydrogen is present in metal lattice as hydrogen atoms rather than H_2 molecule. Thus, due to unpaired electron, metallic hydrides are paramagnetic. Since, unpaired electrons can be bonded to metal, thus less paramagnetism and poor conduction is observed.

Thus, (c) is correct.

14. (a)

	Compounds with H	
(a) N, O, F	NH_3, H_2O, HF	All covalent and H-bonded
(b) Li, Na, K	LiH, NaH, KH	Ionic and no H-bonding
(c) S, Se, Te,	H_2S,	Covalent
	SeH_2	Ionic and no H-bonding
	TeH_2	Ionic
(d) Cl, Br, I	HCl, HBr, HI	Polar covalent, no H-bonding.

15. (c) (a) $ZnCO_3$ (calamine); (b) ($CaCN_2$ + C) (nitrolim)

(c) CaH_2 (hydrolith); (d) Pb_3O_4 (minium)

16. (d) (a) BaH_2 - Ionic; (b) CaH_2 - Ionic

(c) NaH - Ionic; (d) SiH_4 - Covalent

Na^+, Ba^{2+} and Ca^{2+} attain inert gas configuration, thus NaH, BaH_2 and CaH_2 are ionic hydride. Si has four valence electrons. It completes its octet by sharing four electrons with H.

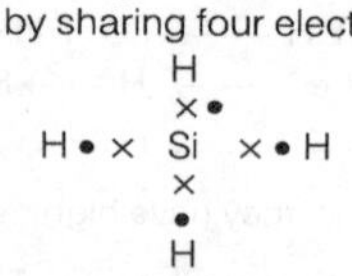

Thus, H and Si form covalent bonds.

17. (b)

I. AlH_3 is electron deficient (octet of Al is incomplete). H^- donates lone pair to AlH_3 hence, H^- is a Lewis base and AlH_3 is a Lewis acid.

$$\underset{\text{Lewis acid}}{AlH_3} + \underset{\text{Lewis base}}{H^-} \longrightarrow AlH_4^-$$

II. H_2O is proton donor. H^- is proton acceptor hence, it is Bronsted base and H_2O is Bronsted acid.

$$\underset{\text{Bronsted acid}_1}{H_2O} + \underset{\text{Bronsted base}_2}{H^-} \longrightarrow \underset{\text{Bronsted acid}_2}{H_2} + \underset{\text{Bronsted base}_1}{OH^-}$$

18. (d)

(a) H—P—H One lone pair. Thus, correct.
    |
    H

(b) H—O—H Two lone pairs. Thus, correct.

(c) H—F : Three lone pairs. Thus, correct.

(d) H—C—H No lone pair. Thus, incorrect. (C bonded to H above and H below)

19. (c) $8LiH + Al_2Cl_6 \longrightarrow 2LiAlH_4 + 6LiCl$

$2LiH + B_2H_6 \longrightarrow 2LiBH_4$

20. (b) CH_4, SiH_4, etc., have zero dipole moments, the only forces of attraction between these molecules are relatively weak van der Waals' forces due to temporary dipoles. Since, these forces result from the mutual repulsion of the electron clouds of neighbouring molecules and they increase as the size of the molecule increases, thus accounting for the steady increase in the boiling point of the hydrides on passing down the group 14.

Thus, Statement I and Statement II are correct but Statement II is not the correct explanation of Statement I.

21. (a)

Group ⟶	14	15	16	17
	C	N	O	F
Hydrides	CH_4	NH_3	H_2O	HF

CH_4 is non-polar hence, there is no dipole moment. Thus, its boiling point is least. As we move left to right in a period, EN increases. Hence, polarity of (M — H) bond increases. This results in the formation of permanent dipole moments in hydrides of group 15, 16 and 17.

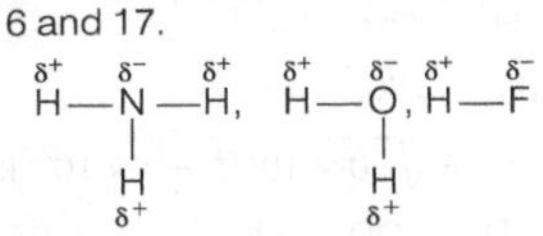

There is association of these molecules together by intermolecular H-bonding. Thus, these hydrides exist as associated molecules. They have higher boiling points than hydrides of group 14.

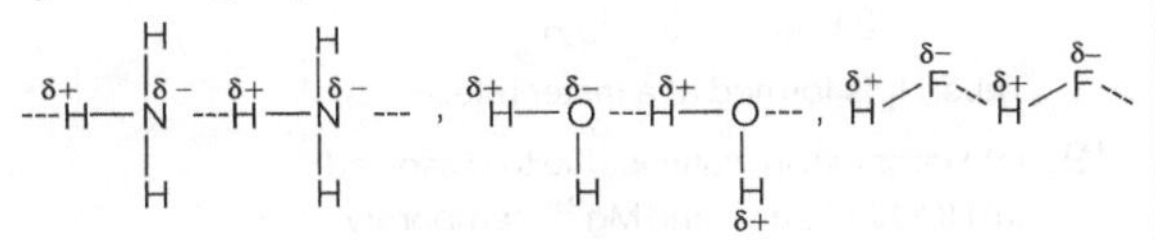

Thus, Statement I and Statement II are correct and Statement II is the correct explanation of Statement I.

22. (c) $LiH + H_2O \longrightarrow LiOH + H_2$

LiH is an ionic hydride. It is decomposed by H_2O forming H_2. All ionic hydrides are decomposed by H_2O forming H_2 and thus, they are instant source of H_2. In all these reactions, H^- is H^+ ion acceptor and is a Lowry-Bronsted base and changes to H_2 (acid). CaH_2 is not a covalent hydride. It is also ionic hydride and decomposed by H_2O.

Thus, Statement I is correct and Statement II is incorrect.

23. (b) Electron deficient hydrides have incomplete octet of the central atom. Hence, they can accept electron-pair from electron rich species. Electron-deficient hydrides are thus Lewis acids.

NH_3 has one extra lone pair hence, it is electron-pair donor (Lewis base).

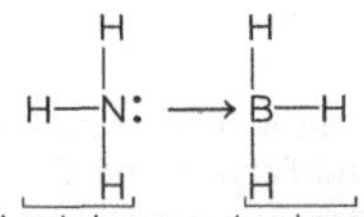

Lewis base   Lewis acid

Thus, Statement I and Statement II are correct but Statement II is not the correct explanation of Statement I.

24. (d) Although, phosphorus exhibits +3 and +5 oxidation states, it cannot form PH_5.

$$H_2(g) \longrightarrow 2H(g), \quad \Delta H = 436.0 \text{ kJ mol}^{-1}$$

Due to high value of dissociation energy of $H_2(g)$ and also high electron gain enthalpy (-71 kJ mol^{-1}), formation of PH_5 is not favoured.

Thus, Statement I is incorrect and Statement II is correct.

25. (a)

(i) B_2H_6 is an electron-deficient hydride and thus, electron-pair acceptor. It is thus a Lewis acid.
Thus, (i) → (p, s)

(ii) CH_4 - Octet of carbon and duplet of hydrogen is complete. Thus, it is an electron-precise hydride.
Thus, (ii) → (u)

(iii) AsH_3 - It has one lone pair. Thus, it is lone-pair donor and a Lewis base. It is electron rich.
Thus, (iii) → (q, t)

(iv) $LiH + H_2O \longrightarrow LiOH + H_2$
On reaction with H_2O, it gives H_2. Thus, it is instant source of H_2.
$$H^- + H_2O \longrightarrow H_2 + OH^-$$
H^- is H^+ ion acceptor (from H_2O). Thus, it is Lowry-Bronsted base.
Thus, (iv) → (r, v)

DPP-3 Soft, Hard and Heavy Water

1. (a) H_2O in the purest form is the universal solvent and is green chemistry solvent.

2. (c) Heavy water is H^2H^2O with isotopic deuterium instead of ordinary hydrogen. Its molar mass is 20 g mol^{-1}.

3. (c) Temporary hardness of water is due to dissolved bicarbonates of calcium and magnesium. CaO (lime) can precipitate bicarbonate as $CaCO_3$.

$$Ca(HCO_3)_2 + CaO \longrightarrow 2CaCO_3 + H_2O$$

Note *Hardness of water is measured in ppm of calcium carbonate.*

4. (b) Ordinary water contains traces of D_2O. D_2O does not undergo electrolysis as early as H_2O. BE of H — O is smaller than that of D — O (about 18 times). When water is electrolysed, H_2 is liberated much faster than D_2 and remaining water thus remains enriched in D_2O.

Note *About 29000 L of water must be electrolysed to give 1L of D_2O.*

5. (c) This represents cation exchange resin.

H^+ ions of resin are replaced by Na^+, Ca^{2+}, Mg^{2+}, Fe^{2+}, etc..., present in hard water.

$$\underset{\text{Cation exchange resin}}{2RH(s)} + M^{2+} \longrightarrow MR_2 + 2H^+ \quad (M = Ca^{2+}, Mg^{2+}, Fe^{2+})$$

Water coming out of the exchange resin is acidic in nature as it has free H^+ ions.

6. (d) Viscosity of water is very low (0.0089 poise), hence this factor does not play a vital role in the biosphere.

7. (d)

(a) Each water molecule is joined to four other water molecules by intermolecular H-bonding, thus correct.

(b) Correct

(c) Correct

8. (b) In H_2O,

$$\theta = 104.5^\circ$$

Bond angle is reduced due to *lp*-*lp*, *lp*-*bp* and *bp*-*bp* repulsion.

Bond length, $x = 95.7$ pm

9. (a) Hard water does not form lather with soap, rather it forms scum/precipitate. Soft water forms lather with soap.

$$\underset{\text{Soap}}{2C_{17}H_{35}COONa(aq)} + \underset{\text{(In hard water)}}{Ca^{2+}} \longrightarrow \underset{\text{Scum}}{(C_{17}H_{35}COO)_2Ca} + 2Na^+$$

10. (c) Temporary hardness is due to dissolved bicarbonates of Ca^{2+} and Mg^{2+}. On addition of lime, $Ca(OH)_2$, $CaCO_3$ and $Mg(OH)_2$ are precipitated which are removed by filtration.

$$Ca(HCO_3)_2 + Ca(OH)_2 \longrightarrow 2CaCO_3 + 2H_2O$$

$$Mg(HCO_3)_2 + 2Ca(OH)_2 \longrightarrow Mg(OH)_2 + 2CaCO_3 + 2H_2O$$

Solubility product of $Mg(OH)_2$ is lower than that of $MgCO_3$, thus $Mg(OH)_2$ and not $MgCO_3$ is precipitated.

11. (d) $Na_6P_6O_{18}$ - Sodium hexametaphosphate $(NaPO_3)_6$. It is called Calgon and this method of removing permanent hardness is called Calgon's method.

$$(NaPO_3)_6 \rightleftharpoons 2Na^+ + Na_4P_6O_{18}^{2-}$$

$$\underset{\substack{(Ca^{2+} \text{ or } Mg^{2+} \\ \text{in hard water})}}{M^{2+}} + Na_4P_6O_{18}^{2-} \longrightarrow \underset{\substack{\text{Soluble complex} \\ \text{(it carries } Ca^{2+} \text{ and } Mg^{2+})}}{Na_2MP_6O_{18}^{2-}} + 2Na^+$$

12. (a) Zeolite is hydrated sodium aluminium silicate. It is used to remove permanent hardness in ion-exchange method (also called Permutit method).

$$NaAlSiO_4 \rightleftharpoons Na^+ + AlSiO_4^-$$

13. (b) Zeolite is $NaAlSiO_4$ represented as $NaZ(s)$. It is packed in ion exchange. Hard water is allowed to pass through ion exchange packed with zeolite. Ca^{2+} and Mg^{2+} ions of hard water are replaced by Na^+ ions of zeolite and water coming out of ion exchange is free from Ca^{2+} and Mg^{2+}.

$$\underset{\substack{\text{Ion exchange} \\ (Z = AlSiO_4)}}{2NaZ(s)} + \underset{\substack{\text{In hard water} \\ M = \text{Ca, Mg}}}{M^{2+}(aq)} \longrightarrow MZ_2 + 2Na^+$$

14. (d) Cation exchange resin contains larger organic molecules with $—SO_3H$. $R—SO_3H$ are water insoluble.

$$RSO_3H + NaCl \longrightarrow \underbrace{RSO_3Na}_{\text{Resin}}$$

Thus, (a) is correct.

When hard water is allowed to pass through synthetic resin, Ca^{2+}/Mg^{2+} ions are replaced by Na^+.

$$2RSO_3Na + Ca^{2+}/Mg^{2+} \longrightarrow (RSO_3)_2Ca/Mg + 2Na^+$$

Thus, (b) is correct.

Resin is said to be exhausted when all Na^+ ions of RSO_3Na are used up. Exhausted resin is regenerated by aqueous NaCl.

$$(RSO_3)_2Ca^{2+}/Mg^{2+} + 2NaCl \longrightarrow 2RSO_3Na + CaCl_2/MgCl_2$$

Thus, (c) is correct.

15. (c) $Cl_2(g) + H_2O(l) \longrightarrow HOCl(aq) + HCl(aq)$

Oxidation states: Cl_2: 0; H_2O: +1, −2; $HOCl$: +1, −2, +1; HCl: +1, −1

Oxidation (0 → +1 in HOCl); Reduction (0 → −1 in HCl)

Cl_2 is oxidised to HOCl. Cl_2 is reduced to HCl.

Thus, Cl_2 disproportionates.

Cl_2 is oxidising agent as well as reducing agent.

HOCl is bleaching agent and has antibacterial action.

16. (a)

(a) Hydrolysis is the interaction of H_2O with cation or anion or both of a salt (or organic ester, etc) to form acidic/basic/neutral solution. Salt of strong acid and strong base is not hydrolysed.

$$NaCl(aq) + H_2O(l) \rightleftharpoons NaOH(aq) + HCl(aq)$$

$$\Updownarrow \qquad\qquad \Updownarrow$$

$$Na^+(aq) + Cl^-(aq) + H_2O(l) \rightleftharpoons Na^+(aq) + OH^-(aq) + H^+(aq) + Cl^-(aq)$$

or $\quad H_2O(l) \rightleftharpoons H^+(aq) + OH^-(aq)$

Thus, NaCl (salt of strong acid (HCl) and strong base (NaOH)) does not react with H_2O.

(b) $CH_3COO^- + H_2O \rightleftharpoons \underset{\text{Weak acid}}{CH_3COOH} + \underset{\text{Basic}}{OH^-}$

(c) $Cu^{2+} + 2H_2O \rightleftharpoons \underset{\text{Weak base}}{Cu(OH)_2} + \underset{\text{Acidic}}{2H^+}$

(d) $CH_3COOC_2H_5 + H_2O \rightleftharpoons CH_3COOH + C_2H_5OH$

$CH_3COOH > C_2H_5OH$ (acid)

Hence, solution is acidic.

17. (d) When there is extensive hydrogen bonding, freezing points and boiling points are higher.

Higher heat is required to convert $H_2O(l)$ to $H_2O(g)$. Thus, heat of vaporisation is high.

Higher heat is required to convert $H_2O(s)$ to $H_2O(l)$. Thus, heat of fusion is high.

18. (a) $H_2O \rightleftharpoons H^+ + OH^-$

$$[H^+]_{H_2O} = \sqrt{K_w}$$

$$= \sqrt{1.0 \times 10^{-14}} = 1 \times 10^{-7} \text{M}$$

$$D_2O \rightleftharpoons D^+ + OD^-$$

$$[D^+]_{D_2O} = \sqrt{K_w} = \sqrt{3.0 \times 10^{-15}}$$

$$= 5.48 \times 10^{-8} \text{ M}$$

$$[H^+]_{H_2O} > [D^+]_{D_2O}$$

Thus, H_2 is formed at a faster rate.

19. (c) Hardness in water is due to dissolved

(a) HCO_3^- of Ca^{2+} and Mg^{2+} - temporary

(b) SO_4^{2-} and Cl^- of Ca^{2+} and Mg^{2+} - permanent

Thus, saturation with $CaSO_4$ will produce hard water.

20. (c) Water has maximum density at 4°C and when ice is formed, volume increases due to air gap. Thus, there is insulation between air and water and this keeps fish and other marine organisms survived for a long period.

21. (d) (a) SO_3 is soluble in H_2O

$$SO_3 + H_2O \rightleftharpoons H_2SO_4$$

(b) F_2 is reduced to HF

$$2H_2O + 2F_2 \longrightarrow 4HF + O_2$$

(c) N_2 is insoluble in H_2O.

(d) Cl_2 disproportionates to HCl (by reduction) and to HClO (by oxidation)

$$Cl_2 + H_2O \longrightarrow HCl + HClO$$

22. (b) H_2O has maximum density at 4°C while D_2O has maximum density at 11.6°C.

23. (a)

(a) Permanent hardness is due to dissolved Cl^- and SO_4^{2-} of Ca^{2+} and Mg^{2+}. They cannot be removed by boiling.

(b) Ion exchange method exchanges Cl^-, SO_4^{2-}, Ca^{2+} and Mg^{2+}

(I) Cation exchange resin

$$RH_2 + Ca^{2+} \longrightarrow RCa + 2H^+$$

(II) Anion exchange resin

$$R(OH)_2 + SO_4^{2-} \longrightarrow RSO_4 + 2OH^-$$

(c) Calgon's method removes permanent hardness by using $(NaPO_3)_6$, Calgon.

$$\underset{(Ca^{2+}/Mg^{2+})}{M^{2+}} + \underset{\text{Calgon}}{Na_4P_6O_{18}^{2-}} \longrightarrow \underset{\text{Soluble}}{[Na_2MP_6O_{18}]^{2-}} + 2Na^+$$

(d) $$\underset{\text{(In hard water)}}{CaSO_4} + \underset{\text{Washing soda}}{Na_2CO_3} \longrightarrow \underset{\text{Insoluble}}{CaCO_3} + Na_2SO_4$$

24. (d) $$\underset{2\text{ mol}}{2C_8H_7SO_3Na} + \underset{1\text{ mol}}{Ca^{2+}} \rightleftharpoons (C_8H_7SO_3)_2Ca + 2Na^+$$

2×206 g resin uptake $Ca^{2+} = 1$ mol

1 g resin uptakes $Ca^{2+} = \frac{1}{412}$ mol

Thus, maximum uptake of Ca^{2+} per gram of resin $= \frac{1}{412}$ mol g^{-1}

25. (a) $$CaSO_4 \rightleftharpoons Ca^{2+} + SO_4^{2-}$$

$$R(OH)_2 + SO_4^{2-} \longrightarrow RSO_4 + 2OH^-$$

$$2OH^- + 2H^+ \longrightarrow 2H_2O$$

SO_4^{2-} in hard water are exchanged by OH^- ions in anion exchange resin. Water coming out of resin is alkaline in nature and neutralised by H_2SO_4.

$$CaSO_4 \equiv SO_4^{2-} \equiv 2OH^- \equiv 2H^+ \equiv H_2SO_4$$

$$M(CaSO_4) \equiv M(H_2SO_4) \equiv 2OH^-$$

$25 \times M$ ($CaSO_4$ in water) $= 20 \times 0.001$ M

$\therefore$ M ($CaSO_4$ in water) $= \frac{20 \times 0.001}{25} = 8 \times 10^{-4}$ M

$\therefore$ $[CaSO_4] = 8 \times 10^{-4}\text{ M} = 8 \times 10^{-4} \times 136\text{ gL}^{-1}$

$CaSO_4$ in 10^6 mL $= \frac{8 \times 10^{-4} \times 136 \times 10^6}{10^3}$ g

$= 108.8$ ppm $CaSO_4$

26. (b) $[MgSO_4] = 0.002$ mol L^{-1}

$= 0.002 \times 120\text{ gL}^{-1}$

$= \frac{0.002 \times 120 \times 10^6}{10^3}$ g in 10^6mL

$= 240$ ppm $MgSO_4$

27. (b) Electronegativity of O is higher than that of S. Thus, there is permanent dipole moment in (H—O) bond in H_2O. Large number of H_2O molecules are attracted by intermolecular H-bonding.

Bond (H—S) is less polar and thus, there is no intermolecular H-bonding in H_2S. Thus, H_2S exists as gas having much lower boiling point than that of H_2O.

Thus, Statement I and Statement II are correct but Statement II is not the correct explanation of Statement I.

28. (d) In water, each water molecule is joined to four other water molecules by intermolecular H-bonding.

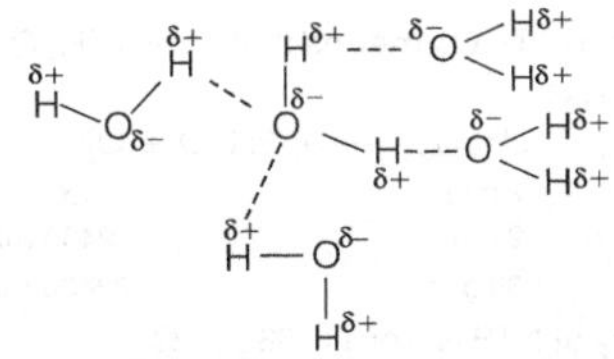

In H—F, individual H—F molecule is joined to two H—F molecules.

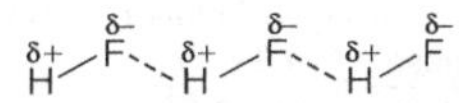

H······F—H H-bonding is stronger but boiling point of H_2O is higher than that of H—F due to more H-bonding (at four points in H_2O and at two points in HF).

Electronegativity of F > O, hence (H—F) bond is more polar than (H—O) bond.

Thus, Statement I is incorrect and Statement II is correct.

29. (b) When ice is formed from liquid water, some air gap is formed (in tetrahedral packing of water molecules). Due to this, volume of ice is greater than that of liquid water. Since, density $\propto \frac{1}{\text{volume}}$, hence density is decreased when ice is formed. Hence, ice floats on water. Reversely, when ice melts, density increases but only upto 4°C, after this, intermolecular H-bonding between H_2O molecules breaks and volume increases, hence density decreases. Thus, Statement I and Statement II are correct but Statement II is not the correct explanation of Statement I.

30. (b)

(i) $NH_3 + H_2O \longrightarrow NH_4OH$

H_2O is H^+ ion donor and NH_3 is H^+ ion acceptor. Thus, H_2O is Lowry-Bronsted acid.

Thus, (i) $\rightarrow$ (t).

(ii) $CH_3COOH + H_2O \longrightarrow CH_3COO^- + H_3O^+$

H_2O is H^+ ion acceptor and CH_3COOH is H^+ ion donor. Thus, H_2O is Lowry-Bronsted base.

Thus, (ii) $\rightarrow$ (u)

(iii) $$\underset{\text{Acid}_1}{H_2O} + \underset{\text{Base}_2}{H_2O} \longrightarrow \underset{\text{Acid}_2}{H_3O^+} + \underset{\text{Base}_1}{OH^-}$$

H_2O is H^+ ion donor and changes to OH^-. H_2O is H^+ ion acceptor and changes to H_3O^+. Thus, it is amphiprotic.

Thus, (iii) $\rightarrow$ (r).

(iv) Fe is oxidised to Fe_3O_4 by H_2O which is reduced to H_2. Thus, H_2O is the oxidising agent.

Thus, (iv) $\rightarrow$ (q).

(v) F_2 is reduced to F^- by H_2O which is oxidised to O_2. Thus, H_2O is the reducing agent.

Thus, (v) $\rightarrow$ (s).

(vi) It is a photochemical reaction. H_2O reduces CO_2 to $C_6H_{12}O_6$ and itself oxidised to O_2. Thus, H_2O is the reducing agent.

Thus, (vi) $\rightarrow$ (p,s)

DPP-4 Hydrogen Peroxide

1. **Plan** "x volume" H_2O_2 means that 1 mL of H_2O_2 at NTP gives 10 mL of O_2 gas.

$$\underset{\substack{2\text{ mol}\\2\text{ mol}\\68\text{ g}}}{2H_2O_2} \longrightarrow 2H_2O + \underset{\substack{1\text{ mol}\\22400\text{ mL at NTP}\\22400\text{ mL at NTP}}}{O_2}$$

22400 mL O_2 at NTP is from = 68g H_2O_2

x mL of O_2 at NTP is from = $\dfrac{68x}{22400}$ g H_2O_2

This amount is in 1 mL H_2O_2 solution of "x volume".

(i) Density = $\dfrac{\text{mass}}{\text{volume}} = \dfrac{68x}{22400}\Big/1$

$= \dfrac{68x}{22400}$ g mL^{-1}

(ii) gL^{-1} = $\dfrac{68x}{22400} \times 1000$ gL^{-1} = $\dfrac{68x}{22.4}$ gL^{-1}

(iii) Percentage = $\dfrac{68x}{22400} \times 100$ g/100 mL = $\dfrac{68x}{224}$ g/100 mL

(iv) Normality = $\dfrac{\text{gL}^{-1}\text{(conc.)}}{\text{equivalent weight}} = \dfrac{68x}{22.4 \times 17} = \dfrac{x}{5.6}$ g equiv L^{-1}

(v) Molarity = $\dfrac{\text{gL}^{-1}\text{ (conc.)}}{\text{molar mass}} = \dfrac{68x}{22.4 \times 34} = \dfrac{x}{11.2}$ mol L^{-1}

2. (d) PbO_2 does not have peroxy linkage hence, H_2O_2 is not obtained by this.

(a) $BaO_2 + H_2SO_4 \longrightarrow BaSO_4 + H_2O_2$

(b) $Na_2O_2 + 2H_2O \longrightarrow 2NaOH + H_2O_2$

(c) $2KO_2 + 2H_2O \longrightarrow 2KOH + H_2O_2$

(d) $PbO_2 + H_2SO_4 \longrightarrow PbSO_4 + H_2O + O_2$

3. (b)

(a) H_2O_2 is oxidised to O_2 thus, it shows reducing action of H_2O_2.

(b) H_2O_2 gains electrons hence, it is reduced. Thus, H_2O_2 is the oxidising agent.

(c) H_2O_2 is oxidised to O_2 and O_3 is reduced to H_2O.

(d) $[Fe(CN)_6]^{3-}$ (ON of Fe = +3) is reduced to $[Fe(CN)_6]^{4-}$ (ON of Fe = +2).

4. (a) $H_2SO_4 \xrightleftharpoons{\text{Current}} H^+ + HSO_4^-$

At anode $\quad 2HSO_4^- \rightleftharpoons \underset{\text{Intermediate}}{H_2S_2O_8} + 2e^-$

$H_2S_2O_8 + H_2O \longrightarrow \underset{\text{Intermediate}}{H_2SO_5} + H_2SO_4$

$H_2SO_5 + H_2O \longrightarrow H_2SO_4 + H_2O_2$

At cathode $\quad 2H^+ + 2e^- \longrightarrow H_2$

Using NH_4HSO_4

$2NH_4HSO_4 \longrightarrow H_2 + \underset{\text{Intermediate}}{(NH_4)_2S_2O_8}$

$(NH_4)_2S_2O_8 + H_2O \longrightarrow 2NH_4HSO_4 + H_2O_2$

5. (d)

(a) $A \xrightarrow{O_2} B + \underset{(C)}{H_2O_2}$

Thus, (a) is correct.

(b) $B \xrightarrow[H_2/\text{Raney Ni}]{H_2/\text{Pd or}} A$

Thus, process is cyclic.

Thus, (b) is correct.

(c) Solution of H_2O_2 is 1% only. It is concentrated under reduced pressure since, it decomposes at its boiling point.

Thus, (c) is correct.

6. (b)

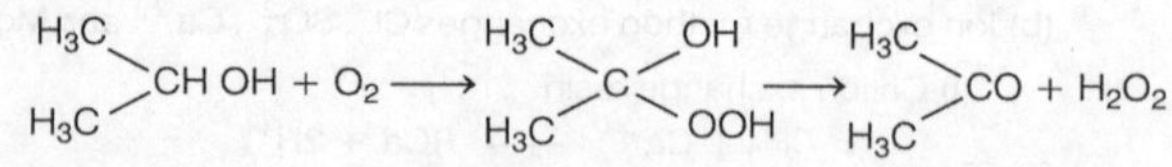

7. (c) H_2O_2 when stored is unstable and decomposes to H_2O and O_2.

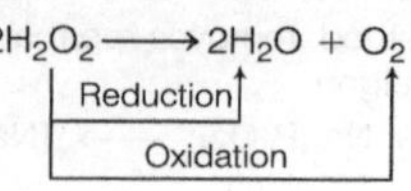

Stabilises as 8-hydroquinoline or sodium stannate can be used as stabiliser to prevent decomposition.

8. (a) $\quad H_2O_2 + 2H^+ + 2e^- \longrightarrow 2H_2O$

At $[H^+] = 1$ M, $E° = 1.77$ V

This indicates that H_2O_2 has been reduced. Thus, H_2O_2 is a strong oxidising agent.

9. (a) $2MnO_4^- + 5H_2O_2 + 6H^+ \xrightarrow{\text{Acidic}} 2Mn^{2+} + 5O_2 + 8H_2O$

$2MnO_4^- + 3H_2O_2 \xrightarrow{\text{Basic}} 2MnO_2 + 3O_2 + 2H_2O + 2OH^-$

10. (d) MnO_2 is reduced to Mn^{2+} by H_2O_2 in acidic medium.

$MnO_2 + H_2O_2 + 2H^+ \longrightarrow Mn^{2+} + O_2 + 2H_2O$

Mn^{2+} is of light pink colour but in dilute solution, it appears colourless.

11. (c) (O—O) bond is stable and is not cleaved in reduction of PbO_2 to PbO. Thus, both ^{18}O of H_2O_2 remain with O_2. Only H—O bond is cleaved.

$PbO_2 + H{\vdots}\overset{18}{O}{-}\overset{18}{O}{\vdots}H \longrightarrow PbO_2 + \overset{18}{O_2} + H_2O$

12. (c) $Ag_2O + H_2O_2 \longrightarrow 2Ag + H_2O + O_2$

$Au_2O + H_2O_2 \longrightarrow 2Au + H_2O + O_2$

CO and H_2 are less active and cannot reduce metal oxides to metal.

13. (b) Oil paints contain Pb^{2+} in the form of $(CH_3COO)_2$ Pb. In contact with atmospheric H_2S, it is blackened due to the formation of PbS(X).

PbS can be cleaned by H_2O_2 which changes PbS to $PbSO_4$ (Y).

$\underset{\text{(In oil paint)}}{(CH_3COO)_2Pb} + H_2S \longrightarrow \underset{\substack{\text{Black}\\(X)}}{PbS} + 2CH_3COOH$

$PbS + 4H_2O_2 \longrightarrow \underset{\substack{\text{White}\\(Y)}}{PbSO_4} + H_2O$

14. (d) $2I^- + H_2O_2 + 2H^+ \longrightarrow \underset{(X)}{I_2} + 2H_2O$

(a) I_2 can be estimated by hypo

$2S_2O_3^{2-} + I_2 \longrightarrow S_4O_6^{2-} + 2I^-$ $\quad$ Thus, correct.

(b) I_2 gives blue colour with starch. Thus, correct.

(c) $2I^- \longrightarrow I_2 + 2e^-$ Thus, correct.

15. (d) In acidic medium, MnO_4^- is reduced to Mn^{2+}.

$2MnO_4^- + 5H_2O_2 + 6H^+ \longrightarrow 2Mn^{2+} + 5O_2$

H_2O_2 oxidised by one mole of MnO_4^- in acidic medium = $\dfrac{5}{2}$

In basic medium, MnO_4^- is reduced to MnO_2.

$2MnO_4^- + 3H_2O_2 \longrightarrow 2MnO_2 + 3O_2 + 2H_2O + 2OH^-$

H_2O_2 oxidised by one mole of MnO_4^- in basic medium = $\dfrac{3}{2}$

Ratio = 5 : 3

16. (c) $Fe^{2+} + H_2O_2 \longrightarrow [FeOH]^{2+} + OH$

$Fe^{2+} + OH \longrightarrow [FeOH]^{2+}$

17. (d) $H_2O_2 \rightleftharpoons H^+ + HO_2^-$

$$pH = 6.0$$

$$\therefore \quad [H^+] = 10^{-pH} = 1 \times 10^{-6} M$$

$$\therefore \quad [HO_2^-] = [H^+] = 1 \times 10^{-6}$$

$$\text{Auto-protolysis constant} = [HO_2^-][H^+] = (1.0 \times 10^{-6})^2 = 1.0 \times 10^{-12}$$

18. (a) H_2O_2 is an oxidising agent as well as a reducing agent.

Oxidising nature $H_2O_2 + 2H^+ + 2e^- \longrightarrow 2H_2O$

Reducing nature $H_2O_2 \longrightarrow 2H^+ + O_2 + 2e^-$

19. (a) Perhydrol is 30% H_2O_2 or "100 volume" H_2O_2.

If it is "100 volume", it means

1 mL of H_2O_2 gives 100 mL of O_2 gas at NTP.

or 1L of H_2O_2 gives 100 L of O_2 gas at NTP.

$$\underset{\substack{2\text{ mol}\\ 2L}}{2SO_2} + \underset{\substack{1\text{ mol}\\ 1L}}{O_2} \longrightarrow 2SO_3$$

$$O_2 \text{ required} = 1\text{ L}$$

$$H_2O_2 \text{ required} = \frac{1}{100}\text{ L} = \frac{1000\text{ mL}}{100} = 10\text{ mL}$$

20. (c) H_2O_2 is oxidised to O_2 and O_3 is reduced to O_2 and H_2O.

$$O_3 \longrightarrow O_2 + O$$

$$H_2O_2 + O \longrightarrow H_2O + O_2$$

O_3 is oxidising agent.

$H_2O_2 \longrightarrow$ diamagnetic, $\quad O_2 \longrightarrow$ paramagnetic

$O_3 \longrightarrow$ diamagnetic, $\quad H_2O \longrightarrow$ diamagnetic

Thus, magnetic behaviour of both changes.

21. (d)

(a) $H_2O_2 + Cl_2 \longrightarrow 2HCl + O_2$

H_2O_2 acts as a reducing agent to counteract Cl_2. Thus, correct.

(b) $2Na_2O_2 + 2CO_2 \longrightarrow 2Na_2CO_3 + O_2$

$4KO_2 + 2CO_2 \longrightarrow 2K_2CO_3 + 3O_2$

Thus, Na_2O_2 and KO_2 both release O_2 by absorbing CO_2. Thus, correct.

Note *This is an application of this type of peroxides in submarines.*

(c) $Ba(OH)_2 + H_2O_2 \longrightarrow BaO_2 + 2H_2O$

In this reaction, H_2O_2 behaves as a dibasic acid. Thus, correct.

22. (d)

(a) $H_2O_2 + 2I^- + 2H^+ \longrightarrow 2H_2O + I_2$

$$I_2 + 2S_2O_3^{2-} \longrightarrow 2I^- + S_4O_6^{2-}$$

$$I_2 \equiv 2S_2O_3^{2-}$$

$$I \equiv S_2O_3^{2-} \equiv \frac{H_2O_2}{2}$$

Thus, equivalent mass of $H_2O_2 = \dfrac{\text{Molar mass}}{2}$

$$\underset{H_2O_2}{N_1V_1} = \underset{\text{Hypo}}{N_2V_2}$$

$$N_1 \times 10 = 1 \times 10$$

$$N_1 (H_2O_2) = 1\text{ N}$$

Thus, (a) is correct.

(b) Volume strength of $H_2O_2 = \dfrac{5.6}{\text{Normality}} = \dfrac{5.6}{1} = 5.6$

Thus, (b) is correct

(c) Concentration in $gL^{-1} = N \times$ equivalent mass $= 1 \times 17 = 17\text{ g L}^{-1}$

Thus, (c) is correct.

23. (a) In H_2O_2,

$$\angle\alpha = 94°48'$$

$$\angle\beta = 111°30'$$

$$(O—O) \text{ bond length} = 148 \text{ pm}$$

24. (b)

(a) $2[\overset{II}{Fe}(CN)_6]^{4-} + H_2O_2 + 2H^+ \xrightarrow{\text{Oxidation}} 2[Fe^{III}(CN)_6]^{3-} + 2H_2O$

$[Fe^{III}(CN)_6]^{3-} + H_2O_2 + 2OH^- \xrightarrow{\text{Reduction}} [Fe^{II}(CN)_6]^{4-} + 2H_2O + O_2$

Thus, correct.

(b) $Mn^{2+} + H_2O_2 + 2OH^- \longrightarrow \underset{\text{(Black)}}{MnO_2} + 2H_2O$

Thus, incorrect.

(c) $2Cr^{3+} + 3H_2O_2 + 10\,OH^- \longrightarrow 2CrO_4^{2-} + 8H_2O$

Thus, correct.

(d) $H_2O_2 + 2H^+ + 2e^- \longrightarrow 2H_2O$

(As an oxidising agent)

Equivalent weight $= \dfrac{M}{2}$

$$H_2O_2 \longrightarrow 2H^+ + O_2 + 2e^-$$

(As a reducing agent)

Equivalent weight $= \dfrac{M}{2}$

Thus, correct.

25. (a)

(a) HCHO is oxidised to HCOOH by H_2O_2 which is reduced to H_2 in the presence of pyrogallol. The reaction in the dark is accompanied by emission of light called chemiluminescence.

(b) This shows acidic behaviour of H_2O_2

$$Na_2CO_3 + H_2O_2 \longrightarrow Na_2O_2 + H_2O + CO_2$$

(c) AsO_3^{3-} is oxidised to AsO_4^{3-} by H_2O_2.

(d) NO_2^- is oxidised to NO_3^- by H_2O_2.

26. (d)

(a) H_2O_2 decomposes carbonates, bicarbonates, etc., and thus behaves as an acid.

$$Na_2CO_3 + H_2O_2 \longrightarrow Na_2O_2 + H_2O + CO_2$$

$$2NaHCO_3 + H_2O_2 \longrightarrow Na_2O_2 + 2H_2O + 2CO_2$$

Thus, correct

(b) H_2O_2 is reduced to H_2O, thus it behaves as an oxidising agent.

$H_2O_2 + 2H^+ + 2e^- \longrightarrow 2H_2O$ Thus, correct.

(c) H_2O_2 is oxidised to O_2, thus it behaves as a reducing agent.

$$H_2O_2 \longrightarrow 2H^+ + O_2 + 2e^-$$

Thus, correct.

Note *(O—O) bond of H_2O_2 remains intact.*

27. (c)

$$\underset{-1}{H_2O_2} \longrightarrow \underset{-2}{H_2O} + \underset{0}{O_2}$$

(ON) $-1 \xrightarrow{\text{Reduction}} -2$; $-1 \xrightarrow{\text{Oxidation}} 0$

H_2O_2 disproportionates to O_2 (by oxidation) and to H_2O (by reduction).

28. (b) H_2O_2 interacts with blood releasing bubbles due to the formation of O_2.

The decomposition is catalysed by an enzyme catalase present in blood.

Thus, Statement I and Statement II are correct but Statement II is not the correct explanation of Statement I.

29. (b) $H_2O_2 \rightleftharpoons HO_2^- + H^+$

K_{a_1} (first ionisation constant) $= \dfrac{[HO_2^-][H^+]}{[H_2O_2]} = 2.0 \times 10^{-12}$

K_{a_2} (second ionisation constant) $<< K_{a_1}$

Thus, HO_2^- is not further ionised. Hence, in alkaline medium, H_2O_2 exists as HO_2^-. Thus, Statement I and Statement II are correct but Statement II is not the correct explanation of Statement I.

30. (a)

(i) $2H_2O + 2F_2 \longrightarrow 4HF + O_2$

H_2O reduces F_2 to HF, thus is reducing agent. H_2O is diamagnetic in nature.

Thus, (i) $\rightarrow$ (p,s)

(ii) $H_2O_2 + 2H^+ + 2e^- \longrightarrow 2H_2O$

H_2O_2 is reduced, thus oxidising agent.

$$H_2O_2 \longrightarrow 2H^+ + O_2 + 2e^-$$

H_2O_2 is oxidised, thus reducing agent. O_2^{2-} is diamagnetic.

$$H_2O_2 \rightleftharpoons H^+ + HO_2^-$$

H_2O_2 is a weak acid. Thus, it turns blue litmus red.

Thus, (ii) $\rightarrow$ (p,q,s,u)

(iii) $Na_2O + H_2O \longrightarrow \underset{\text{Alkaline}}{2NaOH}$

O^{2-} is diamagnetic; it turns red litmus blue being alkaline in nature.

Thus, (iii) $\rightarrow$ (s,t)

(iv) $2KO_2 + 2H_2O \longrightarrow 2KOH + O_2 + H_2O_2$

Since, O_2 and H_2O_2 are formed, hence KO_2 is an oxidising agent.

O_2^- has one unpaired electron, hence paramagnetic.

Being alkaline, it turns red litmus blue.

Thus, (iv) $\rightarrow$ (q,r,t)

(v) $CaH_2 + 2H_2O \longrightarrow Ca(OH)_2 + 2H_2$

It is instant source of H_2 hence, reducing agent.

H^- has paired electron, hence diamagnetic.

Alkaline solution turns red litmus blue.

Thus, (v) $\longrightarrow$ (p,s,t)

Revisal Problems (JEE Main)

1. (d) Ionic hydrides are instant source of H_2 of high purity.

$$2NaH + 2H_2O \longrightarrow 2NaOH + 2H_2$$

2. (d)

(a) Correct

(b) $(CO + H_2)$ is used for the synthesis of methanol, hence mixture is also called synthesis gas

$$CO + 2H_2 \xrightarrow{\text{Cobalt}} CH_3OH$$

(c)

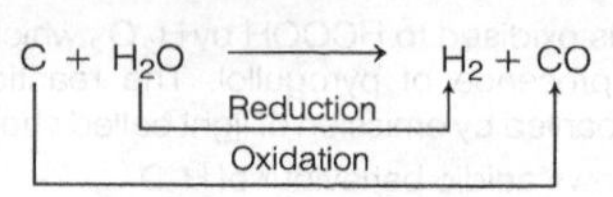

Carbon is oxidised to CO and H_2O is reduced to H_2, thus it is a redox reaction.

3. (c) As we go down the group (top to bottom) in the periodic table, basic nature of hydrides decreases and acidic nature increases. (In a period left to right basic nature increases).

NaH MgH_2 AlH_3 SiH_4 NH_3 ↑ Basic

PH_3 H_2S

AsH_3

SbH_3

$$NH_3 > PH_3 > AsH_3 > SbH_3$$

Tendency to donate lone pair decreases hence, Lewis base nature decreases.

4. (b) "22.4 volume" $H_2O_2 = \dfrac{22.4}{5.6} N = 4\,N = 4 \times 17\ gL^{-1}$

$$\underset{\substack{34\text{ g}\\3.4\text{ g}}}{H_2S} + \underset{\substack{34\text{ g}\\3.4\text{ g}}}{H_2O_2} \longrightarrow 2H_2O + S$$

68 g H_2O_2 is in = 1000 mL

Thus, 3.4 g H_2O_2 is in $= \dfrac{1000}{68} \times 3.4$ mL H_2O_2 = 50 mL

5. (c) $K_2Cr_2O_7 + 2H_2SO_4 \longrightarrow 2KHSO_4 + 2CrO_3 + H_2O$

$$CrO_3 + 2H_2O_2 \longrightarrow CrO_5 + 2H_2O$$

CrO_5 is a diperoxide and gives blue colour in etheral layer.

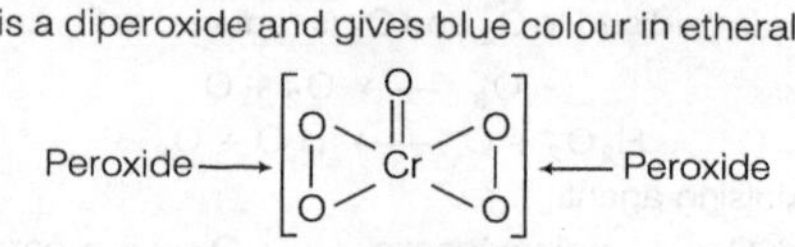

Note *Oxidation number of* Cr *in* $CrO_5 = +6$

Oxidation number of Cr *in* $Cr_2O_7^{2-} = +6$

Hence, this is not a redox reaction.

6. (b) $Cr_2O_7^{2-}$ oxidises SO_3^{2-} to SO_4^{2-} and itself reduced to Cr^{3+} (green) (*X*).

$$3SO_3^{2-} + Cr_2O_7^{2-} + 8H^+ \longrightarrow 3SO_4^{2-} + \underset{\text{Green}}{2Cr^{3+}} + 4H_2O$$

H_2O_2 oxidises SO_3^{2-} to SO_4^{2-}.

$$SO_3^{2-} + H_2O_2 \longrightarrow \underset{\substack{(Y)\\\text{Colourless}}}{SO_4^{2-}} + H_2O$$

Y does not react with $Cr_2O_7^{2-}$.

7. (c) H_2O_2 is reducing agent and itself oxidised to O_2.

$$2H_2O_2 \longrightarrow 2H_2O + O_2$$

or $$H_2O_2 \longrightarrow 2H^+ + O_2 + 2e^-$$

$$IO_4^- \longrightarrow \underset{\text{New compound}}{(X)}$$

IO_4^- and H_2O_2 are consumed in 1 : 1 molar ratio, hence, change in ON in oxidation and reduction reactions are same.

$$\underset{-2}{H_2O_2} \longrightarrow \underset{0}{O_2}$$

(For two O-atoms)

Change in ON = 2 units

$$\underset{+7}{\overset{\uparrow}{IO_4^-}} \longrightarrow \underset{5}{\overset{\uparrow}{(X)}}$$

Change in ON = 2 units ON of I,

$IO_2^- = +3$, $IO^- = +1$

$IO_3^- = +5$, $I_2 = 0$

Thus, X is IO_3^-.

8. (b) 30% H_2O_2 solution is sold in the market under the trade name perhydrol. It is also known as "100 volume H_2O_2".

Molarity of "100 volume" $H_2O_2 = \frac{100}{11.2} = 8.93$ mol L^{-1}

Thus, perhydrol is not 30 volume H_2O_2 solution.

9. (d) $MgSO_4$ = 25 mL of 0.08 N in 100 mL H_2O

$$= 25 \times 0.08 \text{ milliequivalent}$$
$$= \frac{25 \times 0.08}{1000} \text{ equivalent}$$
$$\therefore \quad CaCO_3 = \frac{25 \times 0.08}{1000} \text{ equiv in 100 mL } H_2O$$
$$= 2 \times 10^{-3} \text{ equiv in 100 mL } H_2O$$
$$= 2 \times 10^{-3} \times 50 \text{ g/100 mL}$$
$$= 0.1 \text{ g/100 mL}$$
$$= 0.1 \times 10^6 / 100 \text{ ppm}$$
$$= 1000 \text{ ppm}$$

10. (b) $$Ca^{2+} + \underset{\text{Cation exchange resin}}{RH_2} \longrightarrow RCa + 2H^+$$

$$Ca^{2+} = 80 \text{ ppm} = 80 \text{ g in } 10^6 \text{ mL}$$
$$= \frac{80 \times 10^3}{10^6} \text{ g}L^{-1}$$
$$= \frac{80 \times 10^3}{10^6 \times 40} \text{ mol } L^{-1}$$
$$= 2 \times 10^{-3} \text{ mol } L^{-1}$$
$$[H^+] = 2[Ca^{2+}]$$
$$= 2 \times 2 \times 10^{-3} \text{ mol } L^{-1}$$
$$= 4 \times 10^{-3} \text{ M}$$
$$pH = -\log[H^+]$$
$$= -\log(4 \times 10^{-3})$$
$$= 2.40$$

11. (a) $$H_2R + Ca^{2+} \longrightarrow RCa + 2H^+$$
$$H^+ + OH^- \longrightarrow H_2O$$

Thus, $[H^+] \equiv [OH^-] \equiv [Ca^{2+}]$ in terms of equivalent.

Water coming out of cation exchange resin is acidic in nature and is neutralised by OH^-.

$$\underset{(H^+ \text{ in water})}{N_1V_1} = \underset{\text{NaOH}}{N_2V_2}$$
$$N_1 \times 100 = 0.025 \times 20$$
$$N_1 = \frac{0.025 \times 20}{100} = 5 \times 10^{-3} \text{N}$$

Equivalent of $[H^+]$ = Equivalent of $[Ca^{2+}]$

$$\therefore \quad [Ca^{2+}] = [H^+]$$
$$\Rightarrow \quad 5 \times 10^{-3} \text{N} = 20 \times 5 \times 10^{-3} \text{g}L^{-1}$$
$$= \frac{100 \times 10^{-3}}{10^3} \times 10^6 \text{ g in } 10^6 \text{ mL}$$
$$= 100 \text{ ppm } Ca^{2+}$$

12. (b) In ion-exchange method, sodium zeolite $[NaAlSiO_4]$ is used for removing permanent hardness.

$$2NaZ + M^{2+} \longrightarrow \underset{\text{(Exhausted)}}{MZ_2} + 2Na^+$$

Ca^{2+} or Mg^{2+} ions of hard water are replaced by Na^+.

When all Na^+ ions of zeolite are used up in purification, zeolite is said to be exhausted. It can be put into fresh operation by treated the exhausted zeolite with aqueous NaCl solution.

$$\underset{\text{Exhausted zeolite}}{MZ_2(s)} + 2NaCl \longrightarrow \underset{\text{Regenerated zeolite}}{2NaZ(s)} + MCl_2(aq)$$

MCl_2 (Ca^{2+}, Mg^{2+}) is soluble and is drained out.

Na_2SO_4 can't be used, since $CaSO_4$ and $MgSO_4$ are insoluble.

$AlCl_3$ will form aluminium salt which is not zeolite.

13. (a) Calgon is $[NaPO_3]_6$ sodium hexametaphosphate

$$Na_6P_6O_{18} \rightleftharpoons 2Na^+ + Na_4P_6O_{18}^{2-}$$
$$Na_4P_6O_{18}^{2-} + Ca^{2+} \text{ or } Mg^{2+} \longrightarrow \underset{\text{Soluble complex}}{Na_2[Na_2CaP_6O_{18}]}$$

Similarly with Mg^{2+}.

14. (c) Cation exchange resin is packed with RH. On passing hard water through it, dissolved cations in water are replaced by H^+.

$$2RH(s) + M^{2+}(aq) \longrightarrow MR_2(s) + 2H^+(aq)$$

(M = Ca, Mg, Fe, etc.)

Thus, (a) is correct.

(b) Anion exchange resin is packed with $RNH_3^+ OH^-$. On passing hard water through it, dissolved anions (Cl^-, HCO_3^-, SO_4^{2-}, etc) of water are replaced by OH^-.

$$RNH_3^+OH^-(s) + Cl^-(aq) \longrightarrow RNH_3^+ Cl^-(s) + OH^-(aq)$$
$$2RNH_3^+OH^-(s) + SO_4^{2-}(aq) \longrightarrow (RNH_3)_2SO_4 + 2OH^-(aq)$$

Note *After passing through cation-exchange resin, water coming out has free H^+ ions and after passing through anion exchange resin, water coming out has free OH^- ions. After two stages, water is neutral (pure) due to* $H^+ + OH^- \longrightarrow H_2O$

15. (a)

Isotopes of H, ${}^1_1H$, ${}^2_1H$, ${}^3_1H$

Isotopes of O, ${}^{16}_8O$, ${}^{17}_8O$, ${}^{18}_8O$

$H_2{}^{16}O$, $H_2{}^{17}O$, $H_2{}^{18}O$

$D_2{}^{16}O$, $T_2{}^{16}O$

$D_2{}^{17}O$, $D_2{}^{18}O$

$T_2{}^{17}O$, $T_2{}^{18}O$

$HD{}^{16}O$, $HD{}^{17}O$, $HD{}^{18}O$

$HT{}^{16}O$, $HT{}^{17}O$, $HT{}^{18}O$

$DT{}^{16}O$, $DT{}^{17}O$, $DT{}^{18}O$

Total arrangements = 18

16. (c)

(a) H^+ in aqueous solution exists as H_3O^+. H_2O molecules are associated by H-bonding and exist as $(H_2O)_n$ (n = 2, 3 ···)

$$H_2O + H^+ \rightleftharpoons H_3O^+$$
$$H_3O^+ + H_2O \rightleftharpoons H_5O_2^+$$
$$H_5O_2^+ + H_2O \rightleftharpoons H_7O_3^+$$
$$H_7O_3^+ + H_2O \rightleftharpoons H_9O_4^+ \cdots$$

Thus, correct.

(b) $H + H \longrightarrow H_2$, $\Delta H_f = -436$ kJ mol^{-1}

ΔH_f is very negative thus, H_2 is thermally stable. Thus, correct.

(c) (O—D) BE > (O—H) bond energy. Hence, ionisation of CH_3COOD is slower than that of CH_3COOH.

$$CH_3COOH \rightleftharpoons CH_3COO^- + H^+, \; K_a = 1.8 \times 10^{-5}$$

$$CH_3COOD \rightleftharpoons CH_3COO^- + D^+, \; K_a = 6.0 \times 10^{-6}$$

Thus, kinetic isotopic effect is observed.
Thus, incorrect.

(d) Regular drinking of D_2O instead of H_2O could prove fatal because of slower rate of transfer of D^+ compared with that of H^+ in the acid-base reaction involved in enzyme catalysis. Thus, correct.

17. (d) $8LiH + Al_2Cl_6 \longrightarrow 2LiAlH_4 + 6LiCl$

$LiAlH_4$ is the reducing agent for reducing aldehydes, ketones and acids to alcohols.

18. (c) $n(H_2) = \dfrac{pV}{RT} = \dfrac{150 \times 100}{0.0821 \times 298} = 613.1$ mol

$$\underbrace{CaH_2}_{1\,mol} + 2H_2O \longrightarrow Ca(OH)_2 + \underset{2\,mol}{2H_2}$$

2 mol H_2 is from = 1 mol CaH_2

$\therefore$ 613.1 mol H_2 is from $= \dfrac{613.1}{2}$ mol CaH_2

$= \dfrac{613.1}{2} \times 42$ g CaH_2

$= 12875.1$ g $CaH_2 = 12.875$ kg CaH_2

19. (d)

(a) Larger the ionic character, larger the electrical conductivity, thus, $BeH_2 < CaH_2 < TiH_2$. Thus, correct.

(b) Larger the size of cation, smaller the size of anion, larger the ionic nature, $LiH < NaH < CsH$. Thus, correct.

(c) F—F (158 kJ mol^{-1}) < H—H (436 kJ mol^{-1}) < D—D (443 kJmol^{-1})
Thus, correct.

(d) Larger the ionic nature, greater the formation of H^- ion. Thus, greater the reducing nature.

$$H^- \longrightarrow H + e^-$$

Ionic nature is in order, $AlH_3 < MgH_2 < NaH$
Hence, reducing nature is in order

$$AlH_3 < MgH_2 < NaH$$

Hence, $NaH < MgH_2 < AlH_3$ - reducing nature is incorrect order.

20. (c)

(a) Ionisation constant of 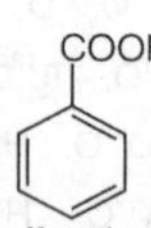 is higher that of 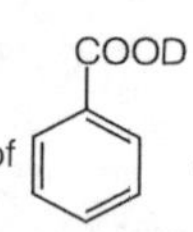

Thus, kinetic isotopic effect is observed.

(b) Rate of nitration of benzene (in which H is replaced by NO_2^+) is higher tha+n that of the deuterio benzene (in which D is replaced by NO_2^+) since, (C—H) bond energy is smaller than that of (C—D) bond. Thus, kinetic isotopic effect is observed.

(c) This reaction explains the role of D^+ in study of mechanism of enolisation of ketone.

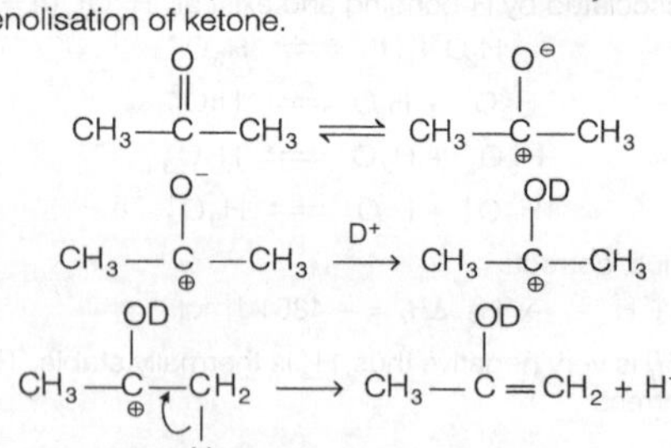

D^+ is attached to O and H^+ is eliminated.

21. (c) Bond length of (H—H), (D—D) and (T—T) bond is constant and is equal to 74 pm.

22. (d) If two nuclei of hydrogen atoms forming H_2 molecule have same spin then it is called *ortho*-hydrogen.

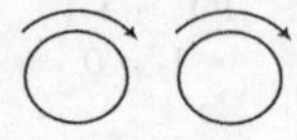

$\text{Sum} = \frac{1}{2} + \frac{1}{2} = 1$

If two nuclei of hydrogen atoms forming H_2 molecule have equal and opposite spin, then it is called *para*-hydrogen.

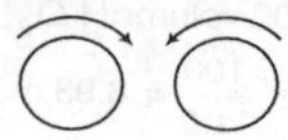

$\text{Sum} = \frac{1}{2} - \frac{1}{2} = 0$

H_2 gas in an equilibrium mixture of *ortho* and *para*-H_2. Ratio varies with temperature. At 300 K, ratio is 3 :1 and remains constant on increasing temperature.

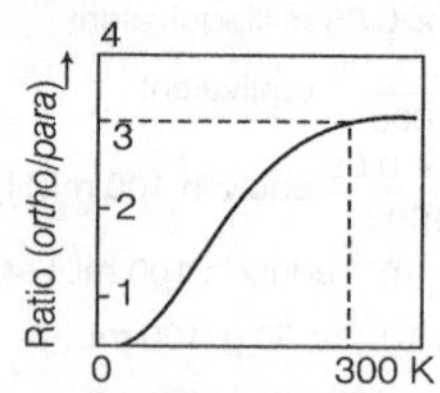

At 0 K, there is 100% *para* but at 300 K and above, there is 75% *ortho* and 25% *para*. Thus, 100% *ortho* form is not isolated.

Note *Ortho and para spin isomerism is also shown by* D_2 *and* T_2.

23. (b) (I) + $\dfrac{(II)}{2} + \dfrac{3}{2}$ (III) gives

$$H_2O \longrightarrow H_2 + \frac{1}{2}O_2$$

24. (d)

(a) HgO is reduced to Hg.
(b) CuO is reduced to Cu.
(c) $AgNO_3$ is reduced to Ag.
(d) Na is oxidised to Na^+ and H is reduced to H^- (in Na^+H^-).

25. (c) When ($CO + H_2$), i.e. water gas is passed over steam in the presence of iron oxide (or cobalt oxide), more H_2 is generated and CO is oxidised to CO_2. This is called water gas shift reaction.

26. (b) In *ortho* isomer, spins of the two nuclei of hydrogen are in same direction. Thus, nuclear spin $= \frac{1}{2} + \frac{1}{2} = 1$

In *para* isomer, spins of two nuclei are in opposite direction, one clockwise and other anticlockwise.

Thus, nuclear spin $= \frac{1}{2} - \frac{1}{2} = 0$

Thus, (b) is incorrect.

27. (b) Hydrogen is economically obtained as a byproduct in the electrolysis of aqueous solution of brine solution in the manufacture of sodium hydroxide.

$$NaCl(aq) \longrightarrow Na^+(aq) + Cl^-(aq)$$

Cl^- is oxidised to Cl_2 and H_2O may be oxidised to O_2. However, in concentrated solution of NaCl, Cl_2 is formed and in dilute solution, O_2 is formed.

$$2H_2O(l) + 2e^- \longrightarrow H_2(g) + 2OH^-(aq) \qquad E° = -0.83\,V$$

$$Na^+(aq) + e^- \longrightarrow Na(s), \qquad E° = -2.7\,V$$

$E°_{red}$ of Na^+ is more negative than H_2O, thus H_2O is reduced more easily as compared to Na^+.

Thus, Statement I and Statement II are correct but Statement II is not the correct explanation of Statement I.

28. (b)

	H	Group 1 (Li)	Group 17 (F)
EC	$1s^1$	(Inert gas) ns^1	(Inert gas) ns^2np^5
(IE)/kJ mol^{-1}	1312	519	1681
Monovalent anion	H^-(He)	Not formed easily	X^- (inert gas)
Monovalent cation	H^+	Na^+	Not formed easily
Electron affinity (kJ mol^{-1})	−72	−52	−348
Structure	H_2	Metallic	F_2

In some respects, H resembles alkali metals and in some, it resembles halogens.

Thus, it is difficult to assign any definite position to hydrogen in the periodic table. It is placed in group 1 as well as in group 17.

Thus, Statement I and Statement II are correct but Statement II is not the correct explanation of Statement I.

29. (c) Hydrogen has $1s^1$ configuration. Like alkali metals with ns^1 configuration, it is placed in group 1 (alkali metals).

H^- has $1s^2$ configuration of helium but it is different from alkali metals in which cations $Li^+, Na^+ \cdots$ have inert gas configuration. Thus, Statement I is correct but Statement II is incorrect.

30. (b) H : $1s^1$

X (halogen) : ns^2np^5

Hydrogen	Halogen
One short of duplet	One short of octet
H^- : $1s^2$ (Helium)	X^- : ns^2np^6 (neon, etc.)
Hydride in NaH	Halide in NaX
H_2	X_2
Covalent compounds CH_4	CCl_4

Thus, hydrogen resembles halogen.

By gain of one electron, both gain inert gas configuration.

Thus, Statement I and Statement II are correct but Statement II is not the correct explanation of Statement I.

31. (b) Non-stoichiometric hydrides are formed between *d* and *f*-block elements and hydrogen atoms. *d* and *f*-block elements have holes and hydrogen atoms occupy these holes. Some holes remain vacant and ratio of metal to hydrogen atoms is fractional and hydride formed is non-stoichiometric. Thus, Statement I and Statement II are correct but Statement II is not the correct explanation of Statement I.

32. (b) Metals (particularly transition metals) have free valencies. They form metal hydrides and can adsorb hydrogen as H-atoms in their lattices. On heating, H_2 and metals are formed. Thus, transition metals (and their alloys) can be used to transport H_2 to be used as fuel. Thus, Statement I and Statement II are correct but Statement II is not the correct explanation of Statement I.

Note *Alloy* $LiNi_5$ *can adsorb seven moles of* H_2 *per mole of alloy at* 2.5 *atm and at room temperature.*

33. (b) Deuterium is heavier than hydrogen hence, its compounds often react more slowly than those of the lighter isotope.

Regular drinking of D_2O instead of H_2O could prove fatal (toxic) because of the slower rate of transfer of D^+ compared to that of H^+ in the acid-base reactions involved in enzyme catalysis. Kinetic isotopic effect (replaced H by D) is also observed in the ionisation of weak acid.

$$CH_3COOH \rightleftharpoons CH_3COO^- + H^+, K_a = 1.8 \times 10^{-5}$$

$$CH_3COOD \rightleftharpoons CH_3COO^- + D^+, K_a' = 6.0 \times 10^{-6}$$

$$K_a' < K_a$$

Thus, Statement I and Statement II are correct but Statement II is not the correct explanation of Statement I.

34. (a) $CuSO_4 \cdot 5H_2O$ is hydrate of copper (II) sulphate and is written as $[Cu(H_2O)_4]SO_4 \cdot H_2O$.

It satisfies coordination number of copper(II) being four. One H_2O molecule is outside the coordinate sphere and bonded to SO_4^{2-} by H-bonding. Thus, Statement I and Statement II are correct and Statement II is the correct explanation of Statement I.

35. (d)

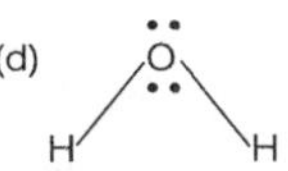

There are two bond pairs and two lone pairs. Thus, oxygen atom is sp^3-hybridised. There are following types of repulsion:

(i) *lp-lp* (ii) *lp-bp* (iii) *bp-bp*

Extent of repulsion of *lp-lp* > *lp-bp* > *bp-bp*. This distorts bond angle to 104.5° and structure is bent (instead of tetrahedral).

Thus, Statement I is incorrect and Statement II is correct.

36. (a) H_2O_2 has six points of H-bonding (four on oxygen and two on hydrogen). H_2O has four points of H-bonding (two on oxygen and two on hydrogen). Thus, H_2O_2 is more H-bonded than H_2O. Hence, boiling point of H_2O_2 is higher than that of H_2O.

Thus, Statement I and Statement II are correct and Statement II is the correct explanation of Statement I.

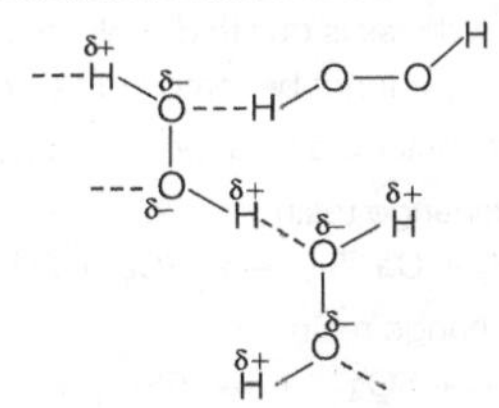

(H-bonding in H_2O_2)

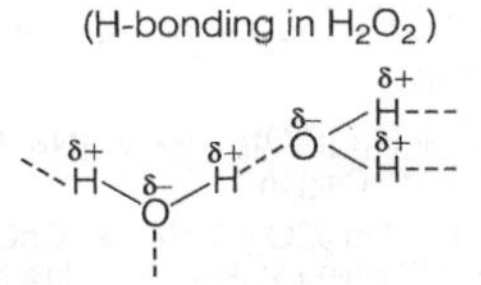

(H-bonding in H_2O)

37. (a) H_2 is formed by sharing of one electron pair between two H-atoms. Duplet of H is complete.

H : H

$$\text{Bond order} = \frac{\text{Electrons shared}}{2}$$

$$= \frac{2}{2} = 1$$

H—H

Also, $H + H \longrightarrow H_2(g), \quad \Delta H = -436 \text{ kJ mol}^{-1}$

ΔH being negative, thus H_2 formed is stable. Hence, hydrogen exists as H_2 molecule. Thus, Statement I and Statement II are correct and Statement II is the correct explanation of Statement I.

38. (a)

(i) ${}_1^2D + {}_1^2D \longrightarrow {}_1^3T + {}_1^1H$

Thus, (i) → (q)

(ii) ${}_3^6Li + {}_0^1n \longrightarrow {}_1^3T + {}_2^4He$

Thus, (ii) → (r)

(iii) ${}_7^{14}N + {}_0^1n \longrightarrow {}_6^{12}C + {}_1^3T,$

Thus, (iii) → (p)

(iv) ${}_1^3T \longrightarrow {}_2^3He + {}_{-1}^0e,$

Thus, (iv) → (s)

39. (b)

(i) $NiH_{0.6}$ - Non-stoichiometric

Thus, (i) → (r)

(ii) KH - Ionic

Thus, (ii) → (q)

(iii) CoH_2-Stoichiometric

Thus, (iii) → (p)

(iv) SiH_4- Covalent

Thus, (iv) → (s)

40. (b)

(i) $LiH \longrightarrow Li^+ + H^-$

H^- is a reducing agent; H^- is a Lewis base.

H^- is a conjugate base of H.

Thus, (i) → (p,q,r)

(ii) $2H_2O + 2F_2 \longrightarrow 4HF + O_2$

H_2O is a reducing agent, H_2O is proton donor as well as proton acceptor.

Thus, Lowry-Bronsted acid and base. H_2O is also a lone pair donor, thus Lewis base.

Thus, (ii) → (p,q,t,u)

(iii) H_3O^+

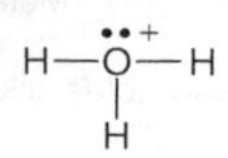

It is lone pair donor thus, Lewis base. It is H^+ ion donor. Thus, Lowry-Bronsted acid. It is conjugated base of H_2O.

Thus, (iii) → (q,r,t)

(iv) BH_3

Electron deficient hence, Lewis acid. Thus, (iv) → (s)

(v) H

Reducing agent. Thus, (v) → (p)

JEE Archive

1. (d) In fuel cell, fuels like CH_4, H_2, etc are combusted producing electrical energy.

$$2H_2(g) + O_2(g) \longrightarrow 2H_2O(l)$$

2. (a)

(a) Permanent hardness is due to dissolved Cl^- and SO_4^{2-} of Ca^{2+} and Mg^{2+}. They cannot be removed by boiling.

(b) Ion exchange method exchanges Cl^-, SO_4^{2-}, Ca^{2+} and Mg^{2+}

(I) Cation exchange resin

$$RH_2 + Ca^{2+} \longrightarrow RCa + 2H^+$$

(II) Anion exchange resin

$$R(OH)_2 + SO_4^{2-} \longrightarrow RSO_4 + 2OH^-$$

(c) Calgon's method removes permanent hardness by using $(NaPO_3)_6$, Calgon.

$$\underset{(Ca^{2+}/Mg^{2+})}{M^{2+}} + \underset{\text{Calgon}}{Na_4P_6O_{18}^{2-}} \longrightarrow \underset{\text{Soluble}}{[Na_2MP_6O_{18}]^{2-}} + 2Na^+$$

(d) $$\underset{\text{(In hard water)}}{CaSO_4} + \underset{\text{Washing soda}}{Na_2CO_3} \longrightarrow \underset{\text{Insoluble}}{CaCO_3} + Na_2SO_4$$

3. (d) $$\underset{2\text{ mol}}{2C_8H_7SO_3Na} + \underset{1\text{ mol}}{Ca^{2+}} \rightleftharpoons (C_8H_7SO_3)_2Ca + 2Na^+$$

2×206 g resin uptake $Ca^{2+} = 1$ mol

1 g resin uptakes $Ca^{2+} = \frac{1}{412}$ mol

Thus, maximum uptake of Ca^{2+} per gram of resin $= \frac{1}{412}\text{mol g}^{-1}$

4. (a) H_2O_2 can act as an oxidising agent as well as a reducing agent also.

5. (d) Release of electron is known as oxidation. So, H_2O_2 acts as reducing agent when it releases electrons.

Here, in reactions (II) and (IV), H_2O_2 releases two electrons hence, reactions (II) and (IV) is known as oxidation.

In reactions (I) and (III), two electrons are being added, so (I) and (III) represents reduction.

6. (d) Hydrides are instant source of hydrogen of higher purity. They react with H_2O forming H_2 gas.

$$CaH_2 + 2H_2O \longrightarrow Ca(OH)_2 + 2H_2$$

7. (d) CO is oxidised to CO_2 with steam in the presence of a catalyst followed by the absorption of CO_2 in alkali.

$$CO + H_2 \xrightarrow[\text{catalyst}]{\text{Steam}} CO_2 + 2H_2$$

$$\downarrow KOH$$

$$K_2CO_3$$

8. (a) H_2O_2 reacts with KIO_4 in the following manner

$$\overset{+7}{KIO_4} + H_2O_2 \longrightarrow \overset{+5}{KIO_3} + H_2O + O_2$$

On reaction of KIO_4 with H_2O_2, oxidation state of I varies from +7 to +5, i.e. decreases, Thus, KIO_4 gets reduced hence, H_2O_2 is a reducing agent here.

With NH_2OH, it gives following reaction

$$\overset{-1}{NH_2OH} + H_2O_2 \longrightarrow \overset{+3}{N_2O_3} + H_2O$$

In the above reaction, oxidation state of N varies from −1 to +3. Here, oxidation number increases hence, H_2O_2 is acting as an oxidising agent here.

9. (b,c,d) $Ca(OH)_2 + Ca(HCO_3)_2 \longrightarrow 2CaCO_3 \downarrow + 2H_2O$ (Clark's method)

$$NaOCl + H_2O \longrightarrow NaOH + HOCl$$

$$HO^- + HCO_3^- \longrightarrow CO_3^{2-} + H_2O$$

$$Ca(HCO_3)_2 + Na_2CO_3 \longrightarrow CaCO_3 \downarrow + 2NaHCO_3$$

3. Alkali Metals (Group 1)

DPP-1 Periodicity in Group 1

1. (b) $Z_{eff} = Z - S$

where, S = Screening constant

Z = atomic number

S = 0.35 for outermost electron (n)

S = 0.85 for (n – 1)th orbit

S = 1.00 for (n – 2), (n – 3) ...

Na(11) = 2, 8, 1

Z_{eff} = 11 – [Screening of valence electrons by 8 electrons in (n – 1)th orbit + 2 electrons in (n – 2)th orbit]

= 11 – (8 × 0.85 + 2)

= 11 – 8.8 = 2.2

K (19) = 2, 8, 8, 1

Z_{eff} = 19 – (8 × 0.85 + 8 × 1.0 + 2 × 1.0)

= 2.2

Rb (37) = 2, 8, 18, 8, 1

Z_{eff} = 2.2

Cs (55) = 2, 8, 18, 18, 8, 1

Z_{eff} = 2.2

Thus, Z_{eff} is same for all elements (Na, K, Rb, Cs). Number of shells increases. Hence, atomic size is dependent on number of shells.

2. (d)

(a) Smaller cation with smaller anion or larger cation with larger anion is more stable.

$$LiOH \longrightarrow Li_2O$$

$$O^{2-} < OH^- \text{ (Size)}$$

Hence, if Li_2O is formed from LiOH, stability increases, hence LiOH decomposes to Li_2O and as ionic size of cation increases, stability increases. Thus, correct.

(b) Li^+F^- is smaller cation with smaller anion, hence is most stable as compared to Na^+F^- with larger cation and smaller anion. Thus, correct.

(c) $2LiNO_3 \longrightarrow Li_2O + 2NO_2 + \frac{1}{2}O_2$

$Li^+NO_3^-$ lattice is with smaller cation and larger anion, while Li_2O lattice is with smaller cation and smaller anion and Li_2O is thus more stable than $LiNO_3$. Stability increases on increasing cationic size. Thus, correct.

3. (a) Smaller the cation,
larger the anion,
larger the charge,
then larger the polarising power,
larger the covalent nature,
then smaller the melting point.
Cation has same charge,
anion is same,
polarising power is dependent on cationic size.

$$\underset{\text{Smallest size}}{Li^+} < Na^+ < K^+ < Rb^+ < Cs^+$$

Li^+ with smallest size has maximum polarising power.
Li_2CO_3 is most covalent.
Cs_2CO_3 is most ionic thus having highest melting point.

$$Li_2CO_3 < Na_2CO_3 < K_2CO_3 < Rb_2CO_3 < Cs_2CO_3$$

4. (a) Li has maximum IE hence, it is least reactive with H_2O. Also, density of Li = 0.53 g/cm^3. It is lighter than water (d = 1.0 g/cm^3) hence, it floats on water.

Density of Na = 0.97 g/cm^3 < 1.0

Density of K = 0.86 g/cm^3 < 1.0

but these are reactive metals.

5. (a) Slow reactivity of Li is due to kinetic factor since, Li in aqueous solution is oxidised to $Li^+(aq)$, and E°_{red} is most negative for Li indicating it is best reducing agent.

6. (b) Solubility of solute also depends on size of cation and anion. Whenever, cation is small and anion is large (as Li^+ and I^-) or cation is large and anion is small (as Cs^+ and F^-), solubility will be larger.

CsF (larger cation and smaller anion)– maximum solubility.

CsI (larger cation and larger anion)– minimum solubility.

Thus, CsF > CsCl > CsBr > CsI

7. (b) Difference in $\Delta G^\circ_{hydration}$ (of Na^+ and K^+) = 71 kJ mol^{-1} and difference in lattice energy of NaCl and KCl = 67 kJ mol^{-1}

Since, $\Delta G^\circ_{hydration} > \Delta U$

Hence, NaCl is more soluble than KCl.

8. (d) Polarising power of

$$Rb^+ < K^+ < Na^+ < Li^+$$

Anion is same (CO_3^{2-}).

Thus, ionic nature of $Rb_2CO_3 > K_2CO_3 > Na_2CO_3 > Li_2CO_3$

Larger the ionic nature, larger the solubility in water.

9. (a) Smaller cations form stable salts with smaller anions and larger cations form stable salts with larger anions.

$$Li_2CO_3 \longrightarrow Li_2O + CO_2$$

O^{2-} is smaller than CO_3^{2-} and Li^+ fits into Li_2O lattice in a better way than Li_2CO_3 lattice.

Na_2CO_3 is stable since, Na^+ fits into Na_2CO_3 lattice in a better way than Na_2O lattice.

10. (a) Smaller the size of cation (anion is same) then larger the hydration and larger the size of hydrated ion. Heavier the ion, smaller the ionic velocity and when an electric field is applied, thus smaller the ionic conductance.

Smaller cation ⇒ Low ionic conductance

$$\underset{\text{Least}}{LiCl} < NaCl < KCl < RbCl < \underset{\text{Maximum}}{CsCl}$$

11. (a)

$$M(s) \longrightarrow M(g), \quad \Delta H_1 = S$$

$$M(g) \longrightarrow M^+(g) + e^-, \quad \Delta H_2 = I$$

$$\frac{1}{2}X_2(g) \longrightarrow X(g), \quad \Delta H_3 = \frac{D}{2}$$

$$X(g) + e^- \longrightarrow X^-(g), \quad \Delta H_4 = -E$$

$$M^+(g) + X^-(g) \longrightarrow M^+X^-(s), \quad \Delta H_5 = -U$$

$$M(s) + \frac{1}{2}X_2(g) \longrightarrow M^+X^-(s),$$

$$\Delta H^\circ_f = S + I + \frac{D}{2} - E - U$$

$$\therefore \quad U = -\Delta H^\circ_f + S + I + \frac{D}{2} - E$$

12. (b) Li is diagonal to Mg. Thus, Li resembles Mg in many properties but differs its own group.

$$2LiNO_3 \xrightarrow{\Delta} Li_2O + 2NO_2 + \frac{1}{2}O_2$$
$$Mg(NO_3)_2 \xrightarrow{\Delta} MgO + 2NO_2 + \frac{1}{2}O_2$$
$$KNO_3 \xrightarrow{\Delta} KNO_2 + \frac{1}{2}O_2$$

13. (a) Smaller the size of cation,
larger the size of anion,
larger the charge,
then larger the polarising power, thus compounds with such ions are covalent

$$Li^+ < Na^+ < K^+ < Rb^+ < Cs^+ \text{ (Size)}$$

Smallest size,
maximum polarising power,
maximum covalent character.

Thus, Statement I and Statement II are correct and Statement II is the correct explanation of Statement I.

14. (b) Li and Mg due to their diagonal relationship resemble each other in many of their properties.

$$Li_2CO_3 \xrightarrow{Heat} Li_2O + CO_2$$

A compound is stable with smaller cation and smaller anion or with larger cation and larger anion.

$$\underset{\substack{\uparrow \\ \text{Smaller cation}}}{Li_2}\ \underset{\substack{\uparrow \\ \text{Larger anion}}}{CO_3} \xrightarrow{\Delta} \underset{\substack{\uparrow \\ \text{Smaller cation}}}{Li_2}\ \underset{\substack{\uparrow \\ \text{Smaller anion}}}{O} + CO_2$$

On heating Li_2CO_3, Li_2O is formed which is stable. Thus, Li_2CO_3 is thermally unstable.

$MgCO_3$ is also thermally unstable.

Thus, Statement I and Statement II are correct but Statement II is not the correct explanation of Statement I.

15. (b) Li^+ (smaller cation) fits with smaller anion N^{3-}, hence Li_3N lattice is formed. K^+ (larger anion) does not fit with smaller anion N^{3-}. Hence, K_3N is not formed.

$$Li_3N + 3H_2O \longrightarrow 3LiOH + NH_3$$

Li_3N is decomposed by H_2O forming NH_3.

Thus, Statement I and Statement II are correct but Statement II is not the correct explanation of Statement I.

16. (b) U (lattice energy) $\propto \frac{1}{r_+ + r_-} \propto$ charge on ion

Charge on cation and anion is same. The variation in U is the greatest when r_- is smallest (for F^-) and least when r_- is largest (for I^-). Energy released when gaseous ions combine to form a lattice is called lattice energy and is thus the driving force for the determination of stability. Thus, Statement I and Statement II are correct but Statement II is not the correct explanation of Statement I.

17. (b) Polarising power of $Li^+ > Na^+$ thus, LiCl is predominantly covalent and NaCl is predominantly ionic. Thus, LiCl is soluble in ether and NaCl is insoluble in it.

Li shows diagonal relationship with Mg and electronegativity values of Li and Mg being same.

Thus, Statement I and Statement II are correct but Statement II is not the correct explanation of Statement I.

18. (b) As we go down the group of alkali metals from Li to Cs, size and repulsion of non-bonded electrons increases. Their cohesive energy and tendency of metallic bonding decreases down the group and thus, softness increases and these metals can be cut with a knife.

As we go down the group, size increases hence, force of attraction between valence electrons and nucleus decreases. Thus, IE decreases. Thus, Statement I and Statement II are correct but Statement II is not the correct explanation of Statement I.

19. (c) In alkali metals, electrons of the noble gas core efficiently shield the lone valence shell electrons from the nuclear charge. Thus, effective nuclear charge felt by the electron in the valence shell of an atom of an alkali metal is least and the valence electron is loosely bound to the nucleus and can move from one part to the other in the lattice. Thus, alkali metals are good conductors of heat and electricity. Thus, Statement I is correct and Statement II is incorrect.

20. (a) Lithium differs in many properties in its compounds.

Li	Na, K...
LiCl – covalent	NaCl – ionic
$Li_2CO_3 \xrightarrow{\Delta} Li_2O + CO_2$	$Na_2CO_3 \xrightarrow{\Delta}$ Stable
	$K_2CO_3 \xrightarrow{\Delta}$ Stable
$LiNO_3 \longrightarrow Li_2O + NO_2 + O_2$	$KNO_3 \xrightarrow{\Delta} KNO_2 + O_2$

Thus, Statement I is correct.

$$\text{Charge density} = \frac{\text{Ionic charge}}{\text{Ionic radius}}$$

Charge is same, ionic radius of Li^+ is least.

Thus, Li^+ has maximum charge density.

Thus, Statement I and Statement II are correct and Statement II is the correct explanation of Statement I.

21. (a,b,c)

(a) $Na_2CO_3 \xrightarrow{\Delta}$ No effect

Na_2CO_3 is thermally stable.

$$Ag_2CO_3 \xrightarrow{\Delta} 2Ag + CO_2 + \frac{1}{2}O_2$$

Thus, correct.

(b) $Ag^+ + 2NH_3 \longrightarrow [Ag(NH_3)_2]^+$

Na^+ does not form complexes due to the absence of d-orbital. Thus, correct.

(c) Polarising power of $Na^+ < Ag^+$ thus, NaCl is ionic hence water soluble. AgCl is covalent thus water insoluble. Thus, correct.

(d) NaCl gives golden-yellow flame while AgCl does not hence, incorrect.

22. (b,c,d)

(a) NaOH is crystalline deliquescent solid hence, it forms hydrated salt.

Thus, it is incorrect.

(b) Correct

(c) Correct

(d) Correct

23. (a,b,c,d)

	Group 1		Group 11
(a)	M^+ : Inert gas electronic configuration $Na^+ : 1s^2 2s^2 2p^6$	(a)	M^+ : Different from inert gas configuration $Cu^+ : [Ar]\,3d^{10}$
(b)	$2Na + 2H_2O \longrightarrow 2NaOH + H_2$	(b)	$Cu + H_2O \longrightarrow$ No reaction
(c)	Na_2SO_4, NaCl, Na_2S } Soluble in water	(c)	CuCl, AgCl, CuS, Ag_2S } Insoluble in water $CuSO_4$ - Less soluble
(d)	$E^\circ_{Li^+/Li} = -3.045$ V	(d)	$Cu^{2+} + 2e^- \longrightarrow Cu$, $E^\circ = 0.34$ V

24. (c,d)

(a) Li (3) : $1s^2 2s^1$ (s-block)

(b) Na (11) : $1s^2 2s^2 2p^6 3s^1$ (s-block)

(c) K(19) : $1s^2 2s^2 2p^6 3s^2 3p^6 3d^1$ (d-block) 3d is being filled in the absence of Aufbau rule.

(d) Rb (37) : $1s^2 2s^2 2p^6 3s^2 3p^6 3d^{10} 4s^2 4p^6 4d^1$ (d-block) 4d is being filled in the absence of Aufbau rule.

K and Rb are s-block elements but in the absence of Aufbau rule, they would be placed in d-block.

25. (a,b,c)

ΔH_f° (kJ mol^{-1})	LiF	LiCl	LiBr	LiI
	−617	−409	−351	−270
NaF	−575	NaCl	−411	
KF	−568	KCl	−436	
RbF	−557	RbCl	−443	
CsF	−555	CsCl	−435	

(a) ΔH_f° are negative values indicating they are stable. Thus, correct.

(b) For MF, ΔH_f° becomes less negative thus, correct.

(c) For a given alkali metal, ΔH_f° becomes less negative thus, correct.

(d) ΔH_f° cannot be measured using Born-Haber cycle.

Explanations for Q. Nos. (26-28)

$$\Delta H^\circ_{hydration}\ (LiCl) = -499 - 382 = -881\ kJ\ mol^{-1}$$

$$\Delta H^\circ_{hydration}\ (NaCl) = -390 - 382 = -772\ kJ\ mol^{-1}$$

$$\Delta H^\circ_{hydration}\ (KCl) = -305 - 382 = -687\ kJ\ mol^{-1}$$

Thus, LiCl has maximum heat of hydration.

LiCl with maximum heat of hydration is best suited to control humidity.

$$\Delta H^\circ_{solution} = \Delta H^\circ_{hydration} - \Delta H^\circ_U$$

For LiCl = − 881 − (− 840)
= − 41 kJ mol^{-1} (exothermic)

For NaCl = − 772 − (− 776)
= + 4 kJ mol^{-1} (endothermic)

For KCl = − 687 − (−703)
= + 16 kJ mol^{-1} (endothermic)

26. (a) **27.** (a) **28.** (c)

Explanations for Q. Nos. (29 and 30)

By Fajan's rule,

Smaller the size of cation, larger the size of anion, larger the charge, then larger the polarising power and thus, larger the covalent nature and larger the lattice energy.

Anion is same with same charge (F^-). Cation has same charge but ionic size of

$$Li^+ < Na^+ < K^+ < Rb^+ < Cs^+$$

Li^+ has maximum polarising power; thus LiF is the most covalent and CsF is most ionic.

Thus, lattice energy of

$$CsF < RbF < KF < NaF < LiF$$

In LiF, LiCl, LiBr and LiI, cation is same.

Charge on anion is same, and ionic size of

$F^- < Cl^- < Br^- < I^-$ (Maximum polarising power)

But polarising power of smallest Li^+ ion predominates the effect of halide ion; thus lattice energy of LiF is highest and that of LiI is least.

$$LiI < LiBr < LiCl < LiF$$

29. (b)

30. (b)

31. (6)

$$\Delta H_{hydration}\ (NaCl) = (-390 - 380)\ kJ\ mol^{-1} = -770\ kJ\ mol^{-1}$$

$$\Delta H_{solution} = \Delta H_{hydration} - \text{Lattice energy} = -770 - (-776) = 6\ kJ\ mol^{-1}$$

Thus, $x = 6$ kJ mol^{-1}

DPP-2 Compounds of Alkali Metals

1. (c) CsI_3 is an ionic polyhalide ionising as

$$CsI_3 \rightleftharpoons Cs^+ + I_3^-$$

These polyhalides are formed when alkali halide reacts with the halogen or interhalogen compounds.

$$KI + I_2 \rightleftharpoons K^+ I_3^-$$

$$KF + BrF_3 \longrightarrow K^+[BrF_4]^-$$

$$KBr + ICl \longrightarrow K^+[BrICl]^-$$

2. (a) $NaHCO_3$ (sodium bicarbonate) is used as a mild antiseptic for skin infections.

3. (b) Li_2CO_3 is used in the treatment of manic-depressive disorders.

4. (b) Alkali metal (group 1) Li forms nitride.

Li resembles Mg diagonally, hence Mg also forms nitride.

$$6Li + N_2 \longrightarrow 2Li_3N$$

$$3Mg + N_2 \longrightarrow Mg_3N_2$$

5. (b) Dimeric structure is formed in $KHCO_3$ while chain-type structure is formed in $NaHCO_3$.

6. (b) Na_2CO_3 is manufactured by Solvay (ammonia-soda) process.

Step I $CaCO_3 \xrightarrow{\Delta} CaO + CO_2$

$CaCO_3$ is required only to start the reaction. Further, recycled CO_2 is used from Step (IV).

Step II

$$NH_3 + H_2O + CO_2 \text{ (from Step I)} \longrightarrow NH_4HCO_3$$

Step III

$$NaCl + NH_4HCO_3 \longrightarrow NaHCO_3 + NH_4Cl$$

Step IV

$$2NaHCO_3 \xrightarrow{\Delta} Na_2CO_3 + H_2O + CO_2$$

CaO of Step I can decompose NH_4Cl of Step III giving NH_3 to be recycled to Step II.

7. (c)

(a) Le-blanc process uses KCl (from carnallite) to prepare K_2CO_3. Thus, correct.

(b) $K_2SO_4 + 2C + CaCO_3 \xrightarrow{\Delta} K_2CO_3 + CaS + 2CO_2$

Thus, correct.

8. (c) $KCl + 3(MgCO_3 \cdot 3H_2O) + CO_2 \longrightarrow$

$$2(KHCO_3 \cdot MgCO_3 \cdot 4H_2O) + MgCl_2$$

$$\downarrow \text{Under pressure } \Delta$$

$$K_2CO_3 + 2MgCO_3 + CO_2 + 9H_2O$$

9. (c) $Na_2SO_4 \cdot 10H_2O \longrightarrow Na_2SO_4 + 10H_2O$

$Na_2CO_3 \cdot 10H_2O \longrightarrow Na_2CO_3 + 10H_2O$

Thus, both effloresce.

$$Na_2SO_4 + BaCl_2 \longrightarrow \underset{\text{(White ppt)}}{BaSO_4} + 2NaCl$$

$$Na_2CO_3 + BaCl_2 \longrightarrow \underset{\text{(White ppt)}}{BaCO_3} + 2NaCl$$

10. (c) Gun powder is a mixture of KNO_3, S and C.

$$2KNO_3(s) + 4C(s) \xrightarrow{\text{Ignition}} K_2CO_3(s) + 3CO(g) + N_2(g)$$

$$2KNO_3(s) + 2S(s) \xrightarrow{\text{Ignition}} K_2SO_4(s) + SO_2(g) + N_2(g)$$

11. (d) Na^+ and K^+ (both alkali metal ions) resemble in many properties but in terms of biological aspect, they differ in

(a) Correct

(b) Correct

(c) Correct

12. (c) Sodium ions are found primarily on the outside of cells being located in

(i) blood plasma

(ii) interstitial fluid.

13. (d) (a) Correct (b) Correct (c) Correct

14. (d) LiCl and $MgCl_2$ are predominantly covalent compounds.

Li^+ and Mg^{2+} resemble in many properties due to diagonal relationship.

LiCl and $MgCl_2$ both are hydrated and reaction is exothermic. They absorb H_2O hence deliquescent.

15. (b) Alkali metals dissolve in liquid NH_3 to give blue solution due to solvated electrons and solvated metal ions.

$$M + (n+m)NH_3 \longrightarrow \underset{\text{Solvated metal ions}}{[M(NH_3)_n]^+} + \underset{\text{Solvated electrons}}{[e(NH_3)_m]^-}$$

Thus, Statement I and Statement II are correct but Statement II is not the correct explanation of Statement I.

16. (a) Na_2CO_3 is prepared by Solvay process through $NaHCO_3$ which on heating forms Na_2CO_3.

$$NH_3 + H_2O + CO_2 \longrightarrow NH_4HCO_3$$

$$\downarrow NaCl$$

$$Na_2CO_3 \xleftarrow{\Delta} NaHCO_3$$

$NaHCO_3$ is less soluble hence separates. Comparatively, $KHCO_3$ is more soluble and does not separate. Thus, K_2CO_3 cannot be prepared by Solvay process.

Thus, Statement I and Statement II are correct and Statement II is the correct explanation of Statement I.

17. (a) $4KO_2 + 2CO_2 \longrightarrow 2K_2CO_3 + 3O_2$

KO_2 absorbs CO_2 and releases O_2.

In space and submarines, there is deficiency of O_2. When we breathe, CO_2 is released. In presence of KO_2, this CO_2 is absorbed and O_2 is released. Thus, KO_2 is used in submarines and space vehicles.

Thus, Statement I and Statement II are correct and Statement II is the correct explanation of Statement I.

18. (a) In the periodic table, size increases on moving down the group. Thus,

$$Na < K \text{ and } Na^+ < K^+$$

K^+ ions with the larger anions tend to be less soluble than Na^+ ions and K^+ ions are more strongly bound to the complex silicates and aluminosilicates in soils.

Thus, Statement I and Statement II are correct and Statement II is the correct explanation of Statement I.

19. (d) K_2O_3 (potassium sesquioxide) is actually peroxodisuperoxide.

O_2^{2-} (peroxo) is diamagnetic but O_2^- (superoxide) is paramagnetic. Hence, sesquioxide is paramagnetic.

Thus, Statement I is incorrect but Statement II is correct.

20. (a,b) Dilute solutions of alkali metals in liquid NH_3 are blue due to solvated electrons and solvated metal ions.

$$M(s) \longrightarrow M^+(g) + e^-$$

$$M^+ + (x+y)NH_3 \longrightarrow \underset{\text{Solvated metal ions}}{[M(NH_3)_x]^+} + \underset{\text{Solvated electrons}}{[e(NH_3)_y^-]}$$

Electrical conductivity is due to solvated electrons.

Note *Values of x and y depend on the extent of solvation by NH_3. Dilute solutions are paramagnetic due to free electrons. In concentrated solutions, solvated electrons $[e(NH_3)_y]^-$ associate to form electron-pairs and their paramagnetic character decreases. Also, in concentrated solutions (above 3 M), metal ion clusters are formed, the solutions are copper coloured with metallic lustre.*

21. (a,b,d) $2NaNO_3 \xrightarrow{500°C} 2NaNO_2 + O_2$

$$2NaNO_3 \xrightarrow{\text{Above } 800°C} N_2 + 2.5O_2 + Na_2O$$

22. (a,b) $4Na + O_2 \longrightarrow 2Na_2O$

$2Na + O_2 \longrightarrow Na_2O_2$

Note Li *forms* LiO_2.

Na *forms* Na_2O, Na_2O_2.

K *forms* K_2O, KO_2, K_2O_2.

23. (a,b,c)

(a) $Na(s) + (x+y)\underset{\text{(Liquid)}}{NH_3} \longrightarrow [Na(NH_3)_x]^+ + [e(NH_3)_y]^-$

Due to solvated (unpaired) electrons, paramagnetic behaviour is shown.

Note *With NH_3 gas, $NaNH_2$ and H_2 are formed which are diamagnetic.*

(b) $K + O_2 \longrightarrow KO_2$

O_2^- is paramagnetic due to one unpaired electron.

(c) $3Cu + 8HNO_3 \longrightarrow 3Cu(NO_3)_2 + 2NO + 4H_2O$

NO has one unpaired electron hence paramagnetic.

$$\dot{\ddot{N}} :: \ddot{O}:$$

Unpaired electron

(d) H_2O_2 is formed which is diamagnetic.

24. (a,c,d) It is called Castner-Kellner cell. Na is formed at cathode by reduction.

$$Na^+(aq) + e^- \longrightarrow Na(s)$$

It combines with Hg forming sodium amalgam and is pumped out. It is allowed to react with H_2O forming NaOH.

$$\underset{\text{Sodium amalgam}}{2Na(Hg)} + 2H_2O \longrightarrow 2NaOH + H_2 + 2Hg$$

There can be side reactions:

$$4OH^- \longrightarrow O_2 + 2H_2O + 2e^-$$

Carbon anodes may be corroded by O_2. Hence, now anodes are made of steel coated with titanium.

25. (a)

(i) $Na_2CO_3 \cdot 10H_2O$ is called washing soda used in soap and detergent.
Thus, **(i) → (t)**

(ii) $Na_2CO_3 \cdot 7H_2O$ is called crystal carbonate used in glass manufacturing.
Thus, **(ii) → (r)**

(iii) $Na_2CO_3 \cdot H_2O$ is called heavy ash.
Thus, **(iii) → (s)**

(iv) $Na_2CO_3 \cdot NaHCO_3 \cdot 2H_2O$ called sesquicarbonate and occurs naturally under the name trona. It is a source of Na_2CO_3 (obtained by heating).

$$2(Na_2CO_3 \cdot NaHCO_3 \cdot 2H_2O) \longrightarrow 3Na_2CO_3 + CO_2 + 5H_2O$$

Thus, **(iv) → (p)**

(v) Na_2CO_3 is called soda ash or light ash used in petroleum refinery and textile industry.
Thus, **(v) → (q)**

26. (a)

(i) Liquid sodium metal is used as coolant in nuclear reactor.
Thus, **(i) → (r)**

(ii) Potassium stearate is a content in soft soap.
Thus, **(ii) → (s)**

(iii) Potassium nitrate along with C and S is a gun powder (explosive). Thus, **(iii) → (q)**

(iv) Potassium superoxide absorbs CO_2 and releases O_2. Thus, it is used in life supporting system.
Thus, **(iv) → (p)**

27. (b)

(i) Na_2CO_3 → (t)
(ii) Na_2SiO_3 → (q)
(iii) NaOH → (s)
(iv) NaOCl → (r)
(v) K_2SO_4 → (p)

Explanations for Q. Nos. (28 and 29)

Oxide *X* turns red litmus blue. Thus, it is alkaline in nature.

$$Na_2O + H_2O \longrightarrow 2NaOH$$

Oxide (*Y*) decolourises red litmus. Thus, it has bleaching action. Thus, it is peroxide.

$$Na_2O_2 + 2H_2O \longrightarrow 2NaOH + H_2O_2$$
$$H_2O_2 \longrightarrow H_2O + O \text{ (bleaching action)}$$

Bleaching action is due to nascent oxygen.

28. (a) **29.** (b)

Explanations for Q. Nos. (30-32)

Electrolysis is carried out in Downs cell using fused 40% NaCl and 60% $CaCl_2$.

$$NaCl \xrightleftharpoons{\text{Current}} Na^+ + Cl^-$$

At anode $2Cl^- \longrightarrow Cl_2 + 2e^-$

At cathode $2Na^+ + 2e^- \longrightarrow 2Na$

Sodium collects at cathode and Cl_2 at anode.

30. (a) **31.** (b) **32.** (a)

Explanations for Q. Nos. (33 and 34)

$$A + NH_4Cl \longrightarrow \underset{(B)}{NH_4^+} + NaCl$$

B is NH_4^+ salt and *A* contains Na^+ ion. Solution *E* is alkaline since, it decomposes *B* forming NH_3 (giving white fumes with HCl).
E is NaOH and *D* is Na_2O.

$$A + Na \longrightarrow Na_2O + C$$

Thus, *A* is $NaNO_2$.

$$2NaNO_2 + 6Na \longrightarrow 4Na_2O + N_2$$

Gas *C* is N_2.

$$B \longrightarrow \underset{(C)}{N_2} + H_2O$$

Thus, (*B*) is NH_4NO_2.

$$\underset{(A)}{NaNO_2} + NH_4Cl \longrightarrow NaCl + \underset{(B)}{NH_4NO_2}$$

$$NH_4NO_2 \xrightarrow{\Delta} \underset{(C)}{N_2} + 2H_2O$$

$$NaNO_2 + 6Na \longrightarrow \underset{(D)}{4Na_2O} + N_2$$

$$\underset{(D)}{Na_2O} + H_2O \longrightarrow \underset{(E)}{2NaOH}$$

33. (c) **34.** (a)

Explanations for Q. Nos. (35 and 36)

$$2KO_2 + S \longrightarrow \underset{(X)}{K_2SO_4}$$

$$K_2SO_4 + Al_2(SO_4)_3 + H_2O \longrightarrow \underbrace{K_2SO_4 \cdot Al_2(SO_4)_3 \cdot 24H_2O}_{(Y)\text{ Alum}}$$

$$K_2SO_4 + 4C \xrightarrow{\Delta} K_2S + 4CO$$

35. (a) **36.** (a)

37. (3) On cooling aqueous solution of $Na_2SO_4(A)$ below 32°C, Glauber's salt is crystallised $Na_2SO_4 \cdot 10H_2O$. Thus, $x = 10$.
On cooling *A* below 12°C, $Na_2SO_4 \cdot 7H_2O$ is formed.
Thus, $y = 7$

$$\therefore \quad (x - y) = 10 - 7 = 3$$

38. (6) $$\underset{\substack{1\text{ mol}\\6\text{ mol}}}{Li_2CO_3} \xrightarrow{\text{Heat}} Li_2O + \underset{\substack{1\text{ mol}\\6\text{ mol}}}{CO_2}$$

$$Na_2CO_3 \xrightarrow{\text{Heat}} \text{No effect}$$

Na_2CO_3 is thermally stable, hence CO_2 is due to Li_2CO_3 only.
Hence, Na_2CO_3 = 6 moles
Six moles CO_2 is due to six moles Li_2CO_3 only.
Being equal moles, Na_2CO_3 is also six moles.

Revisal Problems (JEE Main)

1. (b) Sodium metal exhibits metallic lustre due to oscillation of free electrons.

2. (b)

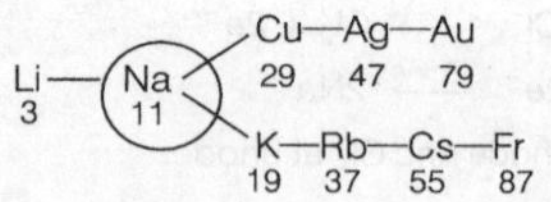

Na links group 1 and group 11 and is called bridge element.

3. (b) Lithium and sodium do not form superoxide.

4. (d)

(a) Correct (b) Correct (c) Correct

5. (b) $2\,Na_2O \longrightarrow Na_2O_2 + 2Na$

Oxidation

Reduction

6. (c) $4KOH + 6NO \longrightarrow 4KNO_2 + N_2 + 2H_2O$

7. (b) $NaCl(molten) \longrightarrow Na^+(l) + Cl^-(l)$

Molten NaCl is conducting due to the presence of ions.

$$2Cl^-(l) \longrightarrow Cl_2(g) + 2e^-$$

$$2Na^+(l) + 2e^- \longrightarrow 2Na(l)$$

8. (c) $\underset{(X)}{2NaHCO_3} \xrightarrow{\Delta} \underset{(Y)}{Na_2CO_3} + CO_2 + H_2O$

$\underset{(Y)}{Na_2CO_3} + H_2O + CO_2 \longrightarrow \underset{(X)}{2NaHCO_3}$

9. (d)

(a) $Na_2CO_3 \xrightarrow{\Delta}$ No effect

(b) $CaCO_3 \xrightarrow{\Delta} \underset{\text{Residue}}{CaO} + CO_2$

CaO (residue) is water soluble but it does not give CO_2 with dilute acid.

(c) $Ca(HCO_3)_2 \longrightarrow \underset{\text{Residue}}{CaCO_3} + CO_2 + H_2O$

$CaCO_3$ (residue) is water insoluble.

(d) $2NaHCO_3 \longrightarrow \underset{\text{Residue}}{Na_2CO_3} + H_2O + CO_2$

Na_2CO_3 (residue) is water soluble; it also gives CO_2 with dil. acids.

10. (b) Oxone is commercial name for Na_2O_2 used in life-supporting system as it absorbs CO_2 and releases O_2.

$$2Na_2O_2 + 2CO_2 \longrightarrow 2Na_2CO_3 + O_2$$

11. (c) Atmospheric CO_2 may react with NaOH forming Na_2CO_3. Silica (SiO_2) of glass can also react with NaOH forming sodium silicate which appears in the form of white deposit.

$$2NaOH + CO_2 \longrightarrow Na_2CO_3 + H_2O$$

$$2NaOH + SiO_2 \longrightarrow Na_2SiO_3 + H_2O$$

Note *Burette in which NaOH is being used as titrant, must be washed after use, otherwise it will jam the stopcock due to* Na_2CO_3 *and* Na_2SiO_3.

12. (d)

(a) H_2O cannot be used as it would further spread fire

$$2Na + 2H_2O \longrightarrow 2NaOH + H_2$$

H_2 is flammable.

(b) N_2 may form nitride.

$$6Li + N_2 \longrightarrow 2Li_3N$$

(c) If metal has converted to oxide or peroxide, CO_2 may react producing O_2 which would help fire.

$$2Na_2O_2 + 2CO_2 \longrightarrow 2Na_2CO_3 + O_2$$

(d) Asbestos blanket is best suited.

13. (d) Solutions of sodium in ether have Na^+ and Na^-. This is confirmed by diamagnetic behaviour for spin-paired Na^- ions in the ground state. In concentrated solutions, Na^+Na^- is formed hence, molar conductivity is decreased.

14. (d) Suitable reactant is that can decompose $NaHCO_3$ releasing CO_2 to extinguish fire.

(a) NaCl solution is neutral.

(b) CH_3COONa solution is basic.

(c) NH_4Cl solution is acidic.

$$NH_4Cl + H_2O \longrightarrow NH_4OH + HCl$$

But at the same time, smell of NH_3 can cause panic.

(d) $Al_2(SO_4)_3 + 6H_2O \longrightarrow 2Al(OH)_3 + 3H_2SO_4$

H_2SO_4 can decompose $NaHCO_3$.

$$2NaHCO_3 + H^+ \longrightarrow Na_2CO_3 + H_2O + CO_2$$

15. (d) NaCl is least soluble.

All nitrates are water soluble.

Solubility of KCl > NaCl.

16. (c) (a), (b) and (d) are regulated by K^+ ions.

(c) Na^+ and K^+ ions responsible for transmission of nerve signals.

17. (c) Baking powder contains

$NaHCO_3$ (baking soda)-30%; Starch-40%

$Ca(H_2PO_4)_2$-10%; $NaAl(SO_4)_2$-20%

Na_2CO_3 and $Mg(HCO_3)_2$ are not contents of baking powder.

18. (b) In Solvay process,

Step I CO_2 and NH_3 are saturated in H_2O.

$$CO_2 + NH_3 + H_2O \longrightarrow NH_4HCO_3$$

CO_2 is obtained by decomposition of limestone.

$$CaCO_3 \xrightarrow{1100°C} CaO + \underset{\text{(Side product)}}{CO_2}$$

Step II $NaCl + NH_4HCO_3 \longrightarrow NaHCO_3 + \underset{\text{(Side product)}}{NH_4Cl}$

Step III $2NaHCO_3 \longrightarrow Na_2CO_3 + H_2O + \underset{\text{(Side product)}}{CO_2}$

CO_2 is recycled in Step I.

CaO of Step I decomposes NH_4Cl of Step II.

$$CaO + 2NH_4Cl \longrightarrow 2NH_3 + CaCl_2 + H_2O$$

19. (c) $Na_2O + H_2O \longrightarrow 2\,NaOH$

$$\underset{+1\;-2}{Na_2O} + \underset{+1\;-2}{H_2O} \longrightarrow 2\,\underset{+1\;-2\;+1}{NaOH}$$

(a) and (d) are incorrect as no change in ON of sodium and oxygen.

(b) Oxide ion has complete octet hence, it cannot accept a pair of electrons. Thus, it is incorrect.

(c) Oxide ion donates a pair of electrons to H in H_2O. Thus, it is correct.

20. (c) (a) $2NaNO_3 \xrightarrow{\Delta} 2NaNO_2 + O_2$

(b) $2KClO_3 \xrightarrow{\Delta} 2KCl + 3O_2$

(c) $CaCO_3 \xrightarrow{\Delta} \underset{\text{Basic}}{CaO} + \underset{\text{Acidic}}{CO_2}$

(d) $NH_4NO_3 \xrightarrow{\Delta} N_2O + 2H_2O$

21. (d) A solution of sodium metal in liquid ammonia is strongly reducing agent due to the presence of solvated electrons.

$$M(s) \longrightarrow M^+(g) + e^-$$

Cation $M^+(g)$ and e^- both are solvated by NH_3 and reducing action is due to solvated electrons.

$$M^+ + (x + y)\,NH_3 \longrightarrow [M(NH_3)_x]^+ + [e(NH_3)_y]^-$$

22. (d)

(a) $MgSO_4 \cdot 7H_2O$ Epsomite
(b) $CuSO_4 \cdot 5H_2O$ Blue vitriol
(c) $FeSO_4 \cdot 7H_2O$ Green vitriol
(d) $Na_2SO_4 \cdot 10H_2O$ Glauber's salt

23. (c) (Na^+ - K^+) pump is operated based on transfer of Na^+ ions from intramolecular fluids to extramolecular fluids in each and every cell of animals. K^+ ions are transferred in reverse way.

24. (c) $LiHCO_3$ is formed only in aqueous solution.

$$Li_2CO_3 + H_2O + CO_2 \longrightarrow LiHCO_3$$

It is not isolated in solid state due to strong polarising power of Li^+ on the HCO_3^- ion of larger size.

25. (c) Na_2CO_3 is thermally stable due to high lattice energy (larger anion fits into larger cation).

$$Na_2CO_3 \xrightarrow{Heat} \text{No effect}$$

$$\underset{\substack{2\text{ mol}\\1\text{ mol}}}{2NaHCO_3} \xrightarrow{Heat} Na_2CO_3 + H_2O + \underset{\substack{1\text{ mol}\\0.5\text{ mol}}}{CO_2}$$

$$\underset{1\text{ mol}}{Li_2CO_3} \xrightarrow{Heat} Li_2O + \underset{1\text{ mol}}{CO_2}$$

Total $CO_2 = 0.5 + 1.0 = 1.5$ moles

26. (a)

(a) $2KNO_3 \longrightarrow 2KNO_2 + O_2$
(b) $2Pb(NO_3)_2 \longrightarrow 2PbO + 4NO_2 + O_2$
(c) $2Cu(NO_3)_2 \longrightarrow 2CuO + 4NO_2 + O_2$
(d) $2AgNO_3 \longrightarrow Ag_2O + 2NO_2 + \frac{1}{2}O_2$

$$Ag_2O \longrightarrow 2Ag + \frac{1}{2}O_2$$

27. (d) Solubility in water is higher if cation is smaller and anion is larger or cation is larger and anion is smaller.

Li^+ is common

$F^- < Cl^- < Br^- < I^-$ (largest in size)

Li^+ is smaller and I^- is larger thus, LiI is soluble in water to maximum extent.

28. (a) Smaller the size of cation, larger the size of anion, larger the charge then
larger the polarising power,
larger the covalent character,
smaller the ionic character,
and thus, smaller the solubility in water.

Fluoride is same, cationic size of

$$\underset{\substack{\uparrow\\\text{Maximum polarising power}}}{Li^+} < Na^+ < K^+ < Rb^+ < Cs^+$$

Thus, CsF is the most ionic and LiF is most covalent.

LiF < NaF < KF < RbF < CsF (maximum solubility)

Note : *In non-polar solvent, solubility of*
LiF > NaF > KF > RbF > CsF
(Maximum)

29. (b) Larger anions O_2^{2-} (peroxide) and O_2^- (superoxide) stabilise larger anion as K^+, Rb^+ through lattice energy effects. Thus, stability of peroxide and superoxide increases down the group.

Li forms Li_2O.
Na forms Na_2O, Na_2O_2.
K forms K_2O, KO_2, K_2O_2.

30. (a) LiI : Smaller cation attached with larger anion thus, unstable.
LiF : Smaller cation attached with smaller anion thus, stable.
Similarly, with KF and KI
Given reaction goes from unstable to stable side, thus $\Delta G° < 0$ (a sign of spontaneity).
$\Delta S° = 0$ since, number of ions remains unchanged.

31. (c) Smaller the size of cations, larger the charge, thus larger the hydration and thus, larger the enthalpy of hydration.

Smaller size, larger charge. $\longrightarrow$

Li^+	Be^{2+}
Na^+	Mg^{2+}
K^+	Ca^{2+}

Ionic size of alkaline earth metal ions are smaller than that of alkali metals. Also charge on alkaline earth metal ions are larger than that of alkali metals. Thus, alkaline earth metal ions are more hydrated than alkali metal ions.
Thus, enthalpy of hydration of alkaline earth metal ions is larger than that of alkali metals.
Thus, Statement I is correct and Statement II is incorrect.

32. (d)

$$Na(g) \longrightarrow Na^+(g) + e^-, \quad (IP)_{Na} = 496 \text{ kJ mol}^{-1}$$
$$Na(g) + e^- \longrightarrow Na^-(g), \quad (EA)_{Na} = -58 \text{ kJ mol}^{-1}$$
$$2Na(g) \longrightarrow Na^+(g) + Na^-(g), \quad \Delta H° = 438 \text{ kJ mol}^{-1}$$

Thus, Statement II is correct.
Since, $\Delta H° > 0$, hence formation of $Na^-(g)$ is not feasible. However, $Na^+(g)$ is formed due to low IP of sodium. Thus, Statement I is incorrect.

33. (c) Smaller the size of cation and larger the size of anion, larger the polarising power, which implies greater covalent nature.
Smaller cation Li^+ polarises larger anion Cl^- and thus, LiCl is predominantly covalent.
Electronegativity of Li = 1.0
Electronegativity of Cl = 3.0
There is large difference in EN of Li and Cl. Thus, Statement I is correct and Statement II is incorrect.

34. (b) Alkali metals have low IE. On heating in Bunsen flame, outermost electrons get excited to higher energy levels.
When these electrons return to the ground state, absorbed energy is released in the form of the light in the visible region of the spectrum.
Colour emission by these elements is the basis of flame photometry, atomic absorption spectroscopy, etc for precise estimation of these elements.
Thus, Statement I and Statement II are correct but Statement II is not the correct explanation of Statement I.

35. (c) In alums,

$$M_2SO_4 \cdot M'_2(SO_4)_3 \cdot 24H_2O$$
$$K_2SO_4 \cdot Al_2(SO_4)_3 \cdot 24\,H_2O$$

(M' is trivalent metal ion. M is monovalent alkali metal ion)
M and M' are coordinated to six separate water molecules

$$[M(H_2O)_6]^+\ [M'(H_2O)_6]^{3+}\ (SO_4^{2-})_2$$

Li^+ does not form alums, probably because the small $[Li(H_2O)_6]^+$ is too small to form stable lattice required for alum formation. Alum is hydrated double salt. Thus, Statement I is correct and Statement II is incorrect.

Revisal Problems (JEE Advanced)

1. (c) Cs with least ionisation energy is used in photo voltaic cell.

2. (d) Al and Zn react with NaOH.

$$Al + NaOH + H_2O \longrightarrow NaAlO_2 + \frac{3}{2}H_2$$
$$Zn + 2NaOH \longrightarrow Na_2ZnO_2 + H_2$$

Glass contains silicate, hence NaOH reacts with silica and silicate.

Poly vessel is best suited.

3. (a) $Na_2O_2 + 2HCl \longrightarrow 2NaCl + H_2O_2$

$$Cr_2O_7^{2-} + 2H^+ \longrightarrow 2CrO_3 + H_2O$$
$$CrO_3 + H_2O_2 \xrightarrow{\text{Ether}} \underset{\text{Violet colour in ether}}{CrO_5} + H_2O$$

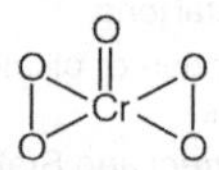

Note *When CrO_3 changes to CrO_5, there is no change in ON of Cr as CrO_5 has two peroxy linkage and one (Cr $=$O) double bond.*

4. (c) $Na_2CO_3 + SO_2 \longrightarrow Na_2SO_3 + CO_2$

5. (d) Dry air contains N_2 and O_2.

Na forms only oxide. It does not form nitride.

Li forms oxide and nitride both.

$$2Li + \frac{1}{2}O_2 \longrightarrow Li_2O$$
$$2Na + \frac{1}{2}O_2 \longrightarrow Na_2O$$
$$6Li + N_2 \longrightarrow 2Li_3N$$

6. (c) (a) $NaHCO_3$ is neutralised by NaOH.

$$NaHCO_3 + NaOH \longrightarrow Na_2CO_3 + H_2O$$

(b) NH_4Cl solution is acidic. $NaHCO_3$ is neutralised by acid.

$$NH_4Cl + H_2O \longrightarrow NH_4OH + HCl$$
$$NaHCO_3 + HCl \longrightarrow NaCl + H_2O + CO_2$$

(c) Na_2CO_3 and $NaHCO_3$ are not reactive with each other, hence can be placed together.

(d) Na_2CO_3 is decomposed by acidic solution of NH_4Cl (as in (b))

$$Na_2CO_3 + 2HCl \longrightarrow 2NaCl + H_2O + CO_2$$

7. (b) NaCl from sea water contains some amount of $MgCl_2$ which being hygroscopic makes NaCl moist. Adding $NaHCO_3$ converts $MgCl_2$ to $MgCO_3$ which is non-hygroscopic salt.

$$MgCl_2(s) + 2NaHCO_3(s) \longrightarrow MgCO_3(s) + 2NaCl(s) + H_2O(l) + CO_2(g)$$

Thus, Statement I and Statement II are correct but Statement II is not the correct explanation of Statement I.

8. (d) $KHCO_3$ (and also NH_4Cl) is highly soluble in water hence, $KHCO_3$ cannot be separated by crystallisation.

Note *Solvay method is not suitable for manufacture of K_2CO_3.*

Thus, Statement I is incorrect and Statement II is correct.

9. (b) In going from Li to Na and then to K there is decrease in screening effect which is in order ($s > p > d > f$). Hence, variation in size is not uniform.

Thus, $(IE)_1$ which depends on force of attraction between valence electrons and nuclear charge, is also dependent on atomic size. Thus, variation in $(IE)_1$ is not regular. As valence electron is easily removed, they are electropositive elements. Thus, Statement I and Statement II are correct but Statement II is not the correct explanation of Statement I.

10. (a) As we go down the group, number of shells increases, hence ionic size increases.

$$Li^+ < Na^+ < K^+ < Rb^+ < Cs^+$$

This increase is rapid upto K^+ due to lack of d- and f-orbitals since, screening effect is $s > p > d > f$.

In Rb^+ and Cs^+, d- and f-electrons are also present. Screening effect is decreased hence, valence electrons are attracted more as compared to in Li^+, Na^+ and K^+, hence variation is small.

Thus, Statement I and Statement II are correct and Statement II is the correct explanation of Statement I.

11. (a,c) $$2MO_2 \longrightarrow \underset{\text{Sesquioxide}}{M_2O_3} + \frac{1}{2}O_2$$

They have formula $(\overset{+}{M})_4\ \underset{\text{Peroxide}}{(O_2^{2-})}\ \underset{\text{Superoxide}}{(O_2^-)_2}$

Due to the presence of O_2^- (superoxide), they are paramagnetic.

O_2^- is a coloured salt.

Thus, option (a) and option (c) are correct.

(b) M_2O_3 is not formed from MO_2 thus, option (b) is incorrect.

(d) They are not diamagnetic, thus option (d) is incorrect.

12. (b,c) It is called Nelson cell.

NaOH obtained is 11% and further concentration (upto 50%) is carried by steam evaporation.

Cathode and anode are separated by asbestos diaphragm, hence, there is no combination of H_2 and Cl_2.

13. (a,b)

(a) $ZnCl_2 + 2NaOH \longrightarrow \underset{\text{White ppt}}{Zn(OH_2)\downarrow} + 2NaCl$

$$Zn(OH)_2 + 2NaOH \longrightarrow \underset{\text{Soluble (sodium zincate)}}{Na_2ZnO_2} + 2H_2O$$

(b) $AlCl_3 + 3NaOH \longrightarrow \underset{\text{White ppt}}{Al(OH)_3\downarrow} + 3NaCl$

$$Al(OH)_3 + NaOH \longrightarrow \underset{\text{Soluble (sodium metaaluminate)}}{NaAlO_2} + 2H_2O$$

Note $NaAlO_2$ *is also written as* $Na[Al(OH)_4]$.

(c) $FeCl_3 + 3NaOH \longrightarrow \underset{\text{Brown ppt}}{Fe(OH)_3\downarrow} + 3NaCl$

$Fe(OH)_3$ is insoluble in excess of NaOH.

(d) $CrCl_3 + 3NaOH \longrightarrow \underset{\text{Green ppt}}{Cr(OH)_3\downarrow} + 3NaCl$

$Cr(OH)_3$ is insoluble in excess of NaOH.

Note $Cr(OH)_3$ *can be made soluble in* NaOH *if* H_2O_2 *is added.*

$$Cr(OH)_3 + 2NaOH + H_2O_2 \longrightarrow \underset{\text{Soluble (yellow)}}{Na_2CrO_4} + H_2O$$

This is a redox reaction.

14. (a,b,c,d) Oxidation takes place

(a) $Al + Na_2O_2 \longrightarrow Na_2O + Al_2O_3$

$\uparrow$ 0 ... $\uparrow$ +3 (Yes)

Na_2O_2 (peroxide) is reduced to Na_2O and Al is oxidised.

(b) $Cr^{3+} + Na_2O_2 + H_2O \longrightarrow CrO_4^{2-}$ (Yes)

$\uparrow$ +3 ... $\uparrow$ +6

Cr^{3+} is oxidised to CrO_4^{2-}.

(c) $Mn(OH)_2 + 2Na_2O_2 \longrightarrow Na_2MnO_4 + 2NaOH$

$\uparrow$ +2 ... $\uparrow$ +6 (Yes)

$Mn(OH)_2$ is oxidised to MnO_4^{2-}.

(d) $CO + Na_2O_2 \longrightarrow Na_2CO_3$

$\uparrow$ +2 ... $\uparrow$ +4 (Yes)

CO is oxidised to CO_3^{2-}.

15. (a,c)

(a) $2CO_2 + 2Na_2O_2 \longrightarrow 2Na_2CO_3 + O_2$

Na_2O_2 is CO_2 absorber and at the same time, O_2 is formed. Thus, Na_2O_2 finds application in submarines, mines and at places where O_2 is insufficient for breathing.

(b) $\underset{\text{dil.}}{H_2SO_4} + 2Na_2O_2 \longrightarrow Na_2SO_4 + H_2O_2$

(c) $\underset{\text{(hot)}}{4HCl} + 2Na_2O_2 \longrightarrow 4NaCl + 2H_2O + O_2$

(d) $CO + Na_2O_2 \longrightarrow Na_2CO_3$

16. (a,b,c,d)

(a) $Li(g) \longrightarrow Li^+(g)$

$(IE)_1$ = energy required to ionise $Li(g)$ to $Li^+(g)$ thus, correct.

(b) $Li(s) \longrightarrow Li(g)$, $\Delta H° = S$ (enthalpy of sublimation)

$Li(g) \longrightarrow Li^+(g) + e^-$, $\Delta H° = I$ (enthalpy of ionisation)

$Li^+(g) + aq \longrightarrow Li^+(aq)$, $\Delta H° = Q$ (enthalpy of hydration)

$$\Delta H°_{oxi} = S + I + Q$$
$$= 155 + 520 - 506$$
$$= 169 \text{ kJ mol}^{-1}$$

(c) $(IE)_1$ of $Li(g)$ is maximum, hence it is difficult to oxidise to $Li^+(g)$.

(d) Thus, $Li(s)$ is easiest to oxidise to $Li^+(aq)$.

17. (a,b)

A solute is soluble in water if

hydration energy > lattice energy

For Na_2SO_4, hydration energy > lattice energy. Hence, it is water soluble.

For $BaSO_4$, hydration energy < lattice energy. Hence, it is water insoluble.

Note *SO_4^{2-} ion is common in Na_2SO_4, and $BaSO_4$. Hence, compare hydration and lattice energy of Na^+ and Ba^{2+}.*

18. (a,b,c)

(a) KO_2 can absorb CO_2 and CO and at the same time releases O_2. Thus, it finds applications in space capsules, submarines and breathing masks (life-supporting systems).

$$CO + 2KO_2 \longrightarrow K_2CO_3 + O_2$$
$$2CO_2 + 4KO_2 \longrightarrow K_2CO_3 + 3O_2$$
$$4CO_2 + 2H_2O + 4KO_2 \longrightarrow 4KHCO_3 + 3O_2$$

Thus, correct.

(b) $2KO_2 + 2H_2O \longrightarrow 2KOH + H_2O_2 + O_2$

$2KO_2 + H_2SO_4 \longrightarrow K_2SO_4 + H_2O_2 + O_2$

In aqueous and acidic solutions, H_2O_2 is formed. H_2O_2 has bleaching properties. H_2O_2 and O_2 has oxidising properties. Thus, correct.

(c) O_2^- has one unpaired electron. Thus, it is paramagnetic. Thus, correct.

(d) O_2^- = 17 electrons

By MO theory, bond order = 1.5

Thus, it is incorrect.

19. (a,b,d)

(a) Li^+ with smallest size (in alkali metal ion) is most polarising hence, Li_2CO_3 is covalent and sparingly soluble in water. $LiHCO_3$ is not isolated. Thus, correct.

(b) $KHCO_3$ is highly soluble in water, hence $KHCO_3$ (intermediate for K_2CO_3) cannot be crystallised to obtain K_2CO_3. Thus, correct

(c) $Li_2CO_3 \xrightarrow{\Delta} Li_2O + CO_2$

$MgCO_3 \xrightarrow{\Delta} MgO + CO_2$

Thus, (c) is incorrect.

(d) Correct.

20. (a,b,d)

(a) $LiNO_3$ and $NaNO_3$ absorb moisture and thus, are deliquescent. Thus, correct.

(b) KNO_3 is not deliquescent hence, $NaNO_3$ is not preferred in gun powder. Thus, correct.

(c) Incorrect

(d) Correct

21. (a,b)

(a) NO_2 with one unpaired electron is paramagnetic.

KO_2 with O_2^- (superoxide) ion is paramagnetic.

(b) NO with one unpaired electron is paramagnetic.

(c) $2NO_2 \rightleftharpoons N_2O_4$

Unpaired electron in NO_2 is used in bonding between N and N. Thus, N_2O_4 is diamagnetic.

O^{2-} has paired electron - diamagnetic.

(d) N_2O_2 - diamagnetic as unpaired electron in NO is used in bonding between N and NO.

O_2^{2-} in K_2O_2 is diamagnetic as all electrons are paired.

22. (a)

(i) KO_2 - Orange (chrome-yellow)

- Paramagnetic due to O_2^- (with one unpaired electron)
- $[2KO_2 + 2H_2O \longrightarrow \underset{\text{aqueous alkaline solution}}{2KOH + H_2O_2} + O_2]$
- $[2KO_2 + CO_2 \longrightarrow K_2CO_3 + 1.5O_2]$

It absorbs CO_2 and O_2 is produced.

Thus, (i) $\rightarrow$ (p,r,s,t)

(ii) K_2O_2 - Colourless in pure form

- Diamagnetic
- $K_2O_2 + 2H_2O \longrightarrow 2KOH + H_2O_2$

Aqueous solution of K_2O_2 is alkaline.

- $K_2O_2 + CO_2 \longrightarrow K_2CO_3 + \frac{1}{2}O_2$

Thus, it is CO_2 absorber and O_2 releasor.

Thus, (ii) $\rightarrow$ (q,r,s)

(iii) K_2O -(Diamagnetic (O^{2-})

- ($K_2O + H_2O \longrightarrow 2KOH$)
Alkaline solution

Thus, (iii) $\rightarrow$ (q,r)

(iv) KO_3 (Potassium ozonide)

- Diamagnetic

- $4KO_3 + 2H_2O \longrightarrow 4KOH + 5O_2$
Alkaline solution

Thus, (iv) $\rightarrow$ (q,r)

(v) K_2O_3 (Potassium sesquioxide)

- Dark coloured

- Paramagnetic as it is actually peroxodisuperoxide.

$(O_2)^{2-}$ $(O_2^-)_2$
Peroxo Disuperoxide

O_2^- is paramagnetic.

- Aqueous solution is alkaline.

- Due to the presence of O_2^-, it is also CO_2 absorber and O_2 releasor.

Thus, (v) $\rightarrow$ (p,r,s,t)

23. (a)

(i) $2Li_3N \xrightarrow{\Delta} 6Li + N_2$

Li_3N on decomposition gives N_2, hence it is used in automobile airbags.

Li_3N is ruby-red in colour.

Thus, (i) $\rightarrow$ (s,t)

(ii) LiCl is covalent and soluble in ether. It absorbs water, thus used to control humidity.

Thus, (ii) $\rightarrow$ (q,r)

(iii) KO_2 is paramagnetic and an orange coloured compound.

Thus, (iii) $\rightarrow$ (p,t)

(iv) Rb_2O_3 is rubidium sesquioxide and exists as Rb_4O_6.

$$Rb_4(O_2^{2-})(O_2^-)_2$$

It contains superoxide (O_2^-) ion hence, paramagnetic.

It is coloured.

Thus, (iv) $\rightarrow$ (p,t)

Explanations for Q. Nos. (24 and 25)

X is a binary salt of alkali metal *M* (say)

$$X + S \longrightarrow Y$$

$Y + BaCl_2 \longrightarrow Z$ (insoluble in conc. HCl)

Thus, *Z* is sulphate (SO_4^{2-}).

Thus, *Y* is M_2SO_4 (sulphate of alkali metal).

Y is formed by oxidation of S by *X*.

X contains oxygen.

X is a binary salt (with one cation and one anion).

Thus, *X* is $MO_2(M^+O_2^-)$

$$\underset{X}{2MO_2} \xrightarrow{S} \underset{Y}{M_2SO_4} \xrightarrow{BaCl_2} \underset{\substack{Z \\ 23.3\text{ g}}}{BaSO_4}$$

$$\frac{2(\text{molar mass of } MO_2)}{14.2} = \frac{\text{molar mass of } BaSO_4}{23.3}$$

$$\frac{2(M+32)}{14.2} = \frac{233}{23.3}$$

$$M = 39$$

Thus, metal is potassium.

X is KO_2.

KO_2 is paramagnetic due to O_2^-. KO_2 is used in submarines since, it is CO_2 absorber and O_2 is releasor.

$$2KO_2 \xrightarrow{S} K_2SO_4 \xrightarrow{BaCl_2} BaSO_4$$

Thus, *Y* is K_2SO_4 and *Z* is $BaSO_4$.

24. (d) **25.** (c)

Explanations for Q. Nos. (26 and 27)

$$NaCl(aq) \underset{}{\overset{\text{Current}}{\rightleftharpoons}} Na^+(aq) + Cl^-(aq)$$

At anode $2Cl^-(aq) \longrightarrow Cl_2(g) + 2e^-$

At cathode $2Na^+(aq) + 2e^- \longrightarrow 2Na(s)$

$$2Na(s) + 2H_2O(l) \longrightarrow 2NaOH(aq) + H_2$$

If cathode and anode are not separated, Cl_2 may react with NaOH at room temperature forming NaCl (by reduction) and NaClO (by oxidation).

$$2NaOH + Cl_2 \longrightarrow NaClO + NaCl + H_2O$$

Thus, Cl_2 disproportionates to NaCl and NaClO.

Note *If* Cl_2 *reacts with boiling* NaOH, *products are* NaCl *and* $NaClO_3$.

$$3Cl_2 + 6NaOH \longrightarrow 5NaCl + NaClO_3 + 3H_2O$$

NaCl solution is neutral being a salt of strong acid and strong base. After electrolysis, *aq*. NaOH is formed which is alkaline in nature. Thus, pH changes from 7.0 to a range of alkaline nature (which is 9.0).

26. (b) **27.** (a)

Explanations for Q. Nos. (28 and 29)

	Na	O
P :	74.19%	25.81%
Moles :	3.226	1.613
Ratio :	2	1

Thus, *P* is Na_2O.

	Na	O
Q :	58.97%	41.03%
Moles :	2.564	2.564
Ratio :	1	1

Thus, *Q* is Na_2O_2 (NaO does not exist).

$$Mn(OH)_2 + Na_2O \xrightarrow{\Delta} \text{No reaction}$$

$$Cr(OH)_3 + Na_2O_2 + H_2O \longrightarrow Na_2CrO_4$$

28. (d) **29.** (a)

Explanations for Q. Nos. (30-32)

$$2NaHCO_3 \longrightarrow Na_2CO_3 + H_2O + CO_2$$

Bread rise due to CO_2.

$$\underset{\text{Acid}}{H_2PO_4^-} \longrightarrow H^+ + HPO_4^{2-}$$

$H_2PO_4^-$ is H^+ ion donor. It decomposes HCO_3^-.

$$HCO_3^- + H^+ \longrightarrow H_2O + CO_2$$

Starch is filler.

$NaAl(SO_4)_2$ slows down the decomposition of $NaHCO_3$.

30. (b) **31.** (a) **32.** (b)

33. (4) Disproportionation is a redox reaction in which same species is oxidised and reduced.

(i) $\underset{+4}{CO_2} + 2NaOH \longrightarrow Na_2\underset{+4}{CO_3} + H_2O$ (No)

(ii) $\underset{+4}{SO_2} + 2NaOH \longrightarrow Na_2\underset{+4}{SO_3} + H_2O$ (No)

(iii) $2NO_2 + 2NaOH \longrightarrow NaNO_2 + NaNO_3$ (Yes)

$\uparrow$ +4 (in NO_2); $\uparrow$ +3 (in $NaNO_2$); $\uparrow$ +5 (in $NaNO_3$)

NO_2 is reduced to NO_2^- and NO_2 is oxidised to NO_3^-.

(iv) $4S + 6NaOH \longrightarrow Na_2S_2O_3 + Na_2S + 3H_2O$ (Yes)

$\uparrow$ 0 (in S); $\uparrow$ 2.0 (in $Na_2S_2O_3$); $\uparrow$ −2 (in Na_2S)

S is oxidised to $S_2O_3^{2-}$ and reduced to S^{2-}.

(v) $Cl_2 + 2NaOH \xrightarrow{\text{Cold}} NaCl + NaClO + H_2O$ (Yes)

$\uparrow$ 0 (in Cl_2); $\uparrow$ −1 (in NaCl); $\uparrow$ +1 (in NaClO)

Cl_2 is oxidised to ClO^- and reduced to Cl^-.

(vi) $\underset{\text{Yellow}}{P_4} + 3NaOH + 3H_2O \longrightarrow 3NaH_2PO_2 + PH_3$ (Yes)

$\uparrow$ 0 (in P_4); $\uparrow$ +1 (in NaH_2PO_2); $\uparrow$ −3 (in PH_3)

P_4 is oxidised to $H_2PO_2^-$ and reduced to PH_3.

(vii) $2B + 2NaOH + 2H_2O \longrightarrow 2NaBO_2 + 3H_2$ (No)

$\uparrow$ 0 (in B); $\uparrow$ +3 (in $NaBO_2$)

B is oxidised and OH^- is reduced.

(viii) $2Al + 2NaOH + 2H_2O \longrightarrow 2NaAlO_2 + 3H_2$ (No)

$\uparrow$ 0 (in Al); $\uparrow$ +3 (in $NaAlO_2$)

As in boron.

(ix) $Zn + 2NaOH \longrightarrow Na_2ZnO_2 + H_2$ (No)

$\uparrow$ 0 (in Zn); $\uparrow$ +2 (in Na_2ZnO_2)

Zn is oxidised to ZnO_2^{2-} and OH^- is reduced.

(x) $CO + NaOH \xrightarrow[100\text{-}200^\circ C]{5\text{-}10\ atm} HCONa$ (with C=O) (No)

$\uparrow$ +2 (in CO); $\uparrow$ +2 (in HCONa)

No change

Only four species undergo disproportionation.

34. (9) $\underset{\substack{\text{Sodium peroxide} \\ 78\ g \\ 6.382\ g}}{Na_2O_2} + O_2 \longrightarrow \underset{\substack{\text{Sodium superoxide} \\ (2 \times 55)\ g \\ ?}}{2NaO_2}$

$$NaO_2 \text{ formed} = \frac{2 \times 55 \times 6.382}{78} = 9.0\ g$$

35. (2) There is one unpaired electron.

Hence, $x = 2$

$(2 - 1) = 1$

JEE Archive

1. (b) $LiNO_3$ behave differently from other alkali metal nitrates.

Lithium nitrates decompose on heating producing NO_2. Whereas other alkali nitrates like $NaNO_3$ produces nitrite and oxygen on decomposition.

$$2LiNO_3 \xrightarrow{\Delta} Li_2O + 2NO_2 + \frac{1}{2}O_2$$

$$NaNO_3 \xrightarrow{\Delta} NaNO_2 + \frac{1}{2}O_2$$

2. (c) $Na_2O + H_2O \longrightarrow 2NaOH$

Oxidation number of Na and oxygen: +1 −2 −2 +1 −2

No change in oxidation number, so option (a) and option (d) are not true.

(b) is also not true.

Oxide ion donates a pair of electrons, thus changes to OH^-.

3. (a) All alkali metals dissolve in liquid ammonia and gives solvated ammoniated electrons.

$M + (x + y)NH_3 \longrightarrow M^+(NH_3)_x + e^-(NH_3)_y$

These electrons are excited to higher energy levels and the absorption of photons occurs in the red region of the spectrum. Thus, solution appears blue.

4. (a) $4KO_2 + 2CO_2 \longrightarrow 2K_2CO_3 + 3O_2$

5. (a,b,c)

(a) Reaction of alkali metals with ammonia depends upon the physical state of ammonia whether it is in gaseous state or liquid state. If ammonia is considered as a gas then reaction will be

$$Na + \underset{\text{(Excess)}}{NH_3} \longrightarrow NaNH_2 + \frac{1}{2}H_2$$

($NaNH_2$ + 1/2 H_2 are diamagnetic)

If ammonia is considered as a liquid then reaction will be

$$M + (x + y)NH_3 \longrightarrow [M(NH_3)_x]^+ + [e(NH_3)_y]^-$$

- ammoniated electron
- blue colour
- paramagnetic
- very strong reducing agent

(b) $K + \underset{\text{(Excess)}}{O_2} \longrightarrow \underset{\substack{\text{Potassium superoxide} \\ \text{Paramagnetic}}}{KO_2(K^+, O_2^-)}$

(c) $3Cu + 8HNO_3 \longrightarrow \underset{\text{Paramagnetic}}{3Cu(NO_3)_2} + \underset{\text{Paramagnetic}}{2NO} + 4H_2O$

(d) 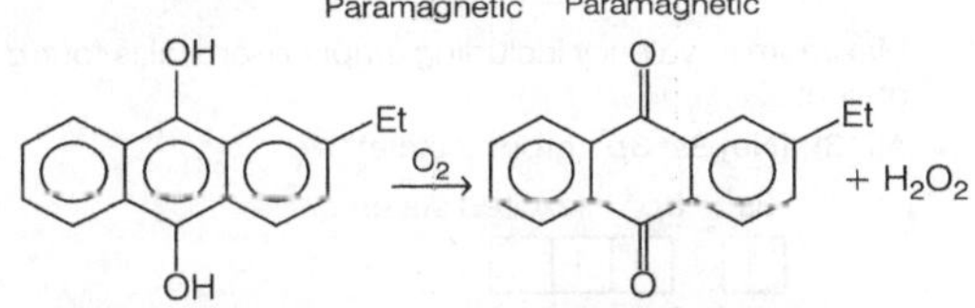

2-ethyl anthraquinol 2-ethyl anthraquinone

H_2O_2 is diamagnetic.

6. (a,b) When sodium metal is burnt in excess of air, mainly sodium peroxide (Na_2O_2) with little sodium oxide (Na_2O) are formed.

7. (b) Both statements are true but blue colour is due to the presence of solvated electrons [$NH_3(e^{-1})$].

4. Alkaline Earth Metals (Group 2)

DPP-1 Periodicity in Group 2

1. (b) Anion is common,
smaller the size of cation,
larger the charge,
then larger the polarising power and smaller the thermal stability.

Size $Be^{2+} < Mg^{2+} < Ca^{2+} < K^{+}$

Thus, thermal stability of

$$BeCO_3 < MgCO_3 < CaCO_3 < K_2CO_3$$

Note $BeCO_3$ *is placed in atmosphere of* CO_2 *to prevent its decomposition.*

2. (a) By Fajan's rule,
smaller the cation,
larger the anion,
larger the charge,
presence of *d*- and *f*-electrons,
then larger the polarisation of the cation by anion and *vice-versa*.
Thus, covalent nature increases, melting point decreases, lattice energy increases and thermal stability increases.

Size $Be^{2+} < Mg^{2+} < Ca^{2+} < Ba^{2+}$ (anion is common)

Thus, $BeCl_2$ has maximum covalent nature and has least melting point.
$BeCl_2$ has maximum lattice energy and has maximum thermal stability.
Thus, (a) is incorrect, (b) is correct, (c) is correct, (d) is correct.

3. (b)

← Basic

Na_2O	MgO ↓
K_2O	CaO
Cs_2O	SrO Basic

Basic nature of oxides decreases left to right and increases top to bottom.

$Na_2O < K_2O < Cs_2O$
$MgO < SrO$
$Cs_2O > SrO$

4. (a) Be(4): $1s^2, 2s^2$ (ground state)

$1s^2$ $2s^1$ $2p^1$ (excited state)
[1] [1][ ][ ]

Maximum covalency including empty p-orbital is four. d-orbital is absent.

Al(13): [Ne] $3s^2 3p^1$ (ground state)

$3s^1$ $3p^2$ (excited state)
[1] [1][1][ ]

In Al, empty d-orbital is also available for bonding. Hence, maximum covalency is six (as in $[AlF_6]^{3-}$).

5. (b) Hydration and lattice energies decrease with increase in size of metal ions. Decreasing lattice energy favours increased solubility while decreasing hydration energies favours decreased solubility. Decrease in hydration energy is more than decrease in lattice energy, hence solubilities of carbonates decrease moving down the group.

$$\underset{\text{Maximum soluble}}{BeCO_3} > MgCO_3 > CaCO_3 > SrCO_3 > BaCO_3$$

6. (b)

(a) Be^{2+}: [He]
Mg^{2+}: [Ne]

Group 2 (alkaline earth) elements form M^{2+} ions with inert gas configuration.

Zn^{2+}: $[Ar]3d^{10}$
Cd^{2+}: $[Kr]4d^{10}$

Zn^{2+} and Cd^{2+} do not form M^{2+} ions with inert gas configuration. Thus (a) is incorrect.

(b) $BeO + 2HCl \longrightarrow BeCl_2 + H_2O$
$BeO + 2NaOH \longrightarrow Na_2ZnO_2 + H_2O$

BeO is amphoteric as it dissolves in both acid and base.

$ZnO + 2HCl \longrightarrow ZnCl_2 + H_2O$
$ZnO + 2NaOH \longrightarrow Na_2ZnO_2 + H_2O$

ZnO is also amphoteric. Thus, (b) is correct.

(c) BeS, MgS ...... are water soluble.
ZnS, CdS, HgS are water insoluble.
Thus, (c) is incorrect.

(d)

$Be(OH)_2$	$Zn(OH)_2$
$Mg(OH)_2$	$Cd(OH)_2$
$Ca(OH)_2$	$Hg(OH)_2$
$Sr(OH)_2$	
Maximum soluble $Ba(OH)_2$	
↓ Solubility increases	↑ Solubility decreases

Thus, (d) is incorrect.

7. (a) Relative size of isoelectronic species is determined by $\frac{e}{Z}$.

Value of e is same, thus size $\propto \frac{1}{Z}$.

Smaller the value of Z, larger the size.
Thus, $F^-(9) > Na^+(11) > Mg^{2+}(12) > Al^{3+}(13) > Si^{4+}(14)$

8. (a)
(a) Correct
(b) $ZnCl_2 + H_2S \longrightarrow \underset{\text{White}}{ZnS\downarrow} + 2H_2O$

$CdCl_2 + H_2S \longrightarrow \underset{\text{Yellow}}{CdS\downarrow} + 2HCl$

$HgCl_2 + H_2S \longrightarrow \underset{\text{Black}}{HgS\downarrow}$

Zn^{2+}, Cd^{2+} and Hg^{2+} are precipitated in basic buffer ($NH_4OH + NH_4Cl$) and not in acidic medium. Be^{2+}, Mg^{2+}, Ca^{2+}, Sr^{2+} and Ba^{2+} are not precipitated as sulphides.

9. (b)
(a) $Be(OH)_2$ least soluble
↓
$Ba(OH)_2$ maximum soluble, thus correct.

(b) Smaller the size of cation, larger the hydration
Be^{2+} (least ionic size, maximum hydration)
↓
Ba^{2+} (maximum size, least hydration) Thus, incorrect.

(c) $BeCO_3 \longrightarrow BeO + CO_2$
$BeCO_3$ decomposes easily as BeO is stabilised by smaller cation and smaller anion. As size of cation increases, thermal stability of MCO_3 increases. Thus, correct.

(d) Be
↓
Ba
As number of shells increases with increase in atomic number from Be to Ba, ionic radius increases.
$Be^{2+} < Mg^{2+} < Ca^{2+} < Sr^{2+} < Ba^{2+}$ Thus, correct.

10. (c) Na or K(Inert gas) $\boxed{1}_{ns^1} \xrightarrow{I_1}$ (Inert gas) $(M^+) \xrightarrow{I_2}$

For Na or K, M^+ ion attains inert gas configuration. Thus, I_2 should be very high.

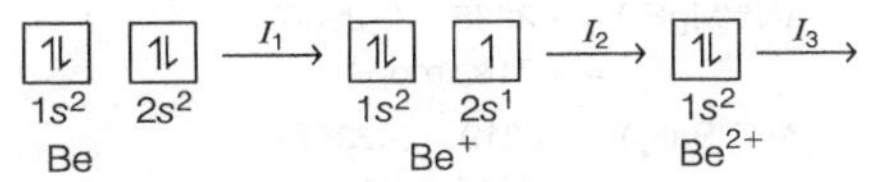

Be^{2+} attains inert gas configuration. Thus, I_3 is very high. It is as per given value.

I_1 of Ne is very high.

11. (b) $BaCO_3 \xrightarrow{1400°C} BaO + CO_2$

Lattice energy of $BaCO_3$ is very high. It decomposes at about 1400°C. In the above reaction, carbon is also added.

$$CO_2 + C \longrightarrow 2CO$$

This converts CO_2 into CO and reaction is made irreversible and BaO is formed easily.

Thus, Statement I and Statement II are correct but Statement II is not the correct explanation of Statement I.

12. (d) Be and Al initially react with HNO_3 but these metals are rendered passive due to the formation of oxide layer. There is no further dissolution due to the oxide layer.

Be and Al have approximately same electronegativity thus, they resemble in many properties. This is called diagonal relationship in terms of position.

Thus, Statement I is incorrect and Statement II is correct.

13. (a) $Ca + 2H_2O \longrightarrow Ca(OH)_2 + H_2$

Similarly, Sr and Ba.

$$M^{2+} + 2e^- \longrightarrow M(s)$$

$E° = -2.87$ V for Ca
$= -2.89$ for Sr
-2.90 for Ba
$= -2.70$ for Na
$= -2.92$ for K

They have high $E°$ values thus are reactive as alkali metals. Thus, Statement I and Statement II are correct and Statement II is the correct explanation of Statement I.

14. (b) Alkaline earth elements are denser than alkali metals because they have two valence electrons per atom (alkali metals have one electron per atom) for bonding the atoms into a metallic lattice and as a result, more mass can be packed into a smaller volume.

Thus, Statement I and Statement II are correct but Statement II is not the correct explanation of Statement I.

15. (b) Decrease in polarising power of the M^{2+} ion and lower lattice energy from BeO to BaO result in increase thermal stability of the hydroxide.

$Be(OH)_2$ is soluble in acid and base. Thus, amphoteric. Thus, Statement I and Statement II are correct but Statement II is not the correct explanation of Statement I.

16. (d) Na^+ attains stable inert gas configuration while Mg^+ has one unpaired electron. Hence, $I_2 > I_2'$ thus, (a) is incorrect and option (c) is also incorrect.

$$I_1(Na) < I_1'(Mg)$$
$$I_1(Na) < I_1''(Al)$$
$$I_1'(Mg) > I_1''(Al)$$

Energy required to remove paired electron is higher than that of unpaired electron.

Thus, $I_1 < I_1''$. Thus, option (b) is incorrect but option (d) is correct.

17. (a,b,c,d)

(a) Be and Al both dissolve in alkali forming H_2. They also dissolve in dil. acids forming H_2 and thus are amphoteric.

$$Be + 2NaOH \longrightarrow \underset{\text{Sodium berylliate}}{Na_2BeO_2} + H_2$$

$$2Al + 2NaOH + 2H_2O \longrightarrow \underset{\text{Sodium metaaluminate}}{2NaAlO_2} + 3H_2$$

(b) Oxides are also soluble in alkali and acid and thus are amphoteric.

$$BeO + 2NaOH \longrightarrow Na_2BeO_2 + H_2O$$
$$Al_2O_3 + 2NaOH \longrightarrow 2NaAlO_2 + H_2O$$

(c) $BeCl_2$ and $AlCl_3$ both are electron-deficient and thus are Lewis acids. They polymerise to complete octet such that metal accepts lone pair and Cl donates lone pair.

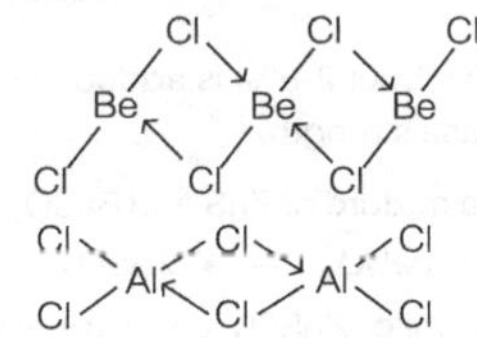

(d) Be^{2+} and Al^{3+} are of comparatively small ionic size and have higher charges. Thus, they are hydrated.

$$[Be(H_2O)_4]^{2+}, [Al(H_2O)_6]^{3+}$$

18. (b,c,d)

(a) Hydration energy decreases more rapidly than decrease in lattice energy, hence solubility of sulphate decreases. Thus, variation is incorrect.

(b) Solubility of hydroxide increases as going down the group. Thus, correct.

(c) Smaller the size of cation, larger the charge then larger the polarising power and thus compounds are predominantly covalent.

$Be^{2+} < Mg^{2+} < Ba^{2+}$ (Size)

$Be^{2+} > Mg^{2+} > Ba^{2+}$ (Polarising power)

$BeCl_2 < MgCl_2 < BaCl_2$ (Ionic)

Thus, correct.

(d) As in (c), polarising power of $Be^{2+} > Li^+ > Mg^{2+} > Na^+$.

Hence, ionic nature of

$$BeCl_2 < LiCl < MgCl_2 < NaCl$$

Thus, correct.

19. (a)

(i) BeO is amphoteric as it dissolves in acid and base both.

Be is diagonally related to Al.

BeO is water insoluble.

Thus, **(i) → (p, q, r)**

(ii) As in (i),

Thus, **(ii) → (p, q, r)**

(iii) Mg is diagonally related to Li. MgO is estimated by EDTA (ethylenediamine tetraacetate).

Thus, **(iv) → (q, s)**

(iv) BaO is maximum soluble in H_2O.

Thus, **(iv) → (t)**

Explanations for Q. Nos. (20 and 21)

Hydration energy

$CaF_2 = Ca^{2+} + 2F^-$

$= -1577 + 2 \times (-457)$

$= -1577 - 914$

$= -2491 \text{ kJ mol}^{-1}$

$MgF_2 = -1921 - 914$

$= -2835 \text{ kJ mol}^{-1}$

$BaF_2 = -1305 - 914$

$= -2219 \text{ kJ mol}^{-1}$

Thus, MgF_2 (with smallest Mg^{2+} ion) has maximum hydration energy.

Standard heat of solution

ΔH° = Hydration energy – lattice energy

$\Delta H^\circ (CaF_2) = -2491 - (-2610)$

$= +119 \text{ kJ mol}^{-1}$

$\Delta H^\circ (MgF_2) = -2835 - (-2906)$

$= +71 \text{ kJ mol}^{-1}$

$\Delta H^\circ (BaF_2) = -2219 - (-2367)$

$= -2219 + 2367$

$= +148 \text{ kJ mol}^{-1}$

Thus, BaF_2 is most soluble (most endothermic).

20. (a)

21. (c)

DPP-2 Compounds of Group 2

1. (a) $CaSO_4 \cdot 2H_2O$ (about 2-5%) is added to slow down the setting process to increase strength.

2. (d) Lithopone is a mixture of ZnS and $BaSO_4$.

$$ZnS + BaSO_4 \longrightarrow ZnSO_4 + BaS$$

ZnS, $BaSO_4$ are white, $ZnSO_4$ and BaS are also white. Hence, lithopone is a better white paint.

3. (c) Magnalium is an alloy of magnesium and aluminium.

4. (b) $Na_2CO_3 + BaSO_4 \longrightarrow BaCO_3 + Na_2SO_4$

$$BaCO_3 \xrightarrow{\Delta} \underset{\text{Residue}}{BaO} + CO_2\uparrow$$

$$Na_2SO_4 \xrightarrow{\Delta} \text{No effect}$$

Na_2SO_4 exists as residue.

5. (c)

(a) NaCl from sea water also contains $MgCl_2$ as impurity. $MgCl_2$ is hygroscopic in nature. It absorbs moisture from atmosphere. Hence, it causes NaCl to clump. Thus, correct.

If $NaHCO_3$ is added, $MgCl_2$ is converted to $MgCO_3$.

$$2NaHCO_3 + MgCl_2 \longrightarrow 2NaCl + MgCO_3 + H_2O + CO_2$$

$MgCO_3$ is non-hygroscopic in nature.

6. (c) NaCl obtained from sea water is also contaminated with small amount of $MgCl_2$ (as impurity) which being hygroscopic makes it (NaCl) moist.

7. (a) $\underset{\text{Dolomite}}{CaCO_3 + MgCO_3} \xrightarrow{\Delta} \underbrace{CaO + MgO}_{\text{Calcined dolomite}}$

$$Ca(OH)_2 \cdot Mg(OH)_2 + \underset{(\text{in } H_2O)}{Mg^{2+}} \longrightarrow Mg(OH)_2 + Ca^{2+}$$

$Mg(OH)_2$ is less soluble than $Ca(OH)_2$. This leads to replacement of Ca^{2+} by Mg^{2+}.

8. (d) $Ca(OH)_2 + Na_2CO_3 \longrightarrow CaCO_3\downarrow + 2NaOH$

Since, $CaCO_3$ is precipitated, hence NaOH solution is concentrated and above reaction proceeds to right side.

(a) Thus, spontaneous.

(b) NaOH being saturated pH > 14.

(c) This provides a cheap way to prepare NaOH.

9. (c) $BeCl_2 + 2H_2O \longrightarrow Be(OH)_2 + 2HCl$

$$AlCl_3 + 3H_2O \longrightarrow Al(OH)_3 + 3HCl$$

Both are hydrolysed in H_2O and HCl is formed; HCl appears in the form of fumes.

10. (b) $BeCl_2$ is electron-deficient in Be and thus is a Lewis acid. To complete its octet, it polymerises to form $(BeCl_2)_4$ in solid state, in which Cl is electron-pair donor and Be is electron-pair acceptor.

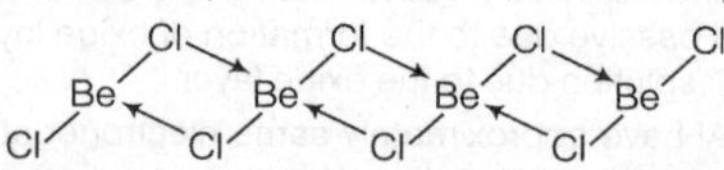

11. (a) $MgO + MgCl_2 \longrightarrow$ Solid mass

Solid mass is called Sorel cement or magnesia cement as it is as hard/stable as cement.

When mixed with saw-dust or cork waste, sorel cement gives a weather-proof composition.

12. (b) Gypsum is $CaSO_4 \cdot 2H_2O$.

Note $CaSO_4 \cdot \frac{1}{2}H_2O$ *is called Plaster of Paris. It sets into hard solid on reaction with water* - $CaSO_4 \cdot 2H_2O$. $CaSO_4$ *is called anhydrite.*

13. (a) Calcium is reactive with H_2O, hence aqueous $CaCl_2$ is not taken. It is obtained by electrolysis of molten $CaCl_2$ when calcium is formed at cathode.

$$CaCl_2(\text{molten}) \rightleftharpoons Ca^{2+} + 2Cl^-$$

At anode $2Cl^- \longrightarrow Cl_2 + 2e^-$

At cathode $Ca^{2+} + 2e^- \longrightarrow Ca$

14. (d)

(a) Barium salt in the form of paste with HCl give green colour in flame- true.

(b) $Ba^{2+} + CrO_4^{2-} \longrightarrow \underset{\text{Yellow ppt}}{BaCrO_4\downarrow}$ Thus, true

(c) $Ba^{2+} + SO_4^{2-} \longrightarrow \underset{\text{White ppt}}{BaSO_4\downarrow}$ Thus, true

15. (a)

(a) MgO dissolves in acid only thus, MgO is basic oxide.

ZnO dissolves in acid and base both thus is an amphoteric oxide.

$$\underset{\text{Basic}}{MgO} + 2HCl \longrightarrow MgCl_2 + H_2O$$

$$\underset{\text{Basic}}{ZnO} + \underset{\text{Acidic}}{2HCl} \longrightarrow ZnCl_2 + H_2O$$

$$\underset{\text{Acidic}}{ZnO} + \underset{\text{Basic}}{2NaOH} \longrightarrow Na_2ZnO_2 + H_2O$$

(b) $MgCO_3 \xrightarrow{\Delta} MgO + CO_2$

$$ZnCO_3 \xrightarrow{\Delta} ZnO + CO_2$$

(c) Mg and Zn are electropositive elements and thus used as electrodes.

(d) $Mg \longrightarrow Mg^{2+} + 2e^-$

$Zn \longrightarrow Zn^{2+} + 2e^-$

Both are reducing agents, thus used to prevent corrosion of metals especially iron.

$$Fe^{2+} + 2e^- \longrightarrow Fe$$

Iron poles are joined to zinc or magnesium metals. If Fe is oxidised, it is reduced by these metals at the same time to iron. It is called **cathodic protection.**

16. (a) When Cl_2 is passed into slaked lime, it disproportionates to Cl^- (by reduction) and to ClO^- (by oxidation).

$$2Ca(OH)_2 + 2Cl_2 \longrightarrow CaCl_2 + \underset{\text{Calcium hypochlorite}}{Ca(ClO)_2} + 2H_2O$$

17. (a)

(a) $Ca(HCO_3)_2 \longrightarrow CaCO_3(Y) + CO_2 + H_2O$

Y is water insoluble.

Y gives CO_2 with dilute acid and also on heating.

$$CaCO_3 \xrightarrow{\Delta} CaO + CO_2$$

$$CaCO_3 + 2HCl \longrightarrow CaCl_2 + H_2O + CO_2$$

(b) $CaCO_3 \xrightarrow{\Delta} \underset{(Y)}{CaO} + CO_2$

CaO does not give CO_2.

(c) $2NaHCO_3 \longrightarrow Na_2CO_3(Y) + H_2O + CO_2$

Na_2CO_3 is water soluble.

(d) $Na_2CO_3 \xrightarrow{\Delta}$ No effect

18. (b)

(a) Be and Al both are made passive by HNO_3 thus, correct.

(b) $BeCl_2$ and $AlCl_3$ are Lewis acids (not Lewis bases) as they are electron deficients thus, incorrect.

(c) $BeCl_2$ and $AlCl_3$ exist as polymeric forms.

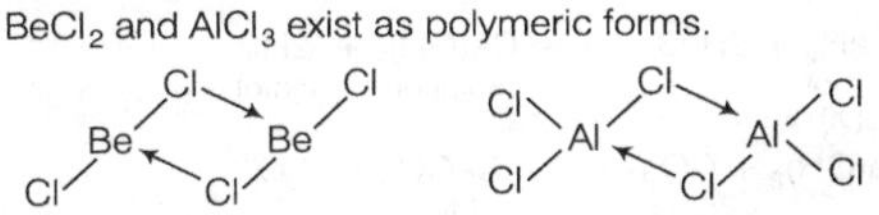

Thus, correct.

(d) $Be(OH)_2$ and $Al(OH)_3$ both are amphoteric as soluble in acid and base both. Thus, correct.

19. (b) $BeCl_2$ has incomplete octet. It can accept lone pair to complete its octet. Thus, $BeCl_2$ is a Lewis acid. $BeCl_2$ accepts lone pair from H_2O and thus, H^+ (of H_2O) and Cl^- (of $BeCl_2$) interact to form $Be(OH)_2$.

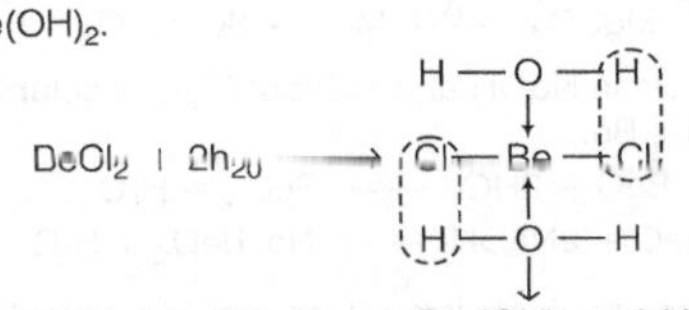

Thus, Statement I and Statement II are correct but Statement II is not the correct explanation of Statement I.

20. (d) Hydrated beryllium chloride is represented as $[Be(H_2O)_4]Cl_2$.

Thus, Statement I is incorrect.

$$[Be(H_2O)_4]^{2+} + H_2O \rightleftharpoons [Be(H_2O)_3(\bar{O}H)]^+ + H_3O^+$$

$[Be(H_2O)_4]^{2+}$ in aqueous solution donates proton.

Thus, $[Be(H_2O)_4]^+$ is a Lowry-Bronsted acid and H_2O is Lowry-Bronsted base. Thus, Statement II is correct.

21. (b,d) On heating $MgCl_2 \cdot 6H_2O$, dehydration does not take place, but partial hydrolysis takes place inernally.

$$MgCl_2 \cdot 6H_2O \longrightarrow MgCl(OH) + HCl + 5H_2O$$

$$MgCl(OH) \longrightarrow MgO + HCl$$

Note $BeCl_2 \cdot 2H_2O$ *on heating results in hydrolysis as in* $MgCl_2 \cdot 6H_2O$.

$$BeCl_2 \cdot 2H_2O \longrightarrow Be(OH)_2 + 2HCl(\text{fumes})$$

22. (b,c,d)

(a) $CaC_2 + H_2O \longrightarrow CH \equiv CH + Ca(OH)_2$

(b) $Al_4C_3 + 12H_2O \longrightarrow 4Al(OH)_3 + 3CH_4$

(c) $Be_2C + 4H_2O \longrightarrow 2Be(OH)_2 + CH_4$

(d) $Mg_2C_3 + 4H_2O \longrightarrow 2Mg(OH)_2 + CH_3C \equiv CH$

23. (a,b,c)

(a) $Ca_3(PO_4)_2$ is a part of bones thus, correct.

(b) $3Ca_3(PO_4)_2 \cdot CaF_2$ is a part of enamel on teeth thus, correct.

(c) Blood clotting involves Ca^{2+} ions thus, correct.

(d) Mg is present in chlorophyll thus, incorrect.

24. (a,b,c,d) Mg^{2+} ions are concentrated more in intracellular than in extracellular fluids in animal bodies. Their presence is also necessary for the activation of phosphate-transfer enzymes.

These enzymes take part in biochemical processes (exothermic) occurring in animal bodies. Mg^{2+} ions are also involved in carbohydrate metabolism.

Thus, options (a), (c) and (d) are correct.

Mg is present in chlorophyll thus, option (b) is also correct.

25. (a,b,c)

(a) Correct

(b) Correct

(c) Correct

(d) Cement is soluble in water and sets as a hard material.

Thus, incorrect.

26. (d)

(i) CaC_2 (calcium carbide) in the presence of moisture liberates ($CH \equiv CH$) which is used in food ripening.

$$CaC_2 + 2H_2O \longrightarrow Ca(OH)_2 + \underset{\text{Acetylene}}{CH \equiv CH}$$

Thus, (i) $\rightarrow$ (q)

(ii) $CaCN_2$ is used as a fertiliser as in contact with H_2O, it forms NH_3.

$$CaCN_2 + 3H_2O \longrightarrow CaCO_3 + 2NH_3$$

Thus, (ii) $\rightarrow$ (s).

(iii) $BaSO_4$ along with ZnS is used as a paint. It is called lithopone.

Thus, (iii) $\rightarrow$ (p).

(iv) Mg_3N_2 on impact releases N_2.

$$Mg_3N_2 \xrightarrow{\text{Impact}} 3Mg + N_2$$

Mg_3N_2 is placed in air-bags of automobiles. In case of accidents, N_2 is formed and N_2 filled air-bags prevent accidents.

Thus, (iv) $\rightarrow$ (r).

27. (b)

(i) CaO - 70% Thus, (i) $\rightarrow$ (t)

(ii) SiO_2 - 20% Thus, (ii) $\rightarrow$ (s)

(iii) Al_2O_3 - 5% Thus, (iii) $\rightarrow$ (r)

(iv) Fe_2O_3 - 3% Thus, (iv) $\rightarrow$ (q)

(v) $CaSO_4 \cdot 2H_2O$ - 2% Thus, (v) $\rightarrow$ (p)

Explanations for Q. Nos. (28 and 29)

$$2BeCl_2 + LiAlH_4 \longrightarrow LiCl + AlCl_3 + 2BeH_2$$

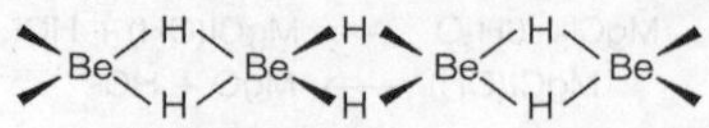

— Bond containing one electron between Be and H.

▬ Bond containing two electrons between Be and H.

Thus, (Be—H—Be) bond is three-centre bond in which banana-shaped molecular orbital covers these atoms (Be — H — Be) and contains two electrons.

28. (c)

29. (d)

Explanations for Q. Nos. (30 and 31)

$BeCl_2$ can't be prepared by

$$Be(OH)_2 + 2SOCl_2 \longrightarrow BeCl_2 + 2SO_2 + 2HCl$$

$BeCl_2$ has polymeric three-centre, two-electron bonds.

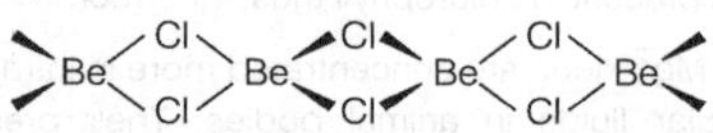

30. (c)

31. (a)

32. (7) Amphoteric species are those which react with acid and base both.

BeO, ZnO and Al_2O_3 are amphoteric.

$$BeO + 2NaOH \longrightarrow Na_2BeO_2 + H_2O \text{ (same for ZnO)}$$
$$Al_2O_3 + 2NaOH \longrightarrow 2NaAlO_2 + H_2O$$
$$BeO + 2HCl \longrightarrow BeCl_2 + H_2O$$
$$Al_2O_3 + 6HCl \longrightarrow 2AlCl_3 + 3H_2O$$

HCO_3^- are decomposed by acid and base both

$$HCO_3^- + OH^- \longrightarrow H_2O + CO_3^{2-}$$
$$HCO_3^- + H^+ \longrightarrow H_2O + CO_2$$

HPO_4^{2-} is also amphoteric.

$$HPO_4^{2-} + OH^- \longrightarrow PO_4^{3-} + H_2O$$
$$HPO_4^{2-} + H^+ \longrightarrow H_2PO_4^-$$

Thus, BeO, $Mg(HCO_3)_2$, $Ca(HCO_3)_2$, $NaHCO_3$, ZnO, Al_2O_3 and MgH_2PO_4 (in all seven) are amphoteric.

33. (6) $Ca_9(Al_6O_{18})$ contains 12-membered rings of Si—O—Si—O made by joining six AlO_4 tetrahedra.

34. (2)Magnesium nitride is Mg_3N_2.

$$Mg_3N_2 + 6H_2O \longrightarrow 3Mg(OH)_2 + 2NH_3 \qquad x = 2$$

35. (7) Epsom salt is $MgSO_4 \cdot 7H_2O$.

36. (6) Epsom salt is $MgSO_4 \cdot 7H_2O$. Kieserite is $MgSO_4 \cdot H_2O$. Thus, difference in water molecules in Epsom salt and Kieserite is 6.

Revisal Problems (JEE Main)

1. (b) $2NaF + BeF_2 \longrightarrow Na_2BeF_4$

$$Na_2BeF_4 \rightleftharpoons \underset{\text{Cation}}{2Na^+} + \underset{\text{Anion}}{BeF_4^{2-}}$$

Be remains with BeF_4^{2-} (anionic part).

Note *BeF_2 can be obtained by*

$$BeO + NH_4F \longrightarrow (NH_4)_2BeF_4 \xrightarrow{\Delta} BeF_2 + 2NH_4F$$

2. (b) If anhydrous $MgCl_2$ is required, it may be obtained by mixing the hydrated salt with thionyl chloride ($SOCl_2$).

$$MgCl_2 \cdot 6H_2O + 6SOCl_2 \longrightarrow MgCl_2 + 6SO_2 + 12HCl$$

3. (d) Plaster of Paris is

$$CaSO_4 \cdot \frac{1}{2}H_2O \text{ or } 2CaSO_4 \cdot H_2O.$$

When it is mixed with quantitative amount of H_2O, it sets as hard mass called gypsum ($CaSO_4 \cdot 2H_2O$).

Note *Due to this property of POP, it is used*
- *for plastering walls.*
- *for making moulds.*
- *for sculptural material.*
- *for encasting broken limbs so that broken bones are set straight.*

4. (d) Mg^{2+} being small in size is hydrated, thus $MgSO_4$ is deliquescent and soluble in water. Thus, correct.

(b) $K_2SO_4 \cdot MgSO_4 \cdot 6H_2O$ is a double salt under the name potash magnesis. Thus, correct.

(c) $K_2SO_4 \cdot MgSO_4 \cdot 6H_2O$ is a fertiliser.

Thus, correct.

5. (b) $Ca^{2+} + H_2Y^{2-} \xrightarrow{\text{Buffer}} [Ca(Y)]^{2-} + 2H^+$

where, (H_2Y^{2-}) is EDTA.

6. (a) Cement contains Ca^{2+} salts. When plaster (on walls) is sprayed with H_2O, hydration takes place hence, heat is liberated. Thus, reaction is exothermic. Cement is thus set up.

7. (a) Calcium cyanamide (nitrolim) is used for fixing of nitrogen.

$$CaCN_2 + H_2O \longrightarrow CaCO_3 + NH_3$$

NH_3 recycles to N_2 by atmospheric oxygen.

8. (b) $\underset{\substack{1\text{ mol}\\(X)}}{CaH_2} + 2H_2O \longrightarrow \underset{\text{Solution}}{Ca(OH)_2} + \underset{2\text{ mol}}{2H_2}$

$$Ca(OH)_2 + CO_2 \longrightarrow \underset{\text{Milky}}{CaCO_3} + H_2O$$

H_2 is readily combustible gas.

9. (c) $Mg(OH)_2$ is used to treat indigestion. It controls acidity.

10. (d) $BeSO_4$ is water soluble. $MgSO_4$ is also water soluble. $CaSO_4$ and $BaSO_4$ are water insoluble.

$$BeSO_4 \xrightarrow{\Delta} BeO + SO_2 + O_2$$
$$MgSO_4 \xrightarrow{\Delta} MgO + SO_2 + O_2$$

BeO is soluble in NaOH and HCl but MgO is soluble in HCl only. Thus, metal is Be.

$$BeO + 2HCl \longrightarrow BeCl_2 + H_2O$$
$$BeO + 2NaOH \longrightarrow Na_2BeO_2 + H_2O$$

11. (a) Tricalcium aluminate is an important constituent of Portland and high alumina cements. Its general formula is $Ca_9(Al_6O_{18})$ because it contains 12-membered rings of (—Si—O—Si—O—) made by joining six AlO_4 tetrahedra.

12. (a) $NaHCO_3$ (sodium bicarbonate) and $Mg(OH)_2$ both are part of antacid medicines.

13. (b) Calcium and barium bicarbonates are stable only in solution and decompose on removal of water to give carbonates. This property of bicarbonate is the reason for **stalacite** and **stalagmite** formation.

Stalacite is the downward growth of $CaCO_3$ formed on the roof of a cave by the trickling of water containing calcium compounds.

Stalagmite is the upward growth of $CaCO_3$ from the floor of a cave as stalacite.

14. (b) Hydration energy of $MgSO_4$ > lattice energy

ΔH(solution) = (Hydration energy – Lattice energy)

Thus, $\Delta H > 0$

Note *If $\Delta H > 0$, then solute is soluble. Since, $MgSO_4$ is water soluble, hence $\Delta H > 0$.*

15. (d) He^+, H, Li^{2+} and Be^{3+} are isoelectronic.

$$(IE)_Z = (13.6)\frac{Z^2}{n^2}\text{ eV}$$

Species	Z	n	$(IE)_Z$
He^+	2	1	54.4 eV
H	1	1	13.6 eV
Li^{2+}	3	1	122.4 eV
Be^{3+}	4	1	217.6 eV

16. (a)

(a) $NaHCO_3$ is more thermally stable than $Ca(HCO_3)_2$.

$$Ca(HCO_3)_2 \longrightarrow CaCO_3 + H_2O + CO_2$$

$$CaCO_3 \xrightarrow{\Delta} CaO + CO_2$$

$$2NaHCO_3 \longrightarrow Na_2CO_3 + CO_2 + H_2O$$

$$Na_2CO_3 \xrightarrow{\Delta} \text{No reaction}$$

(b) $CaCO_3 \xrightarrow{\Delta} CaO + CO_2$,

$Na_2CO_3 \xrightarrow{\Delta}$ No reaction

(c) $BeCO_3$ is so much unstable that it is placed in the atmosphere of CO_2.

(d) As in (a).

17. (c) Li-Mg, Be-Al and B-Si show diagonal relationship.

Na and Mg do not show diagonal relationship.

18. (b) Activity as a reducing agent is based on E°_{red} (standard reduction potential) which is based on

$$M(s) \longrightarrow M(g) \quad \text{(Sublimation)}$$

$$M(g) \longrightarrow M^{2+}(g) + 2e^- \quad \text{(Ionisation)}$$

$$M^{2+}(g) + aq \longrightarrow M^{2+}(aq) \quad \text{(Hydration)}$$

(E°_{red}) Be = −1.85 V

Mg = −2.37 V

Ca = −2.87 V

Sr = −2.89 V

Ba = −2.90 V ↓ Maximum

19. (d)

(a) $Na(s) \longrightarrow Na^+(g) + e^-$

$Na^+(g) + (x+y)NH_3 \longrightarrow [Na(NH_3)_x]^+ + [e^-(NH_3)_y]$ (Solvated electrons)

$Mg(s) \longrightarrow Mg^{2+}(g) + 2e^-$

$Mg^{2+}(g) + (x+y)NH_3 \longrightarrow [Mg(NH_3)_x]^{2+} + [e^-(NH_3)_y]$

(b) Both are blue in colour due to solvated electrons thus, true.

(c) $[Na(NH_3)_x]^+ \xrightarrow{\text{Evaporation}} Na + xNH_3$

$[Mg(NH_3)_x]^{2+} \xrightarrow{\text{Evaporation}} Mg + xNH_3$

Thus, true.

(d) Concentrated solution of $[Na(NH_3)_x]^+$ is of copper coloured (bronze). $[Mg(NH_3)_x]^{2+}$ is also of bronze coloured. Thus, it is incorrect.

20. (b) Size $F^- < Cl^- < Br^- < I^-$

and $Mg^{2+} < Ca^{2+} < Sr^{2+} < Ba^{2+} < Ra^{2+}$

and F^- is smaller as compared to Mg^{2+}, Ca^{2+}, etc. The large cations touch one another without at the same time making contact with the F^- ion. As a result, lattice energies decrease unusually rapidly. For the other halides, the decrease of hydration energy becomes dominant. Thus, Statement I and Statement II are correct but Statement II is not the correct explanation of Statement I.

21. (a) $\Delta H_{solution}$ = Hydration energy – Lattice energy

For $BeSO_4$ and $MgSO_4$,

hydration energy > lattice energy

Hence, $\Delta H_{solution} > 0$

A solute is soluble if dissolution process is endothermic.

Thus, Statement I and Statement II are correct and Statement II is the correct explanation of Statement I.

22. (d)

(i) KCl – Violet Thus, (i) → (s)

(ii) $CaCl_2$ – Brick red. Thus, (ii) → (p)

(iii) $SrCl_2$ – Crimson red. Thus, (iii) → (q)

(iv) $BaCl_2$ – Apple green. Thus, (iv) → (r)

23. (b)

(i) $CaSO_4$ is called anhydrite and is formed when plaster of Paris is heated at 200°C. Thus, (i) → (s).

(ii) $CaSO_4 \cdot \frac{1}{2}H_2O$ is called plaster of Paris. It is formed when gypsum is heated at 150°C.

$$CaSO_4 \cdot 2H_2O \longrightarrow CaSO_4 \cdot \frac{1}{2}H_2O$$

It is also written as

$$2CaSO_4 \cdot 2H_2O \text{ or } (CaSO_4)_2 \cdot H_2O$$

Thus, (ii) → (r)

(iii) $CaSO_4 \cdot H_2O$ is called a labaster, a shiny marble and is used to make ornaments.

Thus, (iii) → (p)

(iv) $CaSO_4 \cdot 2H_2O$ is called gypsum.

Thus, (iv) → (q)

(v) → (r)

24. (c)

(i) $BeCO_3$ decomposes readily (hence placed in the atmosphere of CO_2 to prevent decomposition).

$$BeCO_3 \longrightarrow BeO + CO_2$$

Thus, (i) → (r)

(ii) $BaCl_2$ gives green edged flame.

Thus, (ii) → (s)

(iii) Permanent hardness is due to dissolved SO_4^{2-} and Cl^- of Ca^{2+} and Mg^{2+}.

Thus, (iii) → (q)

(iv) Temporary hardness is due to dissolved HCO_3^- of Ca^{2+} and Mg^{2+}.

Thus, (iv) → (p)

25. (a)

(i) Magnesite - $MgCO_3$. Thus, (i) → (r)

(ii) Dolomite - $MgCO_3 \cdot CaCO_3$. Thus, (ii) → (s)

(iii) Kieserite - $MgSO_4 \cdot H_2O$. Thus, (iii) → (q)

(iv) Carnallite - $KCl \cdot MgCl_2 \cdot 6H_2O$. Thus, (iv) → (p)

Revisal Problems (JEE Advanced)

1. (c) $\underset{\text{Gypsum}}{CaSO_4 \cdot 2H_2O} \xrightarrow{150°C} \underset{\text{Plaster of Paris (POP)}}{CaSO_4 \cdot \frac{1}{2}H_2O} \xrightarrow{200°C} \underset{\text{Anhydrite}}{CaSO_4}$

When POP $\left[CaSO_4 \cdot \frac{1}{2}H_2O \text{ as } (CaSO_4)_2 \cdot H_2O\right]$ is mixed with stoichiometric amount of water, it sets into a solid mass $CaSO_4 \cdot 2H_2O$ (gypsum).

Thus, (a), (b) and (d) are incorrect.

	Formula	Molar Mass	Ca(%)
Gypsum	$CaSO_4 \cdot 2H_2O$	172 g mol^{-1}	23.52
POP	$CaSO_4 \cdot \frac{1}{2}H_2O$	145 g mol^{-1}	27.59

Thus, (c) is correct.

2. (c) Smaller the size of cation, larger the charge, larger the hydration, larger the size of hydrated ion, thus, heavier the ion. Heavier ion moves slowly thus less conducting, lighter ion moves fastly and has more conducting.

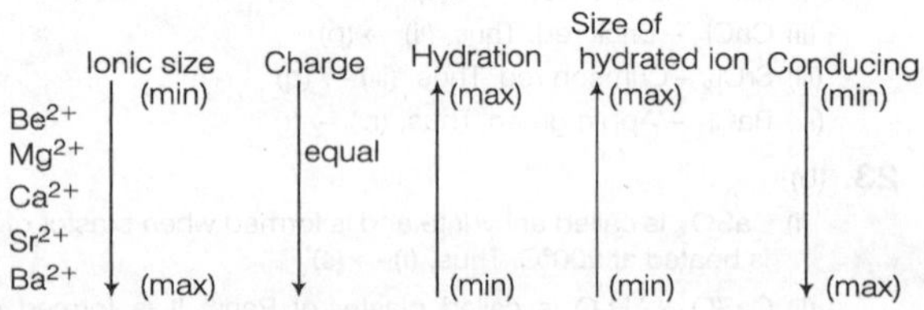

3. (a) Size of cation: $Be^{2+} < Mg^{2+} < Ca^{2+} < Ba^{2+}$

Smaller the size of cation, larger the charge, then larger the polarising power and hence, larger the covalent nature of compounds. Thus, covalent nature is in order

$$BeCl_2 > MgCl_2 > CaCl_2 > BaCl_2$$

4. (a) As cationic size increases, charge increases, polarising power decreases. Hence, lattice energy decreases. Thus,

lattice energy of

$$BeO > MgO > CaO > SrO > BaO$$
$$BeF_2 > MgF_2 > CaF_2 > SrF_2 > BaF_2$$
$$BeCO_3 > MgCO_3 > CaCO_3 > SrCO_3 > BaCO_3$$

5. (a) Smallor the size of the cation, larger the charge, then greater the hydration and thus, larger the hydration energy. Energy released stabilises the hydrated ion.

$$Be^{2+} > Mg^{2+} > Ca^{2+} > Sr^{2+} \text{ (Size)}$$
$$Be^{2+} > Mg^{2+} > Ca^{2+} > Sr^{2+} \text{ (Hydration)}$$
$$[Be(H_2O)_4]^{2+} > [Mg(H_2O)_4]^{2+} > [Ca(H_2O)_4]^{2+} > [Sr(H_2O)_4]^{2+}$$

(Stability of hydrated ion)

6. (b) $2CaCl(s) \longrightarrow CaCl_2(s) + Ca(s)$

$$\Delta H° = [\Delta H_f° (CaCl_2) + \Delta H_f° (Ca)] - 2\Delta H_f° (CaCl)$$
$$= (-795 + 0) - 2(-188) = -795 + 376 = -419 \text{ kJ mol}^{-1}$$

Since, $\Delta H° < 0$ (exothermic), hence reaction is spontaneous, hence $CaCl_2(s)$ exists. For reaction (a), $\Delta H° > 0$ (endothermic).

7. (d) Beryllium is least reducing agent as its $E°_{red}$ is less negative.

$$M^{2+}(aq) + 2e^- \longrightarrow M(s)$$

$$E°_{red} = -1.85 \text{ V for Be}$$
$$= -2.37 \text{ for Mg}$$
$$= -2.87 \text{ for Ca}$$
$$= -2.89 \text{ for Sr}$$
$$= -2.90 \text{ for Ba}$$

The particularly less value for Be arises from the large hydration energy associated with small size of Be^{2+} and relatively with large value of heat of sublimation. Thus, Statement I is incorrect and Statement II is correct.

8. (d) Smaller the size of cation, larger the size of anion, and larger the charge, then larger the polarising power and larger the covalent nature.

Compounds with smaller cation have larger covalent nature and lower melting and boiling points. Thus, Statement I is incorrect and Statement II is correct.

9. (b) Hydration and lattice energies decrease with increase in size of metal ions.

Decreasing lattice energy favours increased solubility, while decreasing hydration energy favours decreased solubility. On moving down the group, the hydration energy decreases more rapidly than the lattice energy, hence solubility of sulphates decreases.

$$\underset{\text{Soluble}}{BeSO_4} > MgSO_4 > CaSO_4 > SrSO_4 > \underset{\text{Insoluble}}{BaSO_4}$$

ΔH (Heat of solution) = Hydration energy – Lattice energy

Thus, Statement I and Statement II are correct but Statement II is not the correct explanation of Statement I.

10. (b) $Mg(s) \longrightarrow Mg^{2+}(aq) + 2e^-$

Mg, Ca, ... have more negative values of $E°_{red}$. This involves

$$M(s) \longrightarrow M(g) \quad \text{(Sublimation)}$$
$$M(g) \longrightarrow M^{2+}(g) + 2e^- \quad \text{(Ionisation)}$$
$$M^{2+}(g) + H_2O \longrightarrow M^{2+}(aq) \quad \text{(Hydration)}$$

This leads to more negative value of $E°_{red}$. Thus, these metals are oxidised and these are reducing agents.

Thus, Statement I and Statement II are correct but Statement II is not the correct explanation of Statement I.

11. (a,b,c,d) As electrons are removed from the atom, effective nuclear charge felt by the electrons is increased, hence ionisation energy increases.

Thus, $I_1 < I_2 < I_3$

$I_1' < I_2' < I_3'$

Thus, (a) and (b) are correct.

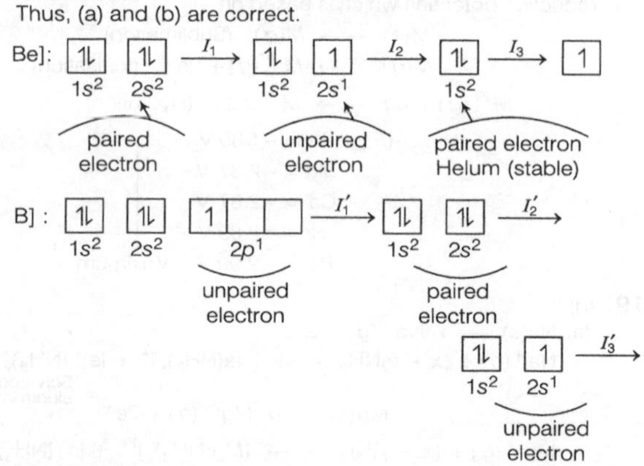

Energy required to remove paired electron is higher than that of unpaired electron. Also (IE) of inert gas is maximum.

12. (a,b,c,d)

(a) On moving from Be to Mg, electronegativity (EN) increases, and on moving from Mg to Al, EN decreases such that EN values of Be and Al are same.

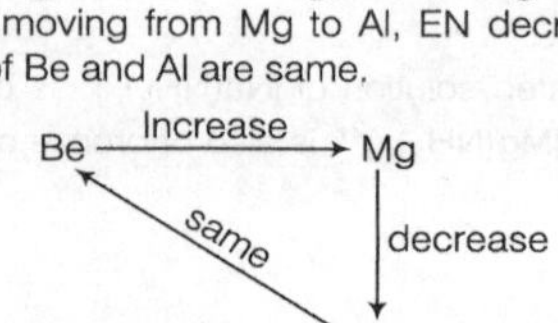

(b) BeO and Al_2O_3 both are soluble in acid as well as in alkali and thus are amphoteric in nature.

$$BeO + 2HCl \longrightarrow BeCl_2 + H_2O$$
$$BeO + 2NaOH \longrightarrow Na_2BeO_2 + H_2O$$
$$Al_2O_3 + 6HCl \longrightarrow 2AlCl_3 + 3H_2O$$
$$Al_2O_3 + 2NaOH \longrightarrow 2NaAlO_2 + H_2O$$

	Effective nuclear charge (Z)	Covalent radius (r)	$\frac{Z}{r}$
Be	1.95	89 pm	0.022
Al	3.50	143 pm	0.024

Thus, ratio of charge to radius ratio is same.

Note *Effective nuclear charge is based on screening effect.*

(d) BeO and Al_2O_3 are water insoluble.

13. (a,b,c,d)

(a) Due to high IE of Be and Mg, they are decomposed by steam and hot water.

$$Be + \underset{\text{Steam}}{H_2O} \longrightarrow BeO + H_2O$$

Ca, Sr and Ba with low IE are decomposed by cold water. Thus, correct.

(b) $3M + N_2 \longrightarrow M_3N_2$

Stability $Be_3N_2 > Mg_3N_2 > Ca_3N_2$. Thus, correct.

(c) Larger cation is stabilised by larger anion hence, BaO_2 is stable. Be, Mg, Ca, Sr form only oxide. Thus, correct.

(d) BeH_2 and MgH_2 are covalent, electron deficient. Other hydrides are ionic. Thus, correct.

14. (a, c, d)

(a) Thermal stability of sulphates and carbonates increases as one moves from Be to Ba.

$BeSO_4 < MgSO_4 < CaSO_4 < SrSO_4 < BaSO_4$

$BeCO_3 < MgCO_3 < CaCO_3 < SrCO_3 > BaCO_3$

Thus, correct.

(b) $2Mg(NO_3)_2 \xrightarrow{\Delta} 2MgO + 4NO_2 + O_2$

$3LiNO_3 \xrightarrow{\Delta} 2Li_2O + 4NO_2 + O_2$

Both give same products. Thus, statement (b) is incorrect.

(c) Ionic size of Group 2 cations is smaller than that of Group 1. Thus, polarising power of M^{2+} (Group 2) < M^+ (Group 1).

Hence, lattice energy of Group 2 salts is higher than that of Group 1 salts.

(d) Being smaller in ionic size of Group 2 than Group 1, hence Group 2 cations are more hydrated than Group 1 cations, hence hydration energies of Group 2 are higher than Group 1. Thus, correct.

15. (b,c,d) Temporary hardness of water is due to dissolved bicarbonates of Ca^{2+} and Mg^{2+}.

(b) $Ca(OH)_2 + Ca(HCO_3)_2 \longrightarrow 2CaCO_3 + 2H_2O$

(c) $NaOCl + H_2O \longrightarrow NaOH + HCl$

NaOH formed decomposes HCO_3^-

$$HCO_3^- + OH^- \longrightarrow H_2O + CO_2$$

(d) $Ca(HCO_3)_2 + Na_2CO_3 \longrightarrow CaCO_3\downarrow + 2NaHCO_3$

16. (a) $MgSO_4 + NH_4OH + Na_2HPO_4 \longrightarrow \underset{\text{(White ppt)}}{Mg(NH_4)PO_4\downarrow} + Na_2SO_4 + H_2O$

17. (a,c,d) $4LiNO_3 \longrightarrow 2Li_2O + 4NO_2 + O_2$

$2Mg(NO_3)_2 \longrightarrow 2MgO + 4NO_2 + O_2$

Li and Mg resemble each other due to diagonal relationship, hence $LiNO_3$ also decomposes to form oxide, NO_2 and O_2.

Ca and Mg are of same group 2, thus $Ca(NO_3)_2$ and $Mg(NO_3)_2$ both decompose to form NO_2, O_2 and oxide.

18. (a,c,d)

(a) $CaCO_3 \xrightarrow{\Delta} \underset{\text{Basic}}{CaO} + \underset{\text{Acidic}}{CO_2}$

(b) $Na_2CO_3 \xrightarrow{\Delta}$ No effect

(c) $2LiNO_3 \xrightarrow{\Delta} \underset{\text{Basic}}{Li_2O} + \underset{\text{Acidic}}{2NO_2} + \frac{1}{2}O_2$

(d) $MgCO_3 \xrightarrow{\Delta} \underset{\text{Basic}}{MgO} + \underset{\text{Acidic}}{CO_2}$

19. (d)

(i) Be shows diagonal relationship with Al.

Be^{2+} with smallest size is maximum hydrated.

Thus, (i) → (p, r)

(ii) Mg shows diagonal relationship with Li.Mg^{2+} ions are involved in transmission of electrical impulses in human body.

Thus, (ii) → (p, t).

(iii) Ca^{2+} ions are also involved in transmission. Thus, (iii) → (t).

(iv) $Be^{2+} + 2e^- \longrightarrow Be, \quad E° = -1.85\,V$

$Mg^{2+} + 2e^- \longrightarrow Mg, \quad E° = -2.37\,V$

$Ca^{2+} + 2e^- \longrightarrow Ca, \quad E° = -2.87\,V$

$Sr^{2+} + 2e^- \longrightarrow Sr, \quad E° = -2.89\,V$

Sr^{2+} has maximum $E°$.

Being largest in ionic size, Sr^{2+} is least hydrated hence has maximum ionic mobility.

Thus, (iv) → (q, s)

20. (b)

(i) Magnesia - MgO. Thus, (i) → (t)

(ii) Epsom salt - $MgSO_4 \cdot 7H_2O$. Thus, (ii) → (r)

(iii) Anhydrone - $Mg(ClO_4)_2$. Thus, (iii) → (u)

(iv) Hydrolith - CaH_2. Thus (iv) → (s)

(v) Beryllia - BeO. Thus, (v) → (p)

(vi) Slaked lime - $Ca(OH)_2$. Thus, (vi) → (q)

21. (c)

(i) Al_2O_3 - 40 Thus, (i) → (p)

(ii) CaO - 40 Thus, (ii) → (p)

(iii) SiO_2 - 10 Thus, (iii) → (q)

(iv) Fe_2O_3 - 10 Thus, (iv) → (q)

Explanations for Q. Nos. (22 and 23)

Beryllium salts are extensively hydrolysed to give a series of hydroxo complexes of the type

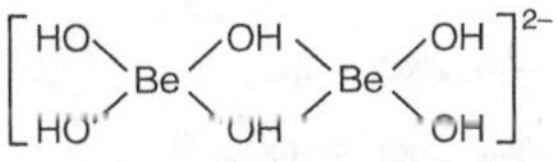

On adding alkali to these solutions, the polymers breakdown to $[Be(OH)_4]^{2-}$ and thus, it assumes tetrahedral structure.

$$[Be_2(OH)_6]^{2-} + 2OH^- \longrightarrow [Be(OH)_4]^{2-}$$

22. (b) **23.** (a)

Explanations for Q. Nos. (24 and 25)

$$Ca + N_2 \longrightarrow Ca_3N_2$$
$$Ca_3N_2 + 6H_2O \longrightarrow \underset{\text{(Alkaline)}}{3Ca(OH)_2} + \underset{(B)}{2NH_3}$$
$$Ca(OH)_2 + \underset{\text{(From air)}}{CO_2} \longrightarrow \underset{(C)}{CaCO_3} + H_2O$$

24. (b)

25. (c)

Explanations for Q. Nos. (26 and 27)

$$CaO + SO_2 \longrightarrow CaSO_3$$
$$Ca(OH)_2 + SO_2 \longrightarrow CaSO_3 + H_2O$$

SO_2 turns acidified $K_2Cr_2O_7$ solution green.

$$Cr_2O_7^{2-} + 3SO_2 + 2H^+ \longrightarrow \underset{\text{Green}}{2Cr^{3+}} + 3SO_4^{2-} + H_2O$$

SO_2 decolourises acidified $KMnO_4$ solution

$$2MnO_4^- + 5SO_2 + 2H_2O \longrightarrow \underset{\text{Colourless}}{2Mn^{2+}} + 5SO_4^{2-} + 4H^+$$

26. (a) **27.** (c)

28. (4) Maximum covalency of Be is 4 since, one 2*s* and three 2*p*- orbitals are available for bonding.

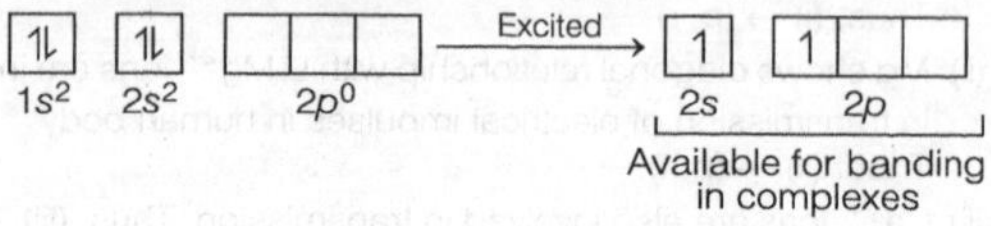

29. (5) Magnesium nitride is Mg_3N_2.

$$(x + y) = 3 + 2 = 5$$

30. (4) Mg is coordinated to four N-atoms in chlorophyll.

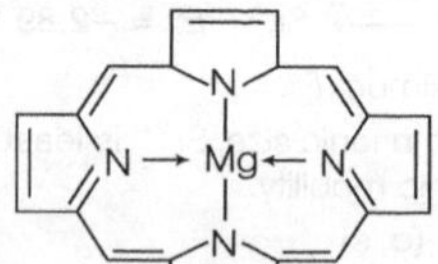

31. (4) There are total eight rings in which four are five-membered and four are six-membered (with Mg as one of the atoms).

32. (5) Magnesium allyide is Mg_2C_3.

$$Mg_2C_3 + 4H_2O \longrightarrow 2Mg(OH)_2 + CH_3C \equiv CH$$

Thus, there are five atoms (two Mg and three C) in one formula unit of magnesium allyide.

$$(Mg^{2+})(C_3^{4-}),\ {}^{3-}C-C\equiv C^-$$

33. (5) $Mg + \frac{1}{2}O_2 \longrightarrow MgO$

$$MgO + H_2O \longrightarrow Mg(OH)_2$$

MgO formed from 25% of Mg $= 96 \times \frac{25}{100} = 24g$

Thus, MgO or $Mg(OH)_2$ formed = 1 mole

$$\underset{\text{1 mole}}{Mg(OH)_2} + \underset{\text{1 mole}}{H_2SO_4} \longrightarrow MgSO_4 + 2H_2O$$

H_2SO_4 used by 1 mole of $Mg(OH)_2$ = 1 mole

Mg left $= 96 - 24 = 72g$

$= \frac{72}{24} = 3$ moles

$$3Mg + N_2 \longrightarrow Mg_3N_2$$

Other residue formed is Mg_3N_2

Mg_3N_2 formed = 1 mole

$$Mg_3N_2 + H_2O \longrightarrow 3Mg(OH)_2 + 2NH_3$$
$$\underset{\substack{\text{1 mole}\\ \text{3 moles}}}{Mg(OH)_2} + \underset{\substack{\text{1 mole}\\ \text{3 moles}}}{H_2SO_4} \longrightarrow MgSO_4 + 2H_2O$$
$$\underset{\text{2 moles}}{2NH_3} + \underset{\text{1 mole}}{H_2SO_4} \longrightarrow (NH_4)_2SO_4$$

Mg_3N_2 on reaction with H_2O gives $Mg(OH)_2$ and NH_3.

H_2SO_4 required by $Mg(OH)_2$ = 3 moles

H_2SO_4 required by NH_3 = 1 mole

Thus, total moles required = 1 + 3 + 1 = 5 moles

34. (4)

Per	cent	Moles	Ratio	Valency
Mg	8.54	0.355	1	+2
Ba	48.75	0.355	1	+2
CO_3	42.71	0.710	2	−2

Thus, $MgBa(CO_3)_2$, molar mass = 281 g mol^{-1}

(+4) valency of Mg^{2+} and Ba^{2+} is satisfied by (−4) of two CO_3^{2-}.

Thus, $(x + y + z) = 1 + 1 + 2 = 4$.

JEE Archive

1. (c)

(a) $2NaNO_3 \xrightarrow{\Delta} 2NaNO_2 + O_2$

(b) $2KClO_3 \xrightarrow{\Delta} 2KCl + 3O_2$

(c) $CaCO_3 \xrightarrow{\Delta} \underset{\text{Basic}}{CaO} + \underset{\text{Acidic}}{CO_2}$

(d) $NH_4NO_3 \xrightarrow{\Delta} N_2O + H_2O$

2. (c) $\underset{\text{1 mol}}{Mg_3N_2(s)} + 6H_2O(l) \longrightarrow 3Mg(OH)_2 + \underset{\text{2 mol}}{2NH_3(g)}$

3. (a) Be (Z = 4) has maximum covalency of 4 while Al (Z = 13) has maximum covalency of 6.

4. (b) The solubilities of carbonates of the alkaline earth metals decrease down the group mainly due to decreasing hydration ethalpies of the cartions from Be^{2+} to Ba^{2+}.

5. (b) Calcined gypsum does not contain $CaCO_3$ as gypsum is $CaSO_4 \cdot 2H_2O$. i.e., it is sulphate.

6. (b) The calcium aluminates and silicates present in cement on hydration gives corresponding hydrated colloidal gel.

$$\underset{\text{Tricalcium aluminate}}{3CaO.Al_2O_3} + 6H_2O \longrightarrow \underset{\text{Hydrated colloidal gel of tricalcium aluminate}}{3CaO \cdot Al_2O_3 \cdot 6H_2O}$$
$$2CaO \cdot SiO_2 + xH_2O \longrightarrow \underset{\text{Hydrated colloidal gel of dicalcium silicate}}{2CaO \cdot SiO_2 \cdot xH_2O}$$

The hydration of calcium aluminates and silicates involved in the setting of cement is an exothermic process, hence cement structures have to be cooled during setting by sprinkling water.

7. (a,b) Both $Be(OH)_2$ and $Mg(OH)_2$ being insoluble are obtained from NaOH solution through precipitation.

8. (b) Mud is colloidal charged solution which on treatment with alum (Al^{3+}) gets neutralised or coagulated.

9. (b) Magnesium-ammonium phosphate is precipitated out

$$MgSO_4 + NH_4OH + Na_2HPO_4 \longrightarrow Mg(NH_4)PO_4 \downarrow + Na_2SO_4$$

5. Group 13 Elements

DPP-1 Periodicity in Group 13

1. (b) Ga(31) : $[Ar]4s^2 3d^{10} 4p^1$

(a) If aufbau rule is not followed, then EC is

$$[Ar]3d^{10}4s^2 4p^1$$

again p-block element.

(b) $1s^3\ 2s^3\ 2p^9\ 3s^3\ 3p^9\ 4s^3\ 3d^1$ -

If each orbital has 3 elements, then last electron goes into $3d$ thus, Ga would be considered as d block element.

2. (a) As we go from B to Tl, basic nature increases, acidic nature decreases.

B_2O_3 - Acidic
Al_2O_3 - Amphoteric
Ga_2O_3 - Amphoteric
In_2O_3 - Basic
Tl_2O_3 - Basic

3. (a) Polarity is in order : $BF_3 > BCl_3 > BBr_3$

Lewis acidity is in order : $BF_3 < BCl_3 < BBr_3$

Thus, stability of adduct is in order

$$(BF_3 \leftarrow L) < (BCl_3 \leftarrow L) < (BBr_3 \leftarrow L)$$

4. (c) $B(g) \longrightarrow B^{3+}(g) + 3e^-, (I_1 + I_2 + I_3)$

Three steps (IE) are very high. Smaller cation with larger charge is highly polarising, hence even water of hydration would be appreciably deformed.

5. (a) B(5) : $1s^2 2s^2 2p^1$, $B^{3+}(1s^2)$: [He]

Al(13) : $1s^2 2s^2 2p^6 3s^2 3p^1$

Al^{3+} : $(1s^2 2s^2 2p^6)$: [Ne]

Ga(31) : $[Ar]3d^{10}4s^2 4p^1$

Ga^{3+} : $[Ar]3d^{10}$

In(49) : $[Kr]4d^{10}5s^2 5p^1$

In^{3+} : $[Kr]4d^{10}$

Tl(81) : $[Xe]4f^{14}5d^{10}6s^2 6p^1$

Tl^{3+} : $[Xe]4f^{14}5d^{10}$

Due to the presence of d-and f-electrons, screening effect decreases. Force of attraction between valence electrons and nucleus increases. Hence, properties of group 13 (Ga, In, Tl) are quite different from group 1 and group 2 elements in which core shells have s^2p^6 configuration (except Li and Be).

6. (a) As we go down the group, there is the occurrence of d and f orbitals. Screening effect due to $s > p > d > f$, d^{10} and f^{14} core electrons decrease screening effect. Hence, valence electrons are tightly bonded resulting in increase in IE.

7. (a) B (5)

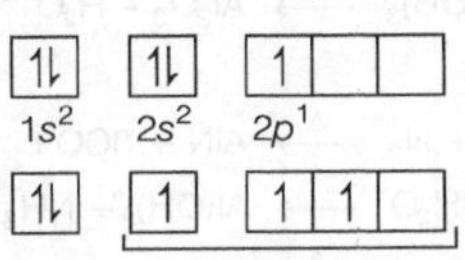

Ground state: $1s^2$ $2s^2$ $2p^1$

First excited state

$2d$ is absent and there are one $2s$ and three $2p$ orbitals - in all four orbitals are available for bonding. In other words, maximum covalency of B is four. Thus,

(a) BF_6^{3-} is not formed.

(b) BH_4^- is formed as $NaBH_4$.

(c) $[B(OH)_4]^-$ is formed as $Na[B(OH)_4]$.

(d) BO_2^- is formed as $NaBO_2$.

8. (c) Lone-pair on F in BF_3 is delocalised hence, (B—F) bond has double bond character. Thus, due to $p\pi$-$p\pi$ interaction between B and F in BF_3 leads to stronger B—F bond in BF_3 as compared to CF_4.

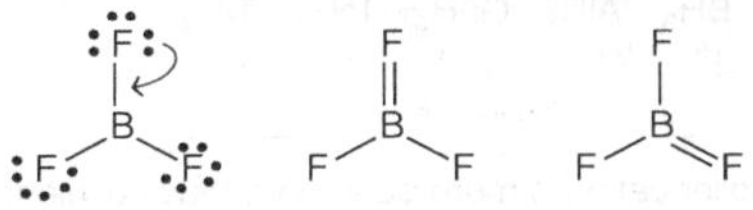

9. (a) B and Al of group 13 form nitrides which are decomposed by H_2O forming NH_3.

$$2B + N_2 \longrightarrow 2BN \xrightarrow{H_2O} NH_3$$

$$2Al + N_2 \longrightarrow 2AlN \xrightarrow{H_2O} NH_3$$

Mg also forms nitride (Mg_3N_2) but Mg belongs to group 2 (alkaline earth metals).

10. (a) As a general rule, electronegativity (EN) decreases down the group as in B to Al but after Al, (EN) increases. In Ga and In, the presence of d^{10} core electron and in Tl presence of d^{10} and f^{14} core electrons causes decrease in screening effect and, thus force of attraction for the added electron is increased. This increases

(i) r_n (ii) (EN) (iii) (IE)

Thus, Statement I and Statement II are correct and Statement II is the correct explanation of Statement I.

11. (c) When we move from Al([Ne] $3s^2 3p^1$) to Ga([Ar] $3d^{10}4s^2 4p^1$) there is increase in electrons in $3d$-orbital. Thus, screening effect decreases ($s > p > d > f$) and force of attraction between valence shell electron and nucleus increases to a greater extent as compared to Al. Thus, atomic radius of Ga is smaller than that of Al.

Thus, Statement I is correct and Statement II is incorrect.

12. (a) Smaller the size of cation, larger the size of anion and larger the charge, then larger the polarising power. This makes the compound containing such ions covalent.

B^{3+} with smaller size and higher charge, makes B^{3+} high polarising. Hence, BCl_3, etc., are covalent in nature.

Thus, Statement I and Statement II are correct and Statement II is the correct explanation of Statement I.

13. (b) As we go down the group, screening effect (SE) decreases. Hence, force of attraction between valence shell electron and nucleus increases. SE is in order

$$s > p > d > f$$

Tendency to unpair ns^2 pair increases.

Group 13 :

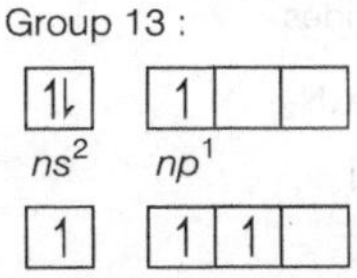

Valency = +1, if ns^2 pair is inert. (ns^2 np^1)

Valency = +3, if ns^2 pair is unpaired

Stability of $Ga^{3+} > In^{3+} > Tl^{3+}$

$$Ga^+ < In^+ < Tl^+$$

$$Ga^+ \longrightarrow Ga^{3+} + 2e^-, \text{ spontaneous}$$

$$Tl^{3+} + 2e^- \longrightarrow Tl^+, \text{ spontaneous}$$

$TlCl_3$ is thus less stable than TlCl.

Thus, Statements I and Statements II are correct but Statement II is not the correct explanation of Statement I.

14. (b,c)

(a) $AlH_3 + H^- \longrightarrow AlH_4^-$

$\uparrow$ sp^2, $\theta = 120°$; $\uparrow$ sp^3, $\theta = 109°28'$

When AlH_3 (with sp^2 hybridised Al) changes to AlH_4^- (with sp^3 hybridised Al), bond angle decreases from 120° to 109° 28′. Thus, it is incorrect.

(b) BH_3 AlH_3 GaH_3 InH_3 TlH_3

B^{3+} Al^{3+} Ga^{3+} In^{3+} Tl^{3+}

Size increases $\longrightarrow$

Smaller cation is more polarising, thus covalent nature of MH_3 decreases, hence stability of hydride decreases moving from B to Tl. Thus, it is correct.

(c) Size of anion

F^-, Cl^-, Br^-, I^- $\downarrow$ Increases

As size of anion increases, polarising power increases. Hence, MF_3 (with common M) is ionic while MCl_3, MBr_3 and MI_3 are covalent.

Also,

Ionic	Covalent
BF_3	BF_3
AlF_3	BCl_3
GaF_3	BBr_3
InF_3	BI_3 $\downarrow$
TlF_3 $\downarrow$	

Thus, it is correct.

(d) $[M(H_2O)_6]^{3+}$ is formed and not $[M(H_2O)_4]^{3+}$. It has octahedral geometry due to sp^3d^2 hybridisation. Thus, it is incorrect.

15. (a,b) As per inert-pair effect, stability of p-block elements decreases for G-state and increases for (G-2) states.

where, G = valence electrons

For group 13, Ga In Tl } Valence electrons $G = 3(ns^2np^1)$

Thus, $Ga^{3+} > In^{3+} > Tl^{3+}$

and $Ga^+ < In^+ < Tl^+$

Thus, $Tl^{3+} + 2e^- \longrightarrow Tl^+$

Tl^{3+} is reduced to Tl^+ (more stable ion) hence Tl^{3+} salts are better oxidising agents.

$Ga^+ \longrightarrow Ga^{3+} + 2e^-$

Ga^+ is oxidised to Ga^{3+} (more stable ion), hence Ga^+ salts are better reducing agents.

16. (a,b,c)

$Al^{3+}(aq) + 3e^- \longrightarrow Al(s)$, $E°_{Al^{3+}/Al} = -1.66V$

$Tl^{3+}(aq) + 3e^- \longrightarrow Tl(s)$, $E°_{Tl^{3+}/Tl} = +1.26V$

$Al(s) \longrightarrow Al^{3+}(aq) + 3e^-$, $E°_{Al/Al^{3+}} = 1.66V$

Since, $E° > 0$, hence Al is oxidised easily thus, Al is an electropositive metal.

$Tl(s) \longrightarrow Tl^{3+}(aq) + 3e^-$, $E°_{Tl/Tl^{3+}} = -1.26V$

Since, $E° < 0$, hence Tl is not oxidised to Tl^{3+}.

Since, Tl^{3+} salts change to Tl^+ easily, hence Tl^{3+} salts are better oxidising agents.

17. (b)

(i) $\rightarrow$ (t)
(ii) $\rightarrow$ (s)
(iii) $\rightarrow$ (r)
(iv) $\rightarrow$ (q)
(v) $\rightarrow$ (p)

Explanations for Q. Nos. (18 and 19)

Donor molecules like C_5H_5N may react with BX_3 adduct to increase the effective size of boron. This is due to mainly by

(i) the reduction of polarising power.
(ii) the compensation of the (IE).

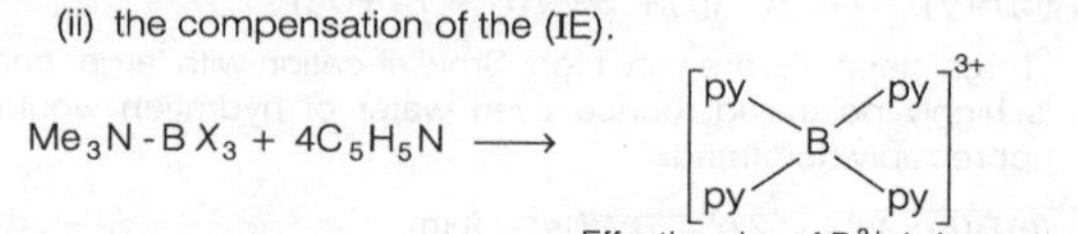

Effective size of B^{3+} is increased.

Note AlH_3, $AlCl_3$ *and* $GaCl_3$ *are Lewis acids (electron-deficients).*

18. (c) **19.** (a)

20. (4)

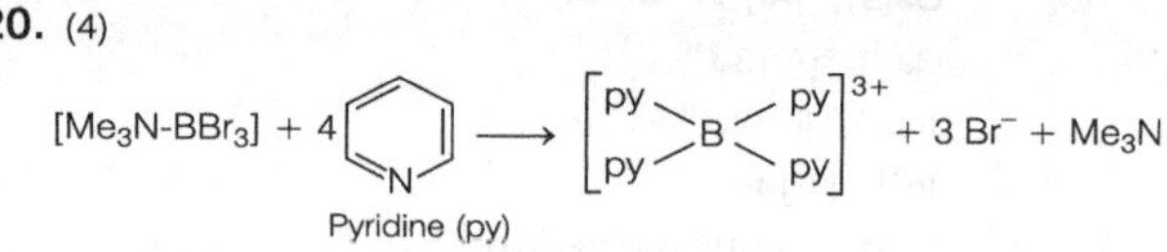

Total ions formed = $[B(py_4)]^{3+}$ - One

= Br^- - Three

= Four

DPP-2 Compounds of Group 13

1. (b) Mg and B react with N_2 forming nitrides.

$$3Mg + N_2 \xrightarrow{\Delta} Mg_3N_2$$

$$2B + N_2 \xrightarrow{\Delta} BN$$

Mg also reacts with B forming borides.

$$Mg + B \longrightarrow Mg_3B_2$$

2. (c)

(a) **Baeyer's process** Fe_2O_3 and TiO_2 are magnetic impurities in alumina. Al_2O_3 is amphoteric and dissolves in either alkali or acid.

$$Al_2O_3 + 2OH^- + 3H_2O \longrightarrow 2[Al(OH)_4]^-$$

$[Al(OH)_4]^-$ can be precipitated as $Al(OH)_3$ by acid or passing CO_2 into it.

$$[Al(OH)_4]^- + H^+ \longrightarrow Al(OH)_3\downarrow + H_2O$$

$$2[Al(OH)_4]^- + CO_2 \longrightarrow 2Al(OH)_3\downarrow + CO_3^{2-} + H_2O$$

$Al(OH)_3$ on heating forms pure Al_2O_3.

$$2Al(OH)_3 \xrightarrow{\Delta} Al_2O_3 + H_2O$$

(b) **Serpeck's process**

$$Al_2O_3 + 3C + N_2 \xrightarrow{\Delta} AlN + 3CO\uparrow$$

$$AlN + 3H_2O \longrightarrow Al(OH)_3 + NH_3$$

$$2Al(OH)_3 \xrightarrow{\Delta} Al_2O_3 + 3H_2O$$

3. (a) Orthoboric acid (H_3BO_3) is a weak monobasic acid due to $p\pi$-$p\pi$ back-bonding between B and O. Direct neutralisation with alkali is not complete. In the presence of *cis*-1, 2-diol, a stable complex is formed and reaction goes to completion.

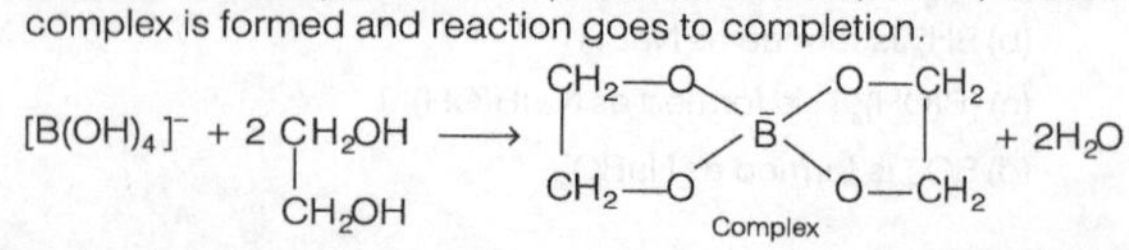

4. (c) $3B(OH)_3 \rightleftharpoons H_3O^+ + [B_3O_3(OH)_4]^- + H_2O$

Orthoboric acid — Polymeric metaborate

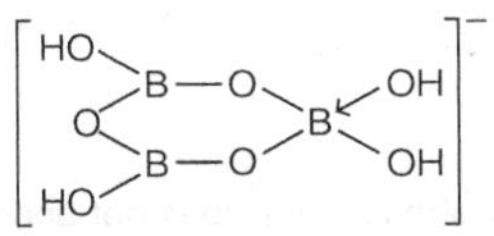

Note *It has total of ten* (B—O) *bonds.*

5. (c)

I. $BCl_3 + 3H_2O \longrightarrow H_3BO_3 + 3HCl$

$SiCl_4 + 4H_2O \longrightarrow H_4SiO_4 + 4HCl$

Thus, Statement I is correct.

II. Mg_3B_2 (magnesium boride; boride as anion) Al_4Si_3 (aluminium silicide, silicide as anion) are binary salts. Thus, Statement II is correct.

III. $B_2O_3 + 2NaOH \longrightarrow 2NaBO_2 + H_2O$

$SiO_2 + 2NaOH \longrightarrow Na_2SiO_3 + H_2O$

Oxides being soluble in NaOH are acidic in nature.

H_3BO_3 and H_4SiO_4 are weak acids.

Thus, III is correct.

IV. B and Si both are non-metals. Thus, IV is incorrect

V. B and Si both are semiconductors. They dissolve in alkali.

$2B + 2NaOH + 2H_2O \longrightarrow 2NaBO_2 + 3H_2$

$Si + 2NaOH + H_2O \longrightarrow Na_2SiO_3 + H_2$

Thus, V is correct.

VI. B and Si exist as crystalline and amorphous forms. Thus, VI is correct.

6. (c) $\Delta H_{\text{hydration}}(AlCl_3) = \Delta H(Al^{3+}) + 3\Delta H(Cl^-)$

$= -4665 + 3(-381) = -4665 - 1143$

$= -5808 \text{ kJ mol}^{-1}$

ΔH (ionisation of Al) $= 5137 \text{ kJ mol}^{-1}$

Hydration energy is larger than ionisation energy, hence $AlCl_3$ ionises and Al^{3+} ions are hydrated at the same time in aqueous solution.

$AlCl_3(aq) \longrightarrow [Al(H_2O)_6]^{3+} + 3Cl^-$

Note *Maximum covalency of* Al = six

7. (a) $B_2O_3 + 6Na \xrightarrow{\Delta} 2B + 3Na_2O$

$B_2O_3 + 3Mg \longrightarrow 2B + 3MgO$

Note *Amorphous boron obtained is 95-98% pure and black in colour.*

B_2O_3 used in reduction, is formed from borax.

$\underset{\text{Borax}}{Na_2B_4O_7 \cdot 10H_2O} + HCl \longrightarrow \underset{\text{Boric acid}}{H_3BO_3} \xrightarrow{\Delta} H_2O + B_2O_3$

8. (c) Boron compounds are covalent and electron deficient. Boron is smaller in size and high charge density, hence due to high polarising power of B^{3+}, compounds are covalent.

Also, $B(g) \longrightarrow B^{3+}(g) + 3e^-$

$\Delta H = I_1 + I_2 + I_3$

$= 801 + 2427 + 3659 = 6887 \text{ kJ mol}^{-1}$

To convert $B(g)$ into $B^{3+}(g)$, very high energy is required. Hence, formation of $B^{3+}(g)$ is difficult. This process is endothermic.

Thus, Statement I is correct but Statement II is incorrect.

9. (a,c,d)

(a) $Na_2B_4O_7 + 2NH_4Cl \longrightarrow 2BN + 4H_2O + 2NaCl + B_2O_3$

Thus, correct.

(b) $B_2H_6 + NH_3 \xrightarrow[\Delta]{\text{High temperature}} BN$

It is not formed at low temperature.

(c) BN (total electrons = 5 + 7 = 12) and C_2 (total electrons = 6 + 6 = 12) are isoelectronic. Thus, $(BN)_x$ has same structure as graphite (C_2) made up of hexagonal rings of alternate B and N atoms joined together. (BN) is called inorganic graphite. Thus, correct.

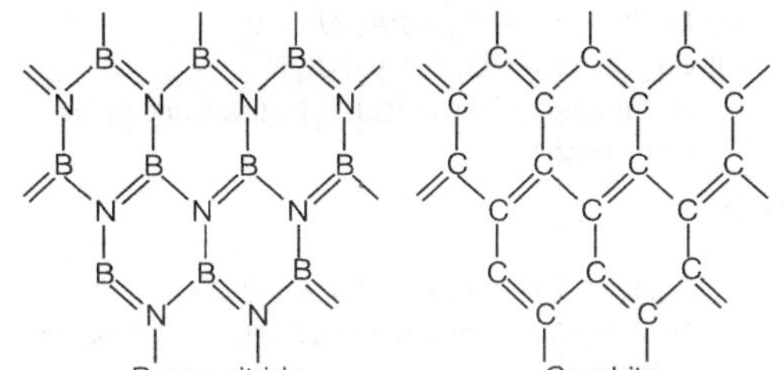

Boron-nitride — Graphite

(d) Graphite is conducting while $(BN)_x$ is non-conducting. Thus, correct.

10. (a,b,d) B_2O_3 forms salts with acids which are stronger than H_3BO_3. Except phenol (C_6H_5OH), all are stronger acids than it.

$B_2O_3 + 2H_3PO_4 \longrightarrow 2BPO_4 + 3H_2O$

$B_2O_3 + 6CH_3COOH \longrightarrow 2B(CH_3COO)_3 + 3H_2O$

$B_2O_3 + 6HNO_3 \longrightarrow 2B(NO_3)_3 + 3H_2O$

11. (a,c,d) A coloured bead is formed from salts which on heating can formed coloured oxide.

Thus, this dry test is only applicable to transition metal salts.

$MnO + B_2O_3 \longrightarrow \underset{\text{Violet}}{Mn(BO_2)_2}$

$CoO + B_2O_3 \longrightarrow \underset{\text{Blue}}{Co(BO_2)_2}$

$Cr_2O_3 + 3B_2O_3 \longrightarrow \underset{\text{Green}}{2Cr(BO_2)_3}$

(Colour in cold oxidising flame)

12. (a,b,c,d)

(a) pH = 13, it means basic medium

$Al(OH)_3 \rightleftharpoons [AlO_2^-] + H_3O^+$

$Al(OH)_3 + OH^- \longrightarrow [Al(OH)_4]^-$

Thus, correct.

(b) Between pH 8 and 12 (which is weakly basic), it exists as octahedrally coordinated aluminium. Thus, correct.

(c) $2[Al(OH)_4]^- \longrightarrow [(HO)_3Al{-}O{-}Al(OH)_3]^{2-}$

Thus, correct.

(d) pH about 2 means acidic medium.

It dissolves forming Al^{3+} and thus hydrated.

$[Al(OH)_4^-] + 4H^+ \longrightarrow Al^{3+} + 4H_2O$

Thus, correct.

13. (a,c,d)

(a) Alum is $M_2SO_4 \cdot M'_2(SO_4)_3 \cdot 24H_2O$.

If $M^+ = K^+$, $(M')^{3+} = Al^{3+}$, then it is

$[K(H_2O)_6]_2SO_4 \cdot [Al(H_2O)_6]_2(SO_4)_3$

Thus, $[K(H_2O)_6]$, $[Al(H_2O)_6]$ and SO_4^{2-} are in 1 : 1 : 2 molar ratio.

Thus, correct.

(b) M^+ (alkali metal) can be ,... Na^+, K^+ but not Li^+, being too small in ionic radius to meet the structural requirement of the crystal. Thus, incorrect.

(c) Correct

(d) The fabric to be dyed is dipped into a solution of alum and heated with steam. Hydrolysis of $[Al(H_2O)_6]^{3+}$ deposits $Al(OH)_3$ into the fibres of the material and the dye is adsorbed on $Al(OH)_3$. Thus, alums are used as mordant in dyes.

Thus, correct.

14. (a,b,c,d)

(a) $2Al + AlCl_3 \xrightarrow{1300\,K} 2AlCl(g) \xrightarrow{Cool} AlCl(s)$

The reaction is spontaneous.

(b) $2Al + 2HCl \xrightarrow{1200\,K} 2AlCl(red) + H_2(g)$

(c) $3AlCl \longrightarrow AlCl_3 + 2Al, \Delta H < 0$

(d) $Ga_2Cl_4 \rightleftharpoons Ga^+ + [GaCl_4]^-$

Ga^+ (monovalent Ga) $[GaCl_4]^-$ (trivalent Ga) are diamagnetic.

15. (a,b,d)

(a) $B_3N_3H_6$: Total electrons $= 3 \times 5 + 3 \times 7 + 6 \times 1 = 42$

C_6H_6 : Total electrons $= 6 \times 6 + 6 = 42$

Thus, $B_3N_3H_6$ (borazine) and C_6H_6 are isoelectronic. Thus, correct.

(b) Due to resonance, (B—N) distances are equal = 144 pm. Thus, correct.

(c) In borazine, nitrogen π-orbitals are much lower in energy than the π-orbitals of boron. Thus, perfect delocalisation of π-electrons does not take place. Thus, incorrect.

(d) Due to availability of π electrons in borazine (as compared to benzene in which π electrons are delocalised), borazine is more reactive than benzene. Thus, correct.

16. (b,c) $Al_2O_3 \longrightarrow 2Al^{3+} + 3O^{2-}$

Anode is a graphite plate.

$$C + 2O^{2-} \longrightarrow CO_2 + 4e^-$$

$$2O^{2-} \longrightarrow O_2 + 4e^-$$

CO_2 and/or O_2 can collect at anode.

Cathode $Al^{3+} + 3e^- \longrightarrow Al$

Al collects at cathode.

17. (b,c,d)

(a) All (B—H) bonds are not equal.

(B—H) bonds formed by two electrons are shorter than that of (B—H) bonds formed by one electron.

Thus, incorrect.

(b) 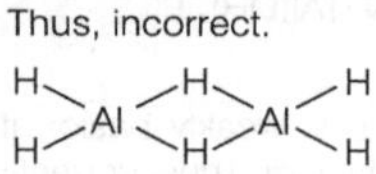 Thus, correct.

(c) $LiAlH_4 \rightleftharpoons Li^+ + AlH_4^-$

Thus, correct.

(d) AlX_3 are all covalent and exist as dimeric (Al_2X_6).

Thus, correct.

18. (a,d) B—H bond is formed by two electrons and B ··· H bond is formed by one electron.

Thus, $l_1 > l_2$ ($l_1 = 1.33$ Å, $l_2 = 1.19$ Å)

Also, $\theta_1 (120°) > \theta_2 (97°)$

19. (a,b,c) Product is formed by unsymmetrical cleavage of B_2H_6 by X^- such that BH_2 goes with X and BH_4 is an anion.

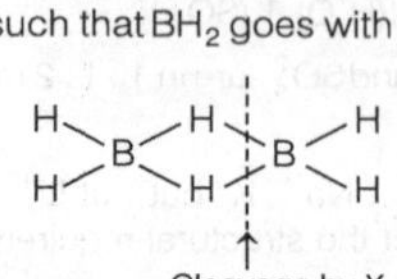

$\xrightarrow{X^-} BH_4^- + [BH_2(X_2)]^+$

This cleavage is possible when X is NH_3, RNH_2 (1° amine) or R_2NH (2° amine).

$$B_2H_6 + 2NH_3 \longrightarrow [BH_2(NH_3)_2]^+ + BH_4^-$$

$$B_2H_6 + 2RNH_2 \longrightarrow [BH_2(RNH_2)_2]^+ + BH_4^-$$

$$B_2H_6 + 2R_2NH \longrightarrow [BH_2(R_2NH)_2]^+ + BH_4^-$$

With 3° amine (R_3N), symmetrical cleavage takes place

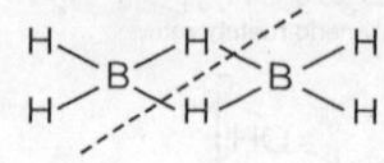 $+ 2R_3N \longrightarrow 2BH_3(R_3N)$

20. (b,d)

(a) H_3BO_3 (orthoboric acid) does not undergo self-ionisation in water but accepts an electron pair from water, thus it is a Lewis monobasic acid.

$$B(OH)_3 + H_2O \rightleftharpoons [B(OH)_4]^- + H^+$$

(H^+ comes from H_2O)

Thus, (a) is incorrect.

(b) In the presence of ethylene glycol (*cis*-diol), a stable complex is formed, thus ionisation of H_3BO_3 is increased. Thus, (b) is correct.

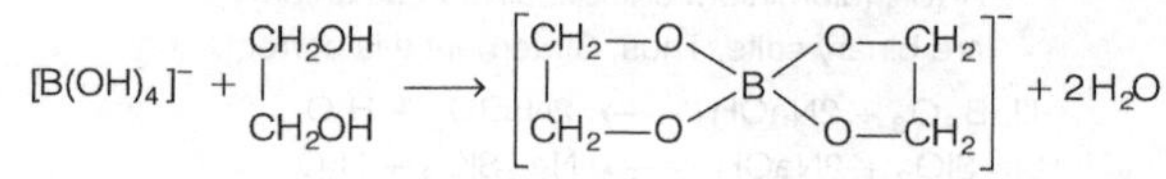

(c) Boric has planar triangular BO_3^{3-} ions which are bonded together through H-bonding in crystalline structure. Thus, (c) is incorrect.

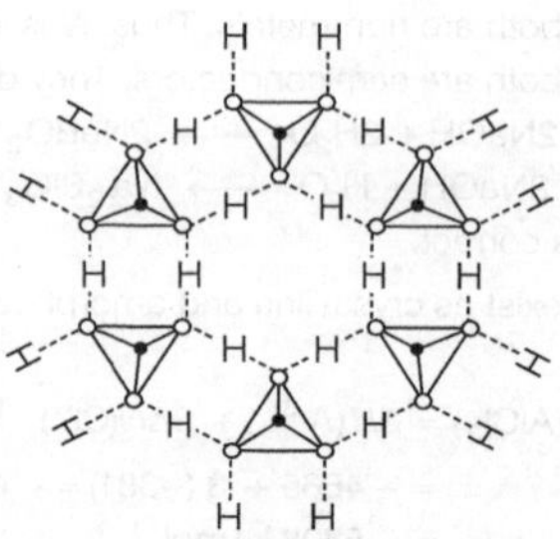

• Boron
∘ Oxygen

Hydrogen bonded structure of orthoboric acid

(d) H_3BO_3 is a weak monobasic Lewis acid. Weak acidic nature is due to $p\pi$-$p\pi$ back bonding as a result of which boron atom is electron-deficient and this decreases Lewis acidic nature of H_3BO_3. Thus, (d) is correct.

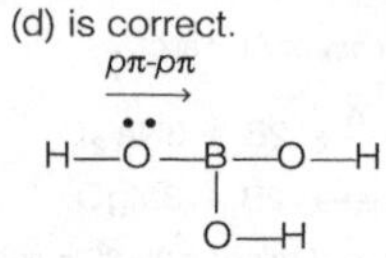

21. (a,b,c)

(a) $B_2H_6 \xrightarrow{\Delta} 2B + 3H_2$

(b) $2BI_3 \xrightarrow{\text{Red hot (W) or (Ta) filament}} 2B + 3I_2$

(c) $BCl_3 + 3H_2 \xrightarrow[\text{(van-Arkel Method)}]{\text{Red hot (W) or (Ta) filament}} 2B + 6HCl$

22. (a)

(i) → (r) Ga melts even on keeping on palm.

(ii) → (t)

(iii) → (p) B_2O_3 is an acidic oxide.

(iv) → (t) $Tl^+ > Tl^{3+}$ ($Tl^+ > In^+ > Ga^+$)

(v) → (p)

(vi) → (p) BF_3

23. (b)

(i) → (u)	(ii) → (s)
(iii) → (r)	(iv) → (t)
(v) → (q)	(vi) → (p)

Explanations for Q. Nos. (24-26)

$$Na_2B_4O_7 \cdot 10H_2O \xrightarrow[740^\circ C]{Heat} \underbrace{NaBO_2 + B_2O_3}_{\text{Glassy transparent bead}}$$

$$CuO + B_2O_3 \xrightarrow{\Delta} Cu(BO_2)_2$$

green - when hot,
blue - when cold
(in oxidising flame)

The bead becomes dull red and opaque when heated in a reducing flame (with C).

$$2Cu(BO_2)_2 + 4NaBO_2 + C \longrightarrow 2Cu + Na_2B_4O_7 + 2C\uparrow$$

24. (c) **25.** (c) **26.** (b)

Explanations for Q. Nos. (27 and 28)

$$B_2H_6 + 2LiH \longrightarrow 2LiBH_4$$

$$\left.\begin{array}{l} RCHO \xrightarrow{LiBH_4} RCH_2OH \\ RCOOH \xrightarrow{LiBH_4} RCH_2OH \\ RCOCl \xrightarrow{LiBH_4} RCH_2OH \end{array}\right\} 1^\circ \text{ alcohol}$$

$\left(\text{—C=C—}\right)$ and (—C≡C—) are not affected.

$$Na + B_2H_6 \longrightarrow NaBH_4 + Na[B_3H_8]$$

27. (a) **28.** (c)

Explanations for Q. Nos. (29 and 30)

$$X \xrightarrow{LiAlH_4} Y \xrightarrow{air} B_2O_3$$

Thus, Y contains boron and hydrogen.

	Percentage	Moles	Ratio
B (11)	78.28	7.11	1
H (1)	21.72	21.72	3

Y (Formula) : BH_3 or B_2H_6

X can be BCl_3 or BF_3.

$$\underset{(X)}{2BCl_3} \xrightarrow{LiAlH_4} \underset{(Y)}{B_2H_6} + LiCl + AlCl_3$$

$$B_2H_6 + 3O_2 \longrightarrow B_2O_3 + 3H_2O$$

29. (a) **30.** (a or c)

31. (2) Borax is actually made of two tetrahedra and two triangular units. It is written as

$$Na_2[B_4O_5(OH)_4] \cdot 8H_2O$$

On hydrolysis, it forms two units each of $B(OH)_3$ and $[B(OH)_4]^-$ per mole of borax.

$$[B_4O_5(OH)_4]^{2-} + 5H_2O \rightleftharpoons 2B(OH)_3 + 2[B(OH)_4]^-$$

$B(OH)_3$ is orthoboric acid. It is not neutralised by HCl. Each unit of $[B(OH)_4]^-$ has only one neutralisable OH^-.

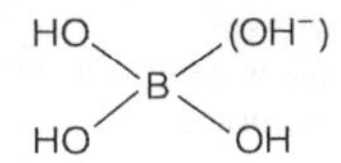

$$[B(OH)_4]^- + HCl \longrightarrow [B(OH)_3] + H_2O + Cl^-$$

Thus, total moles of HCl required for two moles of $[B(OH)_4]^-$ = Two

Note *If dibasic acid is used (as* H_2SO_4*)*

then $2\,[B(OH)_4]^- \equiv 2$ moles $H^+ \equiv 1$ mole H_2SO_4

32. (4) $CH_3Cl + AlCl_3 \longrightarrow AlCl_4^- + CH_3^+$

Aluminium forms four bonds with Cl in $AlCl_4^-$.

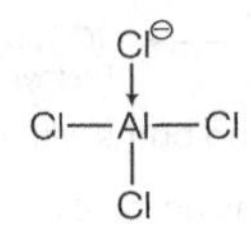

33. (6) Borazine is inorganic benzene ($B_3N_3H_6$).

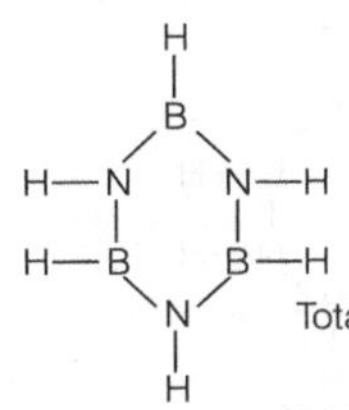

N = 3
B = 3
Total N and B atoms = 6

34. (2) In borazine, each N-atom is bonded to two B-atoms. Thus, $x = 2$.

35. (4) $Na_2B_4O_7 \cdot 10H_2O$ contains the tetranuclear units $[B_4O_5(OH)_4]^{2-}$ and correct formula is

$$Na_2[B_4O_5(OH)_4] \cdot 8H_2O$$

Revisal Problems (JEE Main)

1. (c) Borax is $Na_2B_4O_7 \cdot 10H_2O$. Thus, it is actually a salt of pyroboric acid, $H_2B_4O_7$.
(HBO_2 - metaboric acid; and H_3BO_3 - orthoboric acid are formed on hydrolysis)

$$Na_2B_4O_7 + 2H_2O \longrightarrow H_2B_4O_7 + 2NaOH$$

$$H_2B_4O_7 \longrightarrow 2HBO_2 + B_2O_3$$

2. (d)

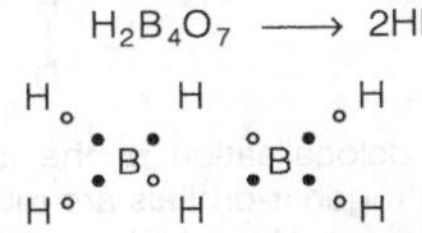

• Boron electron
∘ Hydrogen electron

Bridging groups are electron deficient.
Terminal (B—H) bond ($2c$ - $2e$)
Bridging (B—H—B) bond ($3c$ - $2e$)

3. (c) B_2H_6 uses one s and three p in bonding.

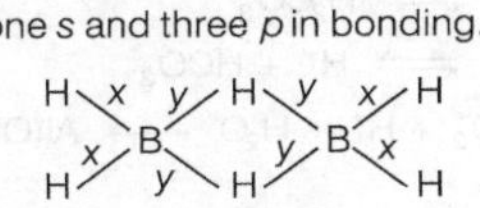

Bond x is formed by sharing two electrons between B and H.
Bond y is formed by sharing one electron between B and H.
Thus, hybridisation of B is sp^3.

4. (b) $AlCl_3$ in aqueous solution exists as $[Al(H_2O)_6]Cl_3$. Thus, it is hydrolysed forming $Al(OH)_3$ and HCl.

$$AlCl_3 + 3H_2O \rightleftharpoons Al(OH)_3 + 3HCl$$

$Al(OH)_3$ on heating to dryness is dehydrated to Al_2O_3.

$$2Al(OH)_3 \longrightarrow Al_2O_3 + 3H_2O$$

5. (b) Mud is colloidal sol which contains charged particles. When alum is added, Al^{3+} ions coagulate oppositely charged particles.

$$3[Fe(OH)_3]OH^- + Al^{3+} \longrightarrow Al(OH)_3 + 3Fe(OH)_3$$

6. (b) $H_2SO_4 \rightleftharpoons 2H^+ + SO_4^{2-}$

$$SO_4^{2-} + H_2O \longrightarrow H_2SO_4 + O^{2-}$$

At anode $O^{2-} \longrightarrow [O] + 2e^-$

$$2Al + 3[O] \longrightarrow \underset{\text{Oxide layer}}{Al_2O_3}$$

At cathode $2H^+ + 2e^- \longrightarrow H_2$

7. (a)

	mp	Physical state
BF_3	$-127.1°C$	Gas
BCl_3	$-107°C$	Gas
BBr_3	$-46°C$	Volatile liquid
BI_3	$50°C$	Solid

Smaller B^{3+} cation and its large charge will polarise X^- to a greater extent. Thus, covalent nature decreases.

8. (c)

(a) When borates are treated with HF (CaF_2 and concentrated H_2SO_4), the volatile compound BF_3 is formed. This BF_3, in Bunsen flame imparts green colour. Thus, borates are identified.

$$\underset{\text{conc.}}{CaF_2 + H_2SO_4} \longrightarrow CaHSO_4 + HF$$

$$H_3BO_3 + 3HF \longrightarrow BF_3 + 3H_2O$$

(b) Borates on heating with conc. H_2SO_4 and ethyl alcohol (C_2H_5OH) and igniting the volatile compound, green edged flames are obtained. This confirms borates.

$$\underset{\text{Borax}}{Na_2B_4O_7} + \underset{\text{Conc.}}{2H_2SO_4} + 5H_2O \xrightarrow{\Delta} 2NaHSO_4 + 4H_3BO_3$$

$$H_3BO_3 + 3C_2H_5OH \longrightarrow \underset{\text{Triethyl borate}}{(C_2H_5O)_3B \text{ or } (C_2H_5)_3BO_3}$$

$(C_2H_5O)_3B$ is volatile and burns with green-edged flame.

9. (c) $B_2H_6 \cdot 2NH_3$ exists as an ionic compound.

$$[H_3N \rightarrow BH_2 \leftarrow NH_3]^+ [BH_4]^-$$

$$B_2H_6 \cdot 2NH_3 \xrightarrow{\Delta}$$

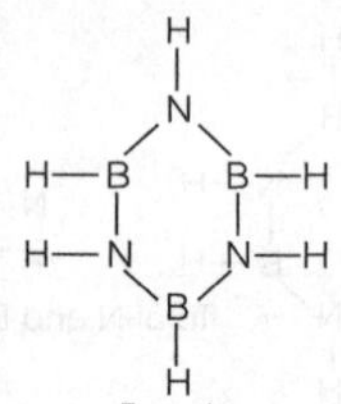

Borazine

Note *Borazine ($B_3N_3H_6$) is called inorganic benzene. It exists as resonance hybrid as*

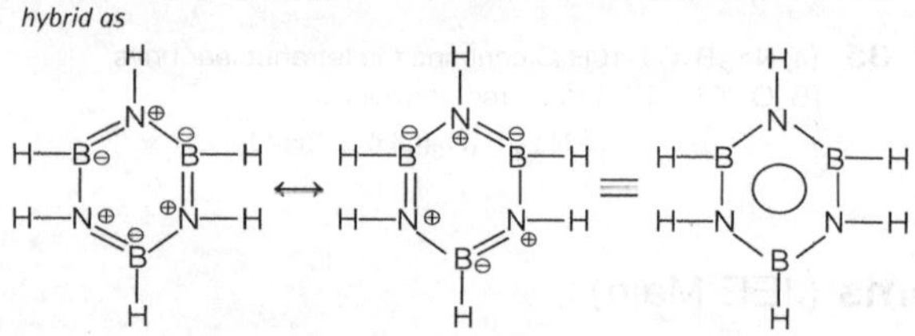

10. (c) $2B(s) + \underset{\text{Hot}}{6HNO_3(aq)} \longrightarrow 2H_3BO_3(aq) + 6NO_2(g)$

$$2Na_2O_2 + 2H_2O \longrightarrow 2NaOH + H_2O_2$$

$$B(s) + 3H_2O_2 \xrightarrow{\Delta} 2H_3BO_3$$

11. (a) Al, Ga, In and Tl form amide $M(NH_2)_3$. Boron forms nitride and H_2.

$$2B + 2NH_3 \longrightarrow 2BN + 3H_2$$

Note *BN is a slippery white solid with layer structure similar to graphite.*

12. (b) $3Mg + 2B \longrightarrow \underset{(X)}{Mg_3B_2}$

$$Mg_3B_2 + 3H_2SO_4 \longrightarrow 3MgSO_4 + \underset{(Y)}{B_2H_6}$$

Note *Borides are hard and refractory.*

13. (b) In BCl_3, B-atom is sp^2-hybridised, thus (Cl—B—Cl) angle is 120°.

In PCl_3, $AsCl_3$ and $BiCl_3$, central atom is sp^3-hybridised, expected angle is 109° 28′, but (lone-pair)-(bond-pair) repulsion decreases this angle. As we go down the group, extent of repulsion increases. Hence, bond angle decreases. Thus,

$$BCl_3 > PCl_3 > AsCl_3 > BiCl_3$$

14. (a) Orthoboric acid (H_3BO_3) is soluble in water and behaves as a weak monobasic acid.

It does not donate protons like other acids but rather accepts OH^- ion. It is therefore a Lewis acid and is better written as $B(OH)_3$.

$$B(OH)_3 + 2H_2O \rightleftharpoons H_3O^+ + [B(OH)_4]^-$$

$p\pi$-$p\pi$ back-bonding between B and O decreases acidic strength thus, it is a weak acid.

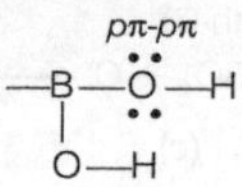

15. (b) When HCl is titrated with borax (alkaline in nature), HCl is neutralised.

$$Na_2B_4O_7 + 7H_2O \longrightarrow 2H_3BO_3 + 2Na[B(OH)_4]$$

$$2Na[B(OH)_4] + 2HCl \longrightarrow 2NaCl + 2H_3BO_3 + 2H_2O$$

Net reaction,

$$Na_2B_4O_7 + 5H_2O + 2HCl \longrightarrow 4H_3BO_3 + 2NaCl$$

One of the products of H_3BO_3 is itself a weak acid. Thus, the indicator used to detect end point of this reaction must be one that is unaffected by H_3BO_3.

Methyl orange (pH range 3.1-4.4) is normally used.

16. (a) Both have same oxidation number of boron = + 3

17. (d) (Ga -As) Gallium arsenides are used in LED, laser diodes and in memory chips of computers.

18. (c)

(a) Boron (in the form of carbide) is a good neutron absorber and is used as shields and control rods in nuclear react.

$${}^{10}_{5}B + {}^{1}_{0}n \longrightarrow {}^{11}_{5}B$$

(b) Boron carbide is hard and used as an abrasive.

19. (a) Soda-free glass is free from Na_2CO_3 and Na_2SiO_3. It is composed of

(i) H_3BO_3

(ii) B_2O_3 and

(iii) $Ca_3(BO_3)_2$

20. (a) In Hall-Heroult process, Al is produced by the electrolysis of Al_2O_3 which is not a good conductor. Na_3AlF_6 (cryolite) improves the electrical conductivity of cell. This also serves as an added impurity and lowers the melting point of the mixture to about 950°C.

21. (d)

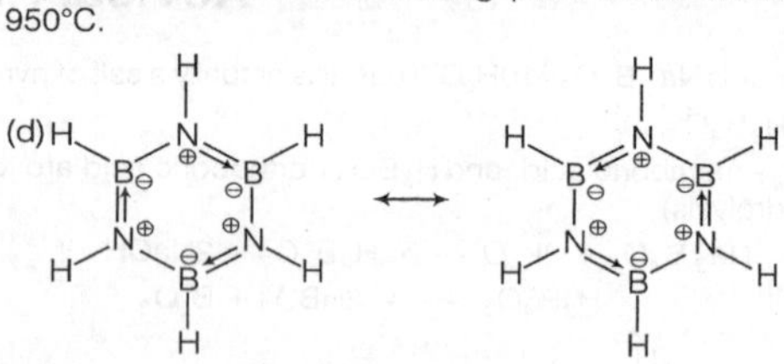

There is no perfect delocalisation of the π-electron cloud in borazine since, the nitrogen π-orbitals are much lower in energy than the π-orbitals of boron (N has higher nuclear charge).

22. (b)

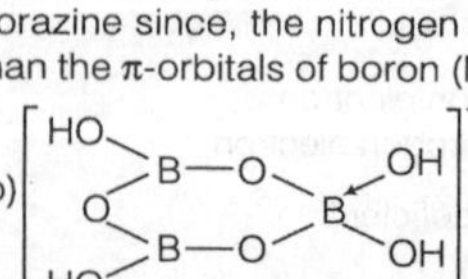

23. (b) $CO_2 + H_2O \rightleftharpoons H_2CO_3$

$$H_2CO_3 \rightleftharpoons H^+ + HCO_3^-$$

$$AlO_2^- + H^+ + H_2O \longrightarrow Al(OH)_3\downarrow$$

24. (d)

(a) $Na_2CO_3 + \underset{\textit{ortho-}\text{boric acid}}{4H_3BO_3} \longrightarrow Na_2B_4O_7 + 6H_2O + CO_2$

(b) $\underset{\text{Colemanite}}{Ca_2B_6O_{11}} + 2Na_2CO_3 \longrightarrow Na_2B_4O_7 + 2CaCO_3 + NaBO_2$

(c) $\underset{\text{Sodium metaborate}}{4NaBO_2} + SO_2 \longrightarrow Na_2B_4O_7 + Na_2SO_3$

25. (b) $\underset{\text{Sodium metaborate}}{4NaBO_2} + CO_2 \longrightarrow \underset{\text{Borax}}{Na_2B_4O_7} + Na_2CO_3$

26. (a) Al^{3+} due to small size and high charge has maximum polarising power. Thus,

$Al(OH)_3 \longrightarrow Al^{3+} + 3\,OH^-$ by (Al—O) bond fission

$Al(OH)_3 \longrightarrow Al(OH)_2(O^-) + H^+$ by (O—H) Bond fission

$\downarrow$

$AlO_2^- + H_2O$

Thus, $Al(OH)_3$ behaves as an acid and is thus soluble in base and it also behaves as a base and is thus soluble in acid.

$$\underset{\text{Acid}}{Al(OH)_3} + \underset{\text{Base}}{NaOH} \longrightarrow NaAlO_2 + 2H_2O$$

$$\underset{\text{Base}}{Al(OH)_3} + \underset{\text{Acid}}{3HCl} \longrightarrow AlCl_3 + 3H_2O$$

Thus, Statement I and Statement II are correct and Statement II is the correct explanation of Statement I.

27. (d) $AlCl_3$ is electron-deficient (octet of Al is incomplete. Thus, $AlCl_3$ is a Lewis acid. Dimer Al_2Cl_6 is formed when Al in one molecule accepts lone pair from Cl of other molecule.

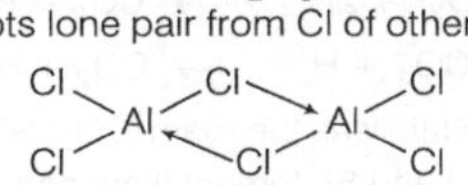

Bridging (Al$\leftarrow$Cl) bonds are different from (Al—Cl) covalent bonds. Thus, Statement I is incorrect Statement II is correct.

28. (c) B atom is electron-deficient. $p\pi$-$p\pi$ back bonding makes B electron-deficient hence, its Lewis acid nature is decreased. In H_2O, it accepts OH^- from it (H_2O), hence H^+ is not released.

$$H_3BO_3 \text{ or } B(OH)_3 + H_2O \rightleftharpoons [B(OH)_4]^- + H^+$$

H^+ is released from H_2O.

Thus, Statement I is correct but Statement II is incorrect.

29. (d) Aluminium reacts with NaOH. In the presence of HNO_3, there is the formation of an oxide layer inside the vessel and metal becomes passive for further attack. Thus, HNO_3 can be stored in aluminium vessel but NaOH can't be stored.

$$2Al + 2NaOH + 2H_2O \longrightarrow 2NaAlO_2 + 3H_2$$

$$Al \xrightarrow{HNO_3} Al_2O_3 \xrightarrow{HNO_3} \text{No reaction}$$

Thus, Statement I is incorrect but Statement II is correct.

30. (a) In BF_3, boron atom is electron-deficient. $p\pi$-$p\pi$ back bonding completes its octet in which F atom donates its lone pair to B and this results in the formation of dative bond. This dative bond may give double bond character and bond-energies of two types of bonds may be different but bonds are equivalent. This is due to delocalisation of dative bond.

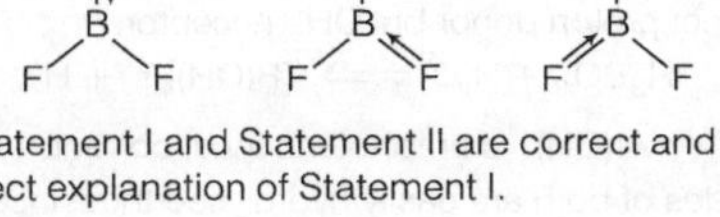

Thus, Statement I and Statement II are correct and Statement II is the correct explanation of Statement I.

31. (d) $2Al(s) + \frac{3}{2}O_2(g) \longrightarrow Al_2O_3(s),\ \Delta H = -1670$ kJ

A large amount of heat is liberated producing a temperature of about more than 1000°C. At this temperature, H_2O is decomposed by Al releasing H_2.

$$2Al + 3H_2O \longrightarrow Al_2O_3 + 3H_2$$

H_2 is highly inflammable, thus it will spread fire. Hence, thermite reaction can't be stopped by H_2O.

Thus, Statement I is incorrect but Statement II is correct.

32. (c) $CO_2 + H_2O \rightleftharpoons H_2CO_3 \rightleftharpoons CO_3^{2-} + 2H^+$

$$2Al^{3+} + 3CO_3^{2-} + 3H_2O \rightleftharpoons 2Al(OH)_3\downarrow + 3CO_2$$

This reaction shows that acidic properties of $Al(OH)_3$ are very weak.

$Al(OH)_3$ dissolves in acid as well as in base, hence it is amphoteric.

Amphiprotic species are proton donor as well as proton acceptor. H_2O is amphiprotic.

$$H_2O \rightleftharpoons H^+ + OH^-$$

$$H_2O + H^+ \rightleftharpoons H_3O^+$$

Thus, Statement I is correct but Statement II is incorrect.

33. (d) $AlCl_3$ and BCl_3 both are electron-deficient and are Lewis acids. $AlCl_3$ completes its octet by exchange of lone pair of Cl and forms dimer.

$$2AlCl_3 \rightleftharpoons Al_2Cl_6$$

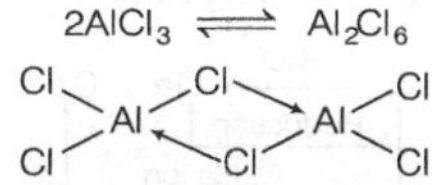

BCl_3 is itself stable due to the stability of (B—Cl)π-bonding. When there is dimer formation, this bonding is disrupted as there would be no vacant orbital for π-bonding. Also boron atom is too small to form the stable 4-membered ring system in such dimers.

Thus, Statement I is incorrect but Statement II is correct.

34. (c)

(i) Prismatic glass is $Na_2B_4O_7 \cdot 10H_2O$.
It has two tetrahedra and two triangular units joined together hence represented as $Na_2[B_4O_5(OH)_4] \cdot 8H_2O$.
Thus, (i) $\rightarrow$ (s)

(ii) Octahedral borax is obtained by crystallising borax above 370 K. It is also called Jeweller's borax.
Thus, (ii) $\rightarrow$ (r), (iv) $\rightarrow$ (r)

(iii) Borax glass is $Na_2B_4O_7$. Thus, (iii) $\rightarrow$ (p)

$$Na_2B_4O_7 \cdot 10H_2O \xrightarrow{\text{Heat above mp}} \underset{\text{Borax glass}}{Na_2B_4O_7}$$

Note *Borax glass absorbs moisture from air and is converted into* $Na_2B_4O_7 \cdot 10H_2O$.

35. (c) $H_3BO_3 \xrightarrow{100°C} HBO_2 + H_2O$

$$4H_3BO_3 \xrightarrow{140°C} H_2B_4O_7 + 5H_2O$$

$$2H_3BO_3 \xrightarrow{\text{Red hot}} B_2O_3 + 3H_2O$$

$$3H_3BO_3(aq) \xrightarrow{\text{Concentrated}} [B_3O_3(OH)_4]^-$$

Thus,

(i) $\rightarrow$ (q) Metaboric acid
(ii) $\rightarrow$ (r) Pyroboric acid
(iii) $\rightarrow$ (s) Boric oxide
(iv) $\rightarrow$ (p) Polymetaborate

Revisal Problems (JEE Advanced)

1. (c) $2B + 2NH_3 \longrightarrow 2BN + 3H_2$

$2Al + 6NH_3 \longrightarrow 2Al(NH_2)_3 + 3H_2$

B forms nitrides and Al forms amides.

2. (a) π-bonding decreases as crowding causes departure from planarity.

3. (a) $[B_4O_5(OH)_4]^{2-} + 5H_2O \rightleftharpoons 2B(OH)_3 + 2[B(OH)_4]^-$

$B(OH)_3$ behaves as a weak monobasic Lewis acid. (It is not a proton donor but OH^- acceptor).

$[B(OH)_4]^-$ is a salt formed from $B(OH)_3$. Thus, it acts as a buffer.

4. (a)

$$2[Al \longrightarrow Al^{3+} + 3e^-, \quad E^\circ_{Al/Al^{3+}} = -E^\circ_{Al^{3+}/Al} = 1.66\text{ V}$$

$$3[2H^+ + 2e^- \longrightarrow H_2 \quad E^\circ_{H^+/H_2/Pt} = 0.00\text{ V}$$

$$2Al + 6H^+ \longrightarrow 2Al^{3+} + 3H_2, \quad E^\circ = 1.66\text{ V}$$

$$2[Al \longrightarrow Al^{3+} + 3e^-, \quad E^\circ = 1.66\text{ V}$$

$$3[SO_4^{2-} + 2e^- + 4H^+ \longrightarrow SO_2 + 2H_2O, \quad E^\circ = 1.17\text{ V}$$

$$2Al + 3SO_4^{2-} + 12H^+ \longrightarrow 2Al^{3+} + 3SO_2 + 6H_2O, \quad E^\circ_{cell} = 2.83\text{ V}$$

For H_2, $E^\circ = 1.17$ V

For SO_2, $E^\circ = 2.83$ V (more spontaneous)

Thus, SO_2 is formed in preference to H_2.

5. (c)

$$Ga^+ \xrightarrow{H_2O} Ga + Ga^{3+}$$

(Reduction: $Ga^+ \to Ga$; Oxidation: $Ga^+ \to Ga^{3+}$)

$Ga^{3+} > Ga^+$ (stability)

Thus, Ga^+ disproportionates to Ga (by reduction) and to Ga^{3+} (by oxidation).

6. (a) $Al \longrightarrow Al^{3+} + 3e^-$, $E^\circ_{Al/Al^{3+}} = 1.66$ V

$E^\circ_{Al/Al^{3+}}$ is maximum, thus $\Delta G^\circ = -nFE^\circ$ is the most negative for ($Al \longrightarrow Al^{3+}$).

Thus, most stable ion in aqueous solution is Al^{3+}.

7. (d)

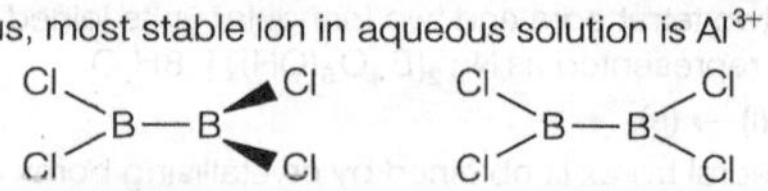

Non-eclipsed Planar

Note $2BCl_3 + 2Hg \xrightarrow[\text{Low pressure}]{\text{Electric discharge}} B_2Cl_4 + Hg_2Cl_2$

8. (c)

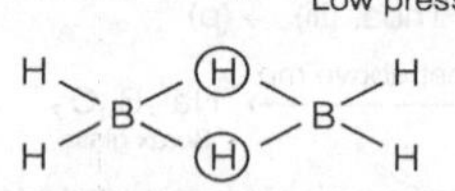

I. Correct

II. Bond angle ≠ 180°

III. H—B bond is different from Ⓗ—B bond.

IV. Total valence electrons for bonding = 6 in H and 6 in B = 12

9. (a) $3BCl_3 + 3NH_4Cl \xrightarrow{C_6H_5Cl}$

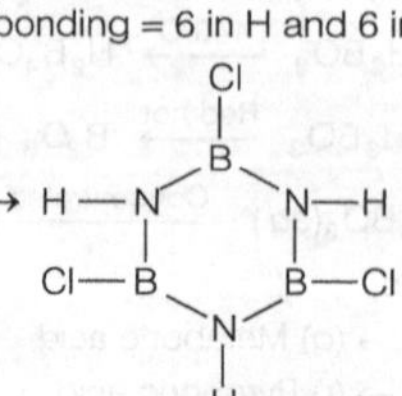

(A)

$$A \xrightarrow{NaBH_4} \underset{(X)}{B_3N_3H_6} \text{ (by reduction)}$$

$$A \xrightarrow{CH_3MgBr} Me_3B_3N_3H_3$$

B-trimethyl borazine (methyl attached to boron) (Y)

10. (b) Al_2O_3 is mixed with Na_3AlF_6 to lower melting point and to increase conductivity of the electrolyte.

Thus, Statement (b) is incorrect.

11. (c) With increase in polarity of (B—X) bond, acidity also increases and should be in the order $BF_3 < BCl_3 < BBr_3$.

There is lateral overlap of the vacant 2*p*-orbital of B with one completely filled orbital of F leading to $p\pi$-$p\pi$ bonds between B and F. This B—F bond thus acquires double bond character. This also leads to compensate electron-deficiency of boron and thus, Lewis acid character of BF_3 is reduced. This $p\pi$-$p\pi$ bonding decreases going from BF_3 to BBr_3 and thus, Lewis acidic nature increases in the order : $BF_3 < BCl_3 < BBr_3$

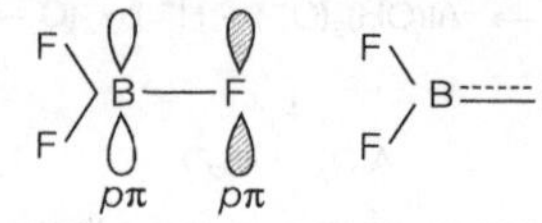

$p\pi$ $p\pi$

Thus, Statement I is correct but Statement II is incorrect.

12. (b) Fire extinguishers can be made that contain $Al_2(SO_4)_3$ together with the usual carbonate or bicarbonate.

Al^{3+} salts are acidic due to hydrolysis.

$$Al^{3+} + 3H_2O \rightleftharpoons Al(OH)_3 + 3H^+$$

H^+ decomposes HCO_3^- and CO_3^{2-} releasing CO_2 that extinguishes fire.

$$CO_3^{2-} + 2H^+ \longrightarrow CO_2 + H_2O$$

$$HCO_3^- + H^+ \longrightarrow CO_2 + H_2O$$

$Al(OH)_3$ formed stabilises the foam (CO_2) which covers the fire.

Thus, Statement I and Statement II are correct but Statement II is not the correct explanation of Statement I.

13. (a,b,c,d) $(M(H_2O)_6)^{3+}$ can be proton donor, thus acts as Bronsted-Lowry acid.

$$[M(H_2O)_6]^{3+} + H_2O \rightleftharpoons [M(H_2O)_5(OH^-)]^{2+} + H_3O^+$$

$$[M(H_2O)_5(OH^-)]^{2+} + H_2O \rightleftharpoons [M(H_2O)_4(OH^-)_2]^+ + H_3O^+$$

$$[M(H_2O)_4(OH^-)_2]^+ + H_2O \rightleftharpoons [M(H_2O)_3(OH^-)_3] + H_3O^+$$

$$[M(H_2O)_3(OH^-)_3] + H_2O \rightleftharpoons [M(H_2O)_2(OH^-)_4]^- + H_3O^+$$

14. (a,b,c)

(a) For this reaction, $\Delta H < 0$ (exothermic) thus, it is correct.

(b) Since, it is exothermic, hence large amount of heat is liberated. It is thus a thermite reaction and used in welding.

$$Fe_2O_3 + 2Al \longrightarrow Al_2O_3 + 2Fe$$

Due to high temperature, Fe melts and welding can be done, thus, correct.

(c) $Cr_2O_3 + 2Al \longrightarrow 2Cr + Al_2O_3$

Thus, correct.

15. (a,b,d)

(a) B_2O_3 – acidic

Al_2O_3 – amphoteric

Thus, it is correct.

(b) H_3BO_3 is a weak monobasic acid due to $p\pi$-$p\pi$ back bonding. It is not proton donor but OH^- acceptor.

$$H_3BO_3 + H_2O \rightleftharpoons [B(OH)_4]^- + H^+$$

$Al(OH)_3$ is amphoteric thus, correct.

(c) Halides of both are easily hydrolysed thus, incorrect.

(d) BH_3 – gaseous compounds

$(AlH_3)_n$–a polymeric solid. Thus, correct.

16. (a, c) $3CH_3CH{=}CH_2 + BH_3 \longrightarrow (CH_3CH_2{-}CH_2)_3B \xrightarrow{H_2O_2/OH^-} \underset{1°\text{alcohol}}{CH_3CH_2CH_2OH} + H_3BO_3$

17. (b,c,d)

(a) Kernite : $Na_2B_4O_7 \cdot 4H_2O$ is an ore of boron thus, incorrect.

(b) $^{10}_{5}B$ (19%) and $^{11}_{5}B$ (81%) are isotopes of boron thus, correct

(c) Bauxite – $Al_2O_3 \cdot 2H_2O$ and
Cryolite – Na_3AlF_6 are minerals of Al thus, correct.

(d) Al (8.3%), O(45.5%) Thus, correct.

18. (b,c,d)

(a) $AlCl_3$ – It has electron-deficient Al hence, it cannot donate lone pair to BF_3.

(b) NH_3 has one lone pair

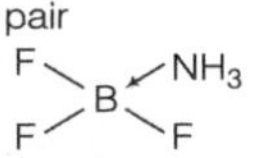

(c) Et_2O can donate lone pair

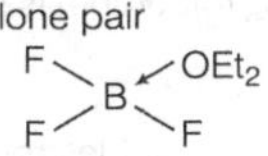

(d) Me_3N can donate lone pair

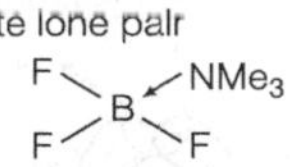

In all cases (b,c,d), sp^3-hybridised tetrahedral structure is attained.

19. (a,b,c)

(a) BF_3 hydrolyses incompletely.

$$BF_3 + 3H_2O \longrightarrow 3HF + H_3BO_3$$

HF first formed combines with H_3BO_3 forming HBF_4.

$$3H_3BO_3 + 12HF \longrightarrow 3HBF_4 + 9H_2O$$

BCl_3 hydrolyses completely to H_3BO_3 and HCl

$$BCl_3 + 3H_2O \longrightarrow 3HCl + H_3BO_3$$

(b) Lewis acid behaviour of $BCl_3 > BF_3$

Due to $p\pi$-$p\pi$ back bonding from F to B, Lewis acid character of BF_3 is decreased.

(c)

	BF_3	BCl_3
mp	– 127.1°C	– 107°C
bp	– 100°C	12.5°C

(d) Both have triangular planar structure.

20. (a,b,c,d)

(a) $Mg_3B_2 + H_3PO_4 \longrightarrow B_4H_{10} \xrightarrow{\Delta} B_2H_6$

(b) $4BF_3 + 3LiAlH_4 \longrightarrow 2B_2H_6 + 3LiF + 3AlF_3$

$2BF_3 + 6NaH \xrightarrow{180°C} B_2H_6 + 6NaF$

(c) $2NaBH_4 + I_2 \xrightarrow{\text{Ether}} B_2H_6 + 2NaI + H_2$

(d) $B_2O_3 + 3H_2 + 2Al \xrightarrow[750\ \text{atm}]{150°C} B_2H_6 + Al_2O_3$

21. (b,d) If $E° > 0$, spontaneous

$E° > 0$, non-spontaneous

$E°_{Tl^{3+}/Tl} = 1.26$ V

I. $Tl^{3+} + 3e^- \longrightarrow Tl$, $E° = 1.26$ V

II. $Tl \longrightarrow Tl^{3+} + 3e^-$, $E° = -1.26$ V $E°_{Tl^+/Tl} = -0.34$ V

III. $Tl^+ + e^- \longrightarrow Tl$, $E° = -0.34$V

IV. $Tl \longrightarrow Tl^+ + e^-$, $E° = +0.34$ V

Thus,

(a) is corect being spontaneous.

(b) is not spontaneous as $Tl^{3+} < Tl^+$ (stability)

(c) is not spontaneous.

(d) is being spontaneous.

22. (b)

(i) Bond length = 206 pm. Thus, (i) → p

(ii) Bond length = 221 pm. Thus, (ii) → r

(iii) Bond angle = 101°. Thus, (iii) → q

(iv) Bond angle = 79°. Thus, (iv) → s

(v) Bond angle = 118°. Thus, (v) → t

Explanations for Q. Nos. (23 and 24)

$$3KF + AlF_3 \longrightarrow \underset{\text{Soluble}}{K_3AlF_6}$$

Maximum covalency of Al is six hence, AlF_6^{3-} is formed.

$$K_3AlF_6 + 3BF_3 \longrightarrow AlF_3 \downarrow + \underset{\text{Soluble}}{3KBF_4}$$

23. (a) **24.** (c)

Explanations for Q. Nos. (25 and 26)

If *A* is Al, molar mass of *B* = 27 + 14 = 41 (AlN)

A is Mg, molar mass of *B* = 72 + 28 = 100 (Mg_3N_2)

A is Li, molar mass of *B* = 21 + 14 = 35 (Li_3N)

Thus, metal *A* is Al.

Composition of *B*

	Percentage	Mole	Ratio
N	34.15	$\frac{34.15}{14} = 2.44$	1
Metal (*A*)	65.85	$\frac{65.85}{27} = 2.44$	1

Thus, *B* is AlN.

$$\underset{(A)}{2\ Al} + N_2 \xrightarrow{\Delta} \underset{(B)}{2AlN}$$

$$AlN + 3H_2O \longrightarrow \underset{(C)}{Al(OH)_3} + \underset{(D)}{NH_3}$$

$Al(OH)_3$ (*C*) is amphoteric as it is soluble in both alkali and acid.

NH_3 (*D*) gives white fumes in HCl.

$$Al(OH)_3 + NaOH \longrightarrow NaAlO_2 + 2H_2O$$

$$Al(OH)_3 + 3HCl \longrightarrow AlCl_3 + 3H_2O$$

$$NH_3 + HCl \longrightarrow NH_4Cl \text{ (white fumes)}$$

25. (b) **26.** (a)

Explanations for Q. Nos. (27 and 28)

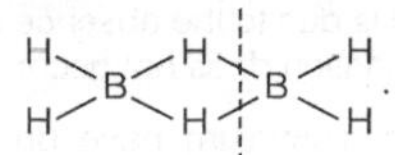

$\cdot\ 2NH_3 \xrightarrow[\text{cleavage}]{\text{Unsymmetrical}} [(BH_2)(NH_3)_2]^+ [BH_4]^-$

$B_2H_6 \cdot 2NH_3 \xrightarrow{\Delta}$

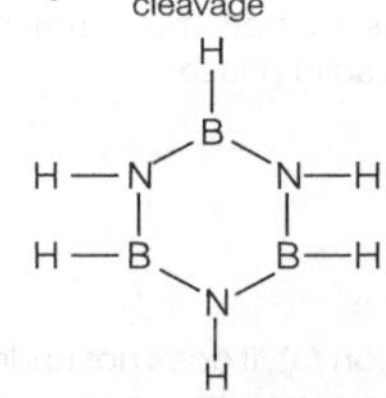

$B_3N_3H_6$ (Borazine)

27. (c) **28.** (b)

29. (6)

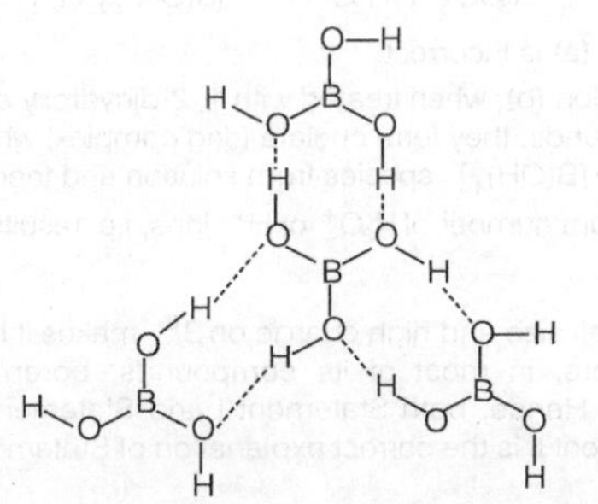

Each unit of H_3BO_3 forms six H-bonds with other units. (Three with H and three with *O*-atoms.)

30. (2) $B_2H_6 + 2Me_3N \longrightarrow \underset{\text{Adduct}}{B_2H_6 \cdot 2Me_3N}$

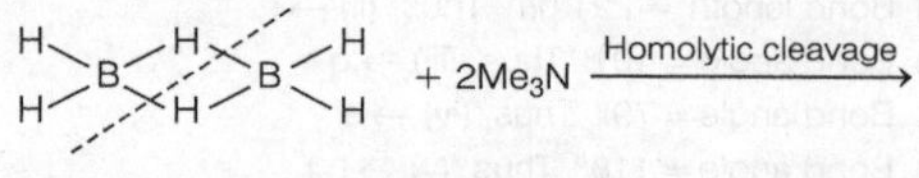

$2[BH_3 \cdot NMe_3]$

By homolytic (symmetric) cleavage, adduct splits into two fragments.

31. (2) $B_2H_6 + 2RNH_2 \longrightarrow [BH_2(RNH_2)_2]^+[BH_4]^-$

Maximum covalency of B is 4, thus $n = 2$.

32. (3) Let formula be $Be_x\ Al_2\ Si_6\ O_{18}$.

On balancing charges, Be = +2, Al = +3

Si = +4, O = −2

$$2x + 6 + 24 - 36 = 0$$
$$x = 3$$

33. (4) There are two tetrahedra and two triangular units joined as

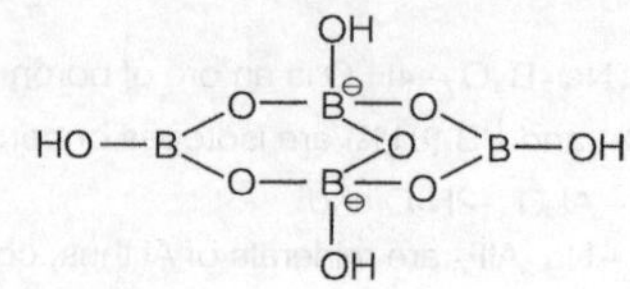

34. (4) There are two tetrahedral units and two triangular units.

B attached to O-atoms in tetrahedral = 4(maximum)

B attached to O-atoms in triangular = 3 (minimum)

JEE Main & AIEEE Archive

1. (a) $_5B$

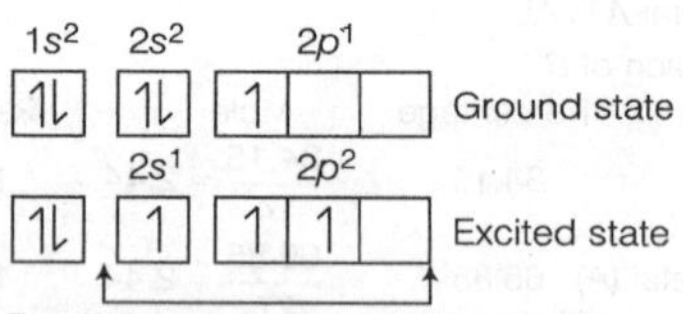

Fourth lone pair is accommodated in this empty orbital

Maximum covalency = 4

Due to the absence of 2*d*-orbital, maximum covalency is four. Thus, BF_6^{3-} is not formed.

Therefore, (a) is not formed while BH_4^- $(BH_3 + H^-)$, $[B(OH)_4^-]$, $(B[OH]_3 + OH^-)$ and BO_2^- are formed.

2. (c) Due to identical sizes of $2p$-orbitals on B and F, back donation of a lone pair of electrons from the filled $2p$-orbital of F to the empty $2p$-orbital of B occurs to a considerable extent.

As a result, B—F bond has some double bond character and hence, bond dissociation energy of B—F bond in BF_3 is much higher that of C—F bond in CF_4. This is due to the absence of empty $2p$-orbital and hence, back donation does not occur.

3. (c) Chlorides of both beryllium and aluminium have bridged structures in solid phase.

4. (c) Boron atom in B_2H_6 (diborane) is sp^3 hybridised.

5. (d) Structure of B_2H_6

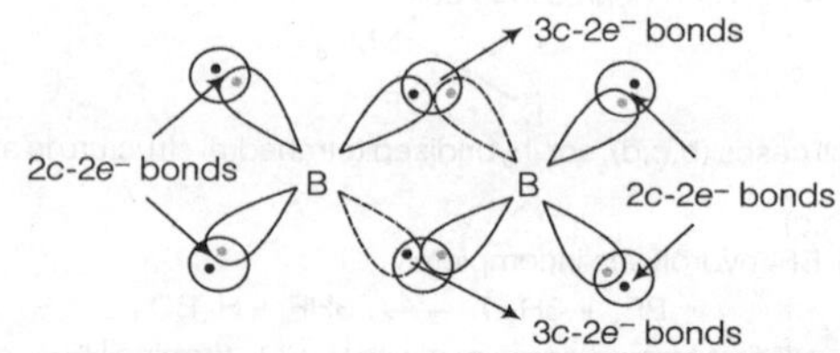

6. (b) Aqueous solution of $AlCl_3$ is acidic, as it gives HCl on hydrolysis.

$$AlCl_3 + 3H_2O \rightleftharpoons Al(OH)_3 + 3HCl$$

On strongly heating, $Al(OH)_3$ is converted into Al_2O_3.

$$2Al(OH)_3 \xrightarrow{\Delta} Al_2O_3 + 3H_2O$$

7. (b) $AlCl_3$ is covalent but in water, it becomes ionic due to large hydration energy of Al^{3+}.

$$AlCl_3 + 6H_2O \rightleftharpoons [Al(H_2O)_6]^{3+} + 3Cl^-$$

JEE Advanced & IIT JEE Archive

1. (b,d) For option (a), it does not undergo self ionisation in water but accepts an electron pair from water, so it behaves as a weak monobasic acid

$$H_3BO_3 + H_2O \rightleftharpoons [B(OH)_4]^- + H^+$$

Hence, (a) is incorrect.

For option (b), when treated with 1, 2-dihydroxy or polyhydroxy compounds, they form chelate (ring complex) which effectively remove $[B(OH)_4]^-$ species from solution and thereby, produce maximum number of H_3O^+ or H^+ ions, i.e. results in increased acidity.

2. (a) Small size and high charge on B^{3+} makes it highly polarising. Therefore, in most of its compounds, boron forms covalent bonds. Hence, both Statement I and Statement II are true and Statement II is the correct explanation of Statement I.

3. (c) Orthoboric acid is a weak monobasic Lewis acid and the poor acidic character is due to $p\pi - p\pi$ back bonding as

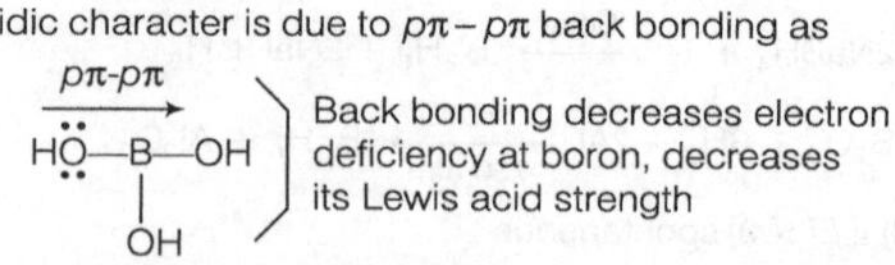

4. (a) Orthoboric acid is a very weak acid, direct neutralisation does not take place. However, addition of *cis*-diol allows the reaction to go to completion by forming a stable complex with $[B(OH)_4]^-$ as

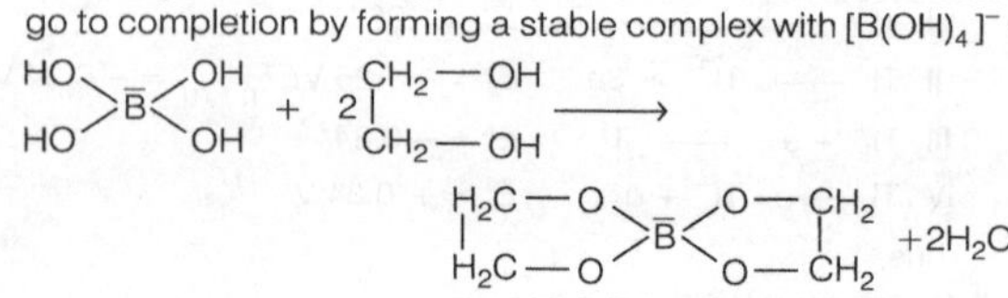

6. **Group 14 Elements** (Carbon Family)

DPP-1 **Periodicity of Group 14**

1. (a) Smaller the size of cation, anion (H^-) is smaller, then larger the polarising power, hence larger the covalent nature. Thus, stability increases in the same order as the covalent nature.

$$\underset{\text{Maximum}}{CH_4} > SiH_4 > GeH_4 > SnH_4 > \underset{\text{Least}}{PbH_4}$$

2. (b) Catenation is a property by virtue of which elements form long chain compounds by single or multiple bond formation between atoms of the same element. Carbon shows this tendency to a remarkable extent because of small size and high bond energy.

Bond	BE (kJ mol^{-1})
C—C	347
Si—Si	226
Ge—Ge	118
Sn—Sn	150

3. (c) Due to inert-pair effect in Group 14, stability of + 2 state increases and that of + 4 state decreases as we go down the group. Thus,

$SiCl_2 < GeCl_2 < SnCl_2 < PbCl_2$ Thus, correct.

$SiCl_4 > GeCl_4 > SnCl_4 > PbCl_4$ Thus, correct.

4. (d)

(a) Stability of +2 oxidation state increases as we go down the group. Thus, PbO_2 is the best oxidising agent and itself reduced to Pb^{2+}.

$$Pb^{2+} > Pb^{4+} \text{ (Stability)}$$

Thus, correct.

(b) As ionic nature of hydride increases, reducing nature also increases.

$CH_4, SiH_4, SnH_4, PbH_4$

$\longrightarrow$ Ionic nature

$\longrightarrow$ Reducing nature

Thus, correct.

(c) $(IE)_1$ of Pb > $(IE)_1$ of Sn.

It is due to poor screening by 4*f*-electrons which increases force of attraction hence increases $(IE)_1$.

Thus, correct.

Note *This effect is also observed in* $(IE)_2, (IE)_3$ *and* $(IE)_4$.

(d) As above thus, incorrect.

5. (b) Silicon exhibits a network structure like diamond–a covalent crystal in solid state.

6. (c) There are three main allotropes of carbon:

(i) Diamond

(ii) Graphite

(iii) Buckminster fullerene (C_{60})

In all cases, there is $p\pi$-$p\pi$ bonding in addition to (sp^2-sp^2) as in graphite and C_{60} and (sp^3-sp^3) bonding as in diamond.

Catenation is the property by which various C-atoms are joined.

7. (b)

(a) $[SiCl_6]^{2-}$: Six large Cl^- ions cannot be accommodated around Si^{4+} due to limitation of its size. Also, interaction between lone pair of Cl^- ions and Si^{4+} is not very strong.

Thus, $[SiCl_6]^{2-}$ does not exist.

(b) $[SiF_6]^{2-}$: F^- ion of smaller size can be accommodated around Si^{4+}

(c) PbI_4 : (Pb —I) bond formed initially does not release enough energy that can excite paired ($6s^2$) electron to (6*p*) to make four unpaired electrons for PbI_4 formation. Also, PbI_2 is more stable than PbI_4 due to inert pair effect.

$$Pb^{4+} + 2e^- \longrightarrow Pb^{2+}$$

8. (c) Smaller the size, larger the charge of cation (anion is same, Cl^-) then larger the polarising power and thus, larger the covalent nature.

M^{4+} is smaller than M^{2+}.

Polarising power of $M^{4+} > M^{2+}$.

Thus, MCl_4 is covalent.

MCl_2 is ionic.

MP and BP of $MCl_4 < MCl_2$.

MCl_4 is volatile, MCl_2 is non-volatile thus, solubility of MCl_4 in ethanol > MCl_2

MCl_4 is more hydrolysed than MCl_2.

Thus, (c) is correct.

9. (a) At very low temperature, β-tin (white tin) changes to α-tin (grey tin), i.e.

$$\underset{\substack{\text{Grey tin} \\ \text{(Diamond structure)}}}{\alpha\text{-Sn}} \underset{}{\overset{13.2^\circ C}{\rightleftharpoons}} \underset{\substack{\text{White tin} \\ \text{(Metallic)}}}{\beta\text{-Sn}}$$

Above equilibrium, shifts in backward side at low temperature.

10. (c) $Sn + \underset{\text{Conc.}}{4HNO_3} \longrightarrow \underset{\text{Metastannic acid}}{H_2SnO_3} + 4NO_2 + H_2O$

11. (a,b) Sn and Pb are slowly attacked by cold alkali and rapidly by hot alkali giving sodium stannate $Na_2[Sn(OH)_6]$ and sodium plumbates $Na_2[Pb(OH)_6]$.

Si reacts slowly with cold aqueous NaOH solution and readily with hot solution forming SiO_4^{2-}.

Ge does not react.

12. (a,b,c,d)

(a) Correct

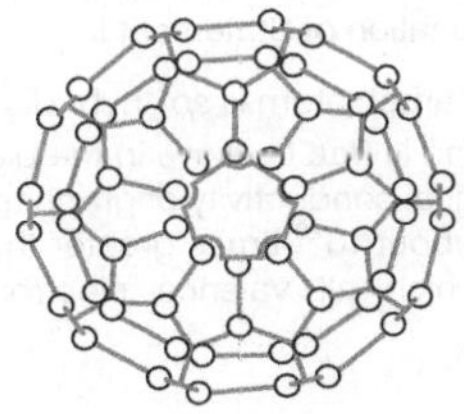

Structure of Buckminster fullerene

(b) Correct

(c) Correct

(d) Correct

13. (b,c,d)

(a) Sn reacts with steam.

(b) $Sn + 2H_2O \text{ (steam)} \longrightarrow SnO_2 + 2H_2$

(c) Oxide layer protects lead from further reaction with H_2O.

(d) C, Si and Ge are unaffected by dilute acids.

14. (a,b) Carbon forms multiple bonds with oxygen, nitrogen and sulphur.

$$>C{=}O, -C{\equiv}N, >C{=}S$$

Silicon also forms multiple bonds with these elements.

$$>Si{=}O$$

15. (a,c) In *p*-block elements, stability of (*G*)-state decreases and that of (*G* − 2) state increases as we go down the group (*G* = valence electrons).

Valence electrons in Group 14 (carbon family)= 4

Thus, stability order is

$$Pb^{2+} > Sn^{2+} > Ge^{2+}$$

and $Pb^{4+} < Sn^{4+} < Ge^{4+}$

$Sn^{4+} > Sn^{2+}$

hence, $Sn^{2+} \longrightarrow Sn^{4+} + 2e^-$

Thus, Sn^{2+} salts are better reducing agents.

$Pb^{4+} < Pb^{2+}$

Thus, $Pb^{4+} + 2e^- \longrightarrow Pb^{2+}$

Thus, Pb^{4+} salts are better oxidising agents.

16. (b) PbO_2 is soluble in concentrated alkali.

$PbO_2 + 2OH^- + 2H_2O \longrightarrow [Pb(OH)_6]^{2-}$

This indicates the acidic nature of PbO_2.

$\underset{\text{IV}}{PbO_2} + 2H_2SO_4 \longrightarrow \underset{\text{II}}{PbSO_4} + 2H_2O + O_2$

PbO_2 liberates O_2 on reaction with H_2SO_4 and $PbSO_4$ is precipitated. Thus, Pb(II) is more stable than Pb (IV) due to inert pair effect. Thus, Statement I and Statement II are correct but Statement II is not the correct explanation of Statement I.

17. (a) As we go down the group from top to bottom, the (IE) values are found to decrease. Thus, (IE) values for Pb (to form Pb^+, Pb^{2+}...) would be expected to be lower than that of Sn (to form Sn^+, Sn^{2+}...). But presence of 4*f*- and 5*d*-electrons decreases screening. Thus, force of attraction increases (screening $s > p > d > f$). Hence, (IE) increases. Thus, Statement I and Statement II are correct and Statement II is the correct explanation of Statement I.

18. (a) Silicon has only one stable allotrope with diamond type structure. It is due to the formation of $p\pi$-$p\pi$ bonds to itself and all the four valence electrons are used in bonding.

Thus, Statement I and Statement II are correct and Statement II is the correct explanation of Statement I.

19. (b) In graphite, carbon atom is sp^2-hybridised and fourth electron is delocalised and is free to move in the plane of the sheet. As a result, the electrical conductivity of graphite in a direction parallel to the sheet is about 10^{20} times greater than the conductivity of diamond (in which all valence electrons are used in sp^3 hybridisation).

The conductivity of graphite perpendicular to the sheets of carbon atom is lower because electron must hop from one sheet to the next.

Thus, Statement I and Statement II are correct but Statement II is not the correct explanation of Statement I.

Note *The substance or crystal (as graphite) that shows different properties in different directions is called anisotropic.*

Explanations for Q. Nos. (20 and 21)

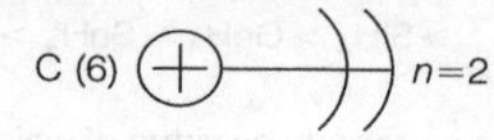

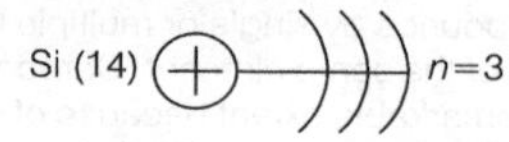

There is increase in size as we move from C to Si due to the introduction of 3rd shell.

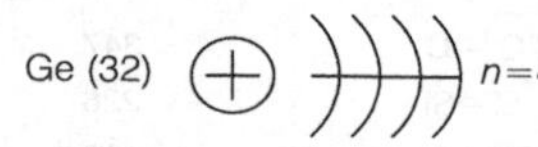

In case of Ge, there is effective nuclear charge due to screening. It balances the effect of addition of 4th shell. Thus, there is only marginal increase in the radius when we move from Si to Ge.

20. (a) **21.** (a)

22. (a) (i) PbO-amphoteric (+2 state) Thus, (i) → (p,r)

(ii) SnO_2 - amphoteric, +4 Thus, (ii) → (q,r)

(iii) PbO_2 - amphoteric, +4 Thus, (iii) → (q,r)

(iv) CO_2 - acidic, +4 Thus, (iv) → (q,t)

(v) GeO - acidic +2 Thus, (v) → (p,t)

23. (3)

Element	Oxidation state	Stability	
		Higher	Lower
B	+3	+3	
Al	+3	+3	
Ga	+3, +1	+3	
Sn	+2, +4	+4	
Ge	+2, +4	+4	
Pb	+2, +4		+2
C	+4	+4	
Tl	+3, +1		+1
Si	+4	+4	
As	+3, +5	+5	
Sb	+3, +5	+5	
Bi	+3, +5		+3

DPP-2 Compounds of Group 14

1. (d) Coal gas (a mixture of CO, H_2, CH_4 and CO_2) has maximum calorific value.

2. (c) CO has total electrons = 14

Thus, it is isoelectronic of N_2.

By MO theory, bond order = 3

$$:C \equiv O:$$

It has one lone pair and is a Lewis base. Thus, bonding between acceptor and itself is dative.

$$M \leftarrow C \equiv O: \text{ (in carbonyl)}$$

3. (b) BH_3 is electron-deficient in B and thus Lewis acid (electron-pair acceptor).

CO is electron rich and thus a Lewis base (electron pair donor).

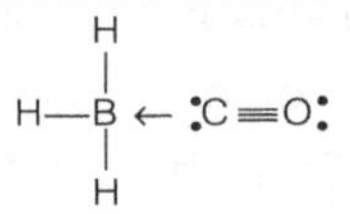

4. (b) Glass is sodium silicate formed when SiO_2 is fused with Na_2CO_3.

$$Na_2CO_3 + SiO_2 \xrightarrow{\Delta} Na_2SiO_3 + CO_2$$

It is called water glass or supercooled liquid.

5. (d) Method *X* is called zone refining.

Silicon is subjected to zone refining to get ultrapure silicon.

6. (d) $4Sn + 10HNO_3 \longrightarrow 4Sn(NO_3)_2 + NH_4NO_3 + 3H_2O$

$$Sn + 2H^+ \longrightarrow Sn^{2+} + 2H$$

Nascent H reduces NO_3^- to NH_3 and NH_3 changes to $NH_4^+NO_3^-$.

7. (b) Estimation of CO is done by I_2O_5 quantitatively.

$$\underset{1\text{ mol}}{I_2O_5} + \underset{5\text{ mol}}{5CO} \longrightarrow 5CO_2 + I_2$$

I_2O_5 is reduced to I_2 and CO is oxidised to CO_2. I_2 formed can be estimated by titration using $Na_2S_2O_3$.

$$I_2 + 2Na_2S_2O_3 \longrightarrow 2NaI + Na_2S_4O_6$$

8. (a,b)

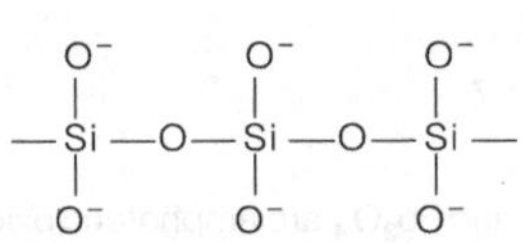
(Pyroxene) silicate

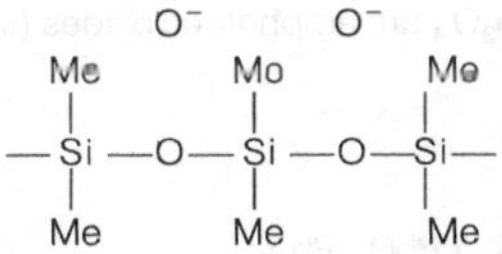
Polydimethyl siloxane (Silicone oil)

(a) O^- has been replaced by $—CH_3$. Thus, correct.

(b) O^-(9 electrons), $—CH_3$ (9 electrons) are isoelectronic, hence silicates and silicones are isoelectronic. Thus, correct.

(c) O (16), $—CH_3$ (15) are of different mass. Thus, (c) is incorrect.

(d) Incorrect based on (b).

9. (c,d) PbO has two forms:

(a) litharge (reddish yellow, tetragonal)

(b) massicot (yellow, orthorhombic)

Being in stable +2 state, they are not reduced. Thus, they are not oxidising agents.

(c) Red lead is $2PbO \cdot PbO_2$ in which PbO_2 is an oxidising agent.

$$PbO_2 + 4HCl \longrightarrow PbCl_2 + 2H_2O + Cl_2$$

(d) Lead dioxide (PbO_2) as in (c).

10. (a,b) Repeating units of silicones are shown as,

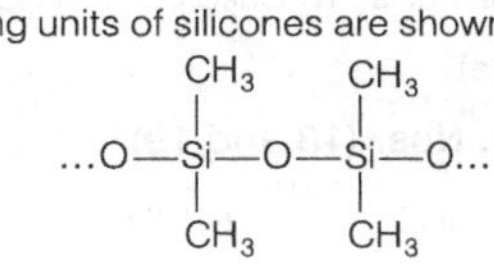

There are (Si—O) and (Si—C) bonds. Thus, thermal and chemical stability of silicones are due to the high stability of (Si— C) and (Si—O) bonds.

11. (a,c,d)

(a) Correct

(b) Incorrect (Structures are of same type)

(c) Correct

(d) Correct

12. (a,b,c,d) $C + H_2SO_4 \longrightarrow H_2O + CO_2 + SO_2$

$$H_2O + CO_2 \longrightarrow H_2CO_3$$

$$H_2O + SO_2 \longrightarrow H_2SO_3$$

13. (a,b)

(a) Correct

(b) Correct

14. (a) Compounds in which the π-electron system of graphite is retained and thus, electrical conductivity is also retained due to the mobile electrons (or even may increase) are called lamellar compounds.

Electrical conductivity is due to transfer of π electrons from metals to π-orbitals of the carbon plane, here they move freely under an electric field. Thus, Statement I and Statement II are correct and Statement II is the correct explanation of Statement I.

15. (b) C : $1s^2 2s^2 2p^2$

Second orbit lacks 2*d*-orbital. Thus, carbon cannot expand its octet beyond eight. Thus, only CF_4 is formed by using four valence electrons.

Si : $1s^2 2s^2 2p^6 3s^2 3p^2$

Empty 3*d* orbital is also available for bonding thus, SiF_6^{2-} is formed in which silicon expands its octet beyond 10 by using 3*d*-orbital.

Glass is a silicate.

$$Na_2SiO_3 + 6HF \longrightarrow Na_2SiF_6 + 3H_2O$$

Thus, HF attacks glass.

Thus, Statement I and Statement II are correct but Statement II is not the correct explanation of Statement I.

Explanations for Q. Nos. (16 and 17)

Silicate is in the chain form having three SiO_4.

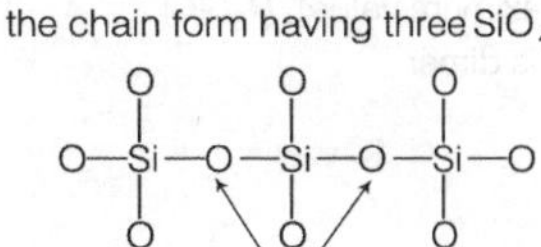

Thus, anion is Si_3O_{10}.

$$\text{Charge} = 4 \times 3 + (-2) \times 10 = -8$$

–8 charge is balanced by Ca^{2+} and Cu^{2+}.

Being in 1 : 1 ratio

$$Cu^{2+} = 2$$
$$Ca^{2+} = 2$$
$$H_2O = 2$$

Thus, silicate mineral is $Ca_2Cu_2Si_3O_{10} \cdot 2H_2O$.

16. (b) **17.** (a)

Explanations for Q. Nos. (18 and 19)

$$Ni + 4CO \xrightarrow{301\,K} Ni(CO)_4$$

$$Ni(CO)_4 \xrightarrow{503\,K} \underset{(Pure)}{Ni} + 4CO \uparrow$$

This process is called Mond's process.

18. (a) **19.** (b)

20. (4)

$$\underset{\substack{5\text{ mol}\\10\text{ mol}}}{5CO} + I_2O_5 \longrightarrow 5CO_2 + \underset{\substack{1\text{ mol}\\2\text{ mol}}}{I_2}$$

$$\underset{\substack{1\text{ mol}\\2\text{ mol}}}{I_2} + \underset{\substack{2\text{ mol}\\4\text{ mol}}}{2Na_2S_2O_3} \longrightarrow Na_2S_4O_6 + 2NaI$$

21. (4)

SiO_4^{4-} anion is in

(1) Orthosilicate

(7) Single chain silicate

(8) Double chain silicate

(9) Sheet like silicate

Pyrosilicate has $Si_2O_7^{6-}$ and $Si_6O_{18}^{12-}$ anions.

(Si—O) is in

(3) Pyroxene (Si—O chains)

(4) Amphibole (Si—O double chains)

(5) Mica (Si—O sheets)

(6) Zeolite (Si—O 3-D network)

22. (6)

Each SiO_4 tetrahedron shares one oxygen atom with each of the two neighbouring tetrahedra forming a ring anion. In three tetrahedra,

$(SiO_4)_3 = Si_3O_{12}$, charge $= -12$

Oxygen common = 3, charge $= -6$

Thus, formula is Si_3O_9, Net charge $= -12 - (-6)$
$= -6$

Thus, anion is $(Si_3O_9)^{6-}$.

23. (3) R_nSiCl_{4-n}

When $n = 1$, $RSiCl_3$

Silicon formed is cross-linked type.

When $n = 2$, R_2SiCl_2

Silicon formed is repeated chain type.

When $n = 3$ (maximum value), R_3SiCl

Silicon formed is dimer.

24. (2)

Liquefied Petroleum Gas	Mixture of *n*-butane and propane
Water gas	$(CO + H_2)$
Producer gas	$(CO + N_2)$
Coal gas	CO, CH_4, H_2, CO_2
CNG (Compressed Natural Gas)	Mainly CH_4

25. (4)

		σ	π	
C_2^{2-}	$^-C{\equiv}C^-$	1	2	
C_3^{4-}	$^-C{\equiv}C{-}C^{3-}$	2	2	Equal
CN_2^{2-}	$^-N{=}C{=}N^-$	2	2	Equal
CO_2	$O{=}C{=}O$	2	2	Equal
C_2H_4	H—C(H)=C(H)—H	5	1	
CN^-	$^-C{\equiv}N$	1	2	
C_3O_2	$O{=}C{=}C{=}C{=}O$	4	4	Equal

26. (2)

Si and Ge are used as semiconductors in Group 14.

27. (5)

$$\underset{\text{Malonic acid}}{CH_2(COOH)_2} \longrightarrow \underset{\substack{\text{Carbon}\\\text{suboxide}}}{C_3O_2} + 2H_2O$$

$$O{=}C{=}C{=}C{=}O$$

Total atoms = 5

28. (2)

C = + 4

Si = + 4

Ge = + 2, + 4,

Sn = + 2, + 4,

Pb = + 2, + 4

S = + 2, + 4, + 6

Cl = − 1, + 1, + 3, + 5, + 7

29. (5)

SnO, SnO_2, PbO, PbO_2 and Pb_3O_4 are amphoteric oxides (soluble in alkali and acid both).

Note *Acidic* : CO, CO_2, SiO_2, GeO

Basic : GeO_2

Amphoteric : $SnO, SnO_2, PbO, PbO_2, Pb_3O_4$

30. (0) Lead pencil contains graphite and lead content in zero percentage.

Revisal Problems (JEE Main)

1. (a)

(a) $SnCl_2$ dissolves in KCl forming $K[SnCl_3]$.

$$SnCl_2 + KCl \longrightarrow K[SnCl_3]$$

(b) $SnCl_2 + H_2O \longrightarrow Sn(OH)Cl$

$SnCl_2$ is partially hydrolysed in H_2O.

(c) $SnCl_2$ reduces $FeCl_3$ to $FeCl_2$.

$$2FeCl_3 + SnCl_2 \longrightarrow 2FeCl_2 + SnCl_2$$

(d) $SnCl_2$ reduces $HgCl_2$ to Hg.

$$HgCl_2 + SnCl_2 \longrightarrow SnCl_4 + Hg$$

Note SnF_2 *dissolves in* KF *forming* $K[SnF_3]$.

2. (b)

Graphite + HNO_3 (Conc. and hot) $\longrightarrow$

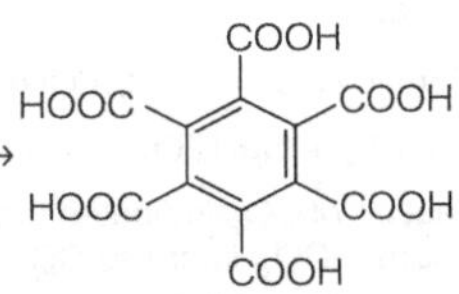

Mellitic acid

Note *Graphite* $\xrightarrow[HNO_3/HF]{\text{Hot conc.}}$ *Graphite oxide*

3. (a) About one-half of the anticipated greenhouse effect is due to CO_2 and the remainder due to other gases as CH_4, O_3, N_2O, CFC (chlorofluorocarbons). These gases are even stronger IR absorber than CO_2 and thus cause global warming.

4. (c)

(a) $FeCl_3$ is reduced to $FeCl_2$.

$$2FeCl_3 + SnCl_2 \longrightarrow SnCl_4 + 2FeCl_2$$

(b) $CuCl_2$ is reduced to CuCl.

$$2CuCl_2 + SnCl_2 \longrightarrow SnCl_4 + 2CuCl$$

(c) $HgCl_2$ is first reduced to Hg_2Cl_2 and then to Hg.

$$2HgCl_2 + SnCl_2 \longrightarrow Hg_2Cl_2 + SnCl_4$$

$$Hg_2Cl_2 + SnCl_2 \longrightarrow 2Hg + SnCl_4$$

(d) $FeCl_2$ is not reduced.

5. (d) SiC has diamond type structure and is also called artificial diamond. It has tetrahedral structure in which each C-atom is bonded to four Si-atoms and *vice-versa*. C-atom is sp^3 hybridised and Si-atom is also sp^3-hybridised.

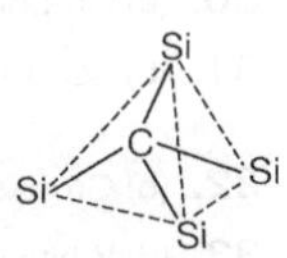

Being hard, it is used in cutting tools and also as an abrasive. It is also called carborundum.

6. (d)

(a) CO is a reducing agent and reduces metal oxides to metals (except alkali and alkaline earth metals like Al).

$$Fe_2O_3 + 3CO \longrightarrow 2Fe + 3CO_2$$

$$ZnO + CO \longrightarrow Zn + CO_2$$

(b) CO is a main content of water gas (synthesis gas), producer gas thus, used as a fuel.

(c) CO makes carbonyl with metals, e.g. $Ni(CO)_4$, $Fe(CO)_5$, etc.

(d) CO does not act as an oxidising agent.

7. (b) Carbides which on decomposition with H_2O form methane, are called methanides.

$$CaC_2 + 2H_2O \longrightarrow Ca(OH)_2 + \underset{\text{Acetylene}}{CH{\equiv}CH}$$

$$Mg_2C_3 + 4H_2O \longrightarrow 2Mg(OH)_2 + \underset{\text{Propyne}}{CH_3{-}C{\equiv}CH}$$

$$Al_4C_3 + 12H_2O \longrightarrow 4Al(OH)_3 + 3CH_4$$

$$Be_2C + 4H_2O \longrightarrow 2Be(OH)_2 + CH_4$$

Thus, Al_4C_3 and Be_2C are methanides.

8. (d) $O \underset{\pi}{\overset{\sigma}{=}} \underset{\underset{sp}{\uparrow}}{C} \underset{\pi}{\overset{\sigma}{=}} O$

Out of four valence electrons of carbon, two are involved in *sp* hybridisation. These *sp* hybrid orbitals form σ-bonds with *p*-orbitals of O-atom. Two electrons of C-atoms are involved in *pπ*-*pπ* bonding with O-atoms.

Shape is linear (θ = 180°)

Due to resonance, bond length are equal.

$$:\ddot{O}{=}C{=}\ddot{O}: \rightleftharpoons :\ddot{O}^{-}{-}C{\equiv}\overset{\oplus}{O}: \rightleftharpoons :\overset{\oplus}{O}{\equiv}C{-}\ddot{O}^{-}:$$

9. (b) H_2CO_3 (weak acid) and its conjugate base HCO_3^- is used to control pH of blood.

$$pH = pK_a + \log \frac{[HCO_3^-]}{[H_2CO_3]}$$

10. (c)

(a) Carbonic acid is aqueous CO_2.

$$CO_2 + H_2O \rightleftharpoons H_2CO_3$$

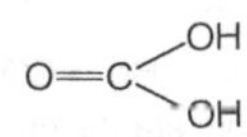

It behaves as a weak dibasic acid. The solubility of CO_2 in water increases with pressure.

(b) Chloral hydrate is stable due to intramolecular H-bonding.

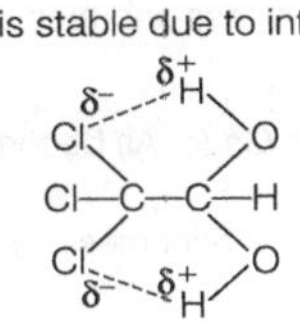

11. (c)

(a) MnO_2/CuO catalytically oxidises CO to CO_2 in the presence of air.

$$CO + \frac{1}{2}O_2 \xrightarrow{MnO_2/CuO} CO_2$$

(b) K_2O_2 reacts with CO forming K_2CO_3.

$$K_2O_2 + CO \longrightarrow K_2CO_3$$

Thus, its (CO) fatal effect is minimised.

12. (b) $:\bar{C}\vdots\vdots\bar{C}:$ Valence electrons in C = 4

Electron gained by each C = 1

Total electrons on each C = 5

Electrons shared between two C-atoms = 6

$$\text{Bond order} = \frac{6}{2} = 3$$

$:\bar{C}{\equiv}\bar{C}:$

13. (d) In silicon dioxide, each silicon is bonded to four oxygen atoms in tetrahedral structure such that Si is at the centre and four oxygen atoms are at the four corners, in each case bonding is covalent.

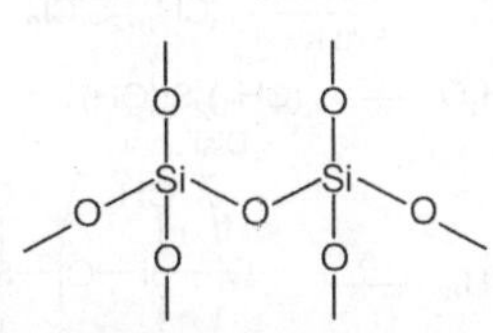

14. (b)

(a) $6CO_2 + 6H_2O \xrightarrow{\text{Sunlight}} 6O_2 + C_6H_{12}O_6$

This process is called photosynthesis.

(b) Reverse of reaction (a) is called respiration in which animals and plants release energy.

$$C_6H_{12}O_6 + 6O_2 \longrightarrow 6CO_2 + 6H_2O + \text{Energy}$$

15. (c) $CO_2(g) \xrightarrow{-57°C} CO_2(l)$

$$CO_2(s) \xrightarrow[\text{Atmospheric pressure}]{-78°C} CO_2(g)$$

Solid CO_2 is called dry ice. It is also called cardice. It is used to preserve meat, frozen foods and as a coolant in the laboratory.

16. (b) In sheet silicate, three oxygen atoms of SiO_4^{4-} are shared in repeating structures.

17. (d)

(a) Water gas is an equimolar mixture of CO and H_2.

$$C + H_2O \xrightarrow{\text{Red hot}} CO + H_2$$

(b) Producer gas is a mixture of CO and N_2 in 1 : 2 molar ratio.

$$\underset{\text{(Red)}}{C} + \underbrace{O_2 + 4N_2}_{\text{(Air)}} \longrightarrow CO_2 + 4N_2$$

$$\downarrow \text{Red hot C}$$

$$2CO + 4N_2$$

Producer gas is less efficient than water gas.

(c) Coal gas is a mixture of CO, CH_4, H_2 and CO_2.

18. (b) CO is toxic because it forms complex with haemoglobin in the blood.

$$Hb—O_2 + CO \rightleftharpoons Hb—CO + O_2$$

This complex (Hb—CO) is more stable than oxyhaemoglobin. This prevents the haemoglobin in the red blood corpuscles from carrying oxygen round the body. This causes oxygen deficiency leading to unconsciousness and then death.

19. (d)

I. Silicon doped with Ga (or Al) becomes *p*-type semiconductor. Thus, correct.

II. Silicon doped with P becomes *n*-type semiconductor. Thus, correct

III. Hyperfine is the purest form of silicon and is used in the production of transistors. Thus, correct.

20. (c) There are in all six tetrahedra with SiO_4^{4-} in which one oxygen shared between two tetrahedra.

Charge on each $SiO_4 = -4$

Charge on six $SiO_4 = -24$

Formula of six SiO_4 (if no oxygen shared) = Si_6O_{24}

Oxygen shared = 6

Thus, formula = $[Si_6O_{24} - 6(O)] = Si_6O_{18}$

Charge = $(+4 \times 6) + (-2) \times 18 = -12$

Thus, silicate is $Si_6O_{18}^{12-}$.

21. (c) Slaked lime is $Ca(OH)_2$. Only CO_2 is absorbed by $Ca(OH)_2$ making it milky.

$$Ca(OH)_2 + CO_2 \longrightarrow \underset{\text{Milky}}{CaCO_3} + H_2O$$

22. (c) $\underset{\text{Magnesium silicide}}{Mg_2Si} + 4NH_4Br \longrightarrow \underset{\text{Silane}}{SiH_4} + 2MgBr_2 + \underset{\text{Ammonia}}{4NH_3}$

23. (b) $2CH_3Cl + Si \xrightarrow[570\ K]{\text{Cu powder}} (CH_3)_2SiCl_2$

$$(CH_3)_2SiCl_2 + H_2O \longrightarrow \underset{\text{Disilanol}}{(CH_3)_2Si(OH)_2}$$

$n(CH_3)_2Si(OH)_2 \xrightarrow{\Delta}$

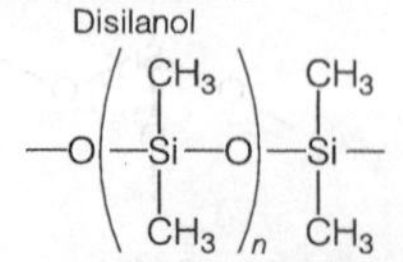

Silicones

24. (b)

(a) Quartz is extensively used as a piezoelectric material. When polar crystals is subjected to mechanical stress, electricity is produced (piezoelectricity).

(b) Kieselguhr, an amorphous form of silica, is used in filtration plants.

(c) Silane – SiH_4

(d) Silicones – R_2SiO

25. (c) Silicones are hydrophobic (water repellant) and not hydrophilic.

I, II, III and IV are associated with silicones.

26. (c) ZSM-5, $H_x[AlO_2]_x[SiO_2]_{96-x} \cdot 16H_2O$, converts alcohols to gasoline (petrol) by dehydrating the alcohols.

Hydrated zeolites are used as ion exchangers in softening of hard water.

27. (c) $SnO + H_2O \longrightarrow Sn(OH)_2$

$$Sn(OH)_2 + NaHCO_3 \longrightarrow CO_2 + H_2O + Sn^{2+} + Na^+$$

Formation of CO_2 indicates that SnO is an acidic oxide (in this reaction). HCO_3^- liberates CO_2 with acid.

$$HCO_3^- + H^+ \longrightarrow H_2O + CO_2$$

Note SnO *and* SnO_2 *both are amphoteric oxide.*

28. (c) If CaO in glass is replaced by PbO, lead glass has a higher refractive index used in making optical parts and glass ornaments.

Note *Normal glass – Sodium calcium silicate*

Heat resistant glass (pyrex) – Borosilicate glass

Opal glass – CaF_2 *is added*

Bubble free glass – $NaNO_3$ *or* As_2O_3 *is added*

Coloured glass – by coloured metal oxide

Green	– Fe_2O_3/CuO
Yellow	– UO_2
Blue	– CuO/CoO
Red	– *Colloidal* Cu, Au

29. (c) CO and CO_2 are stable oxides of carbon.

30. (a) Graphite is the most stable form of carbon thermodynamically.

31. (b) $CO + Cl_2 \longrightarrow \underset{\text{Phosgene}}{COCl_2}$

32. (b) Crooks glass is used to cut ultraviolet rays.

33. (d) White lead is basic lead carbonate $Pb(OH)_2 \cdot PbCO_3$.

34. (c) Graphite consists of planar hexagonal layers of carbon atoms held together by relatively weak van der Waals', forces. Within each layer, a carbon atom is bonded to three carbon atoms by sp^2 hybrid bonds and remaining fourth *p*-electron is delocalised in extended π-bonding. Thus, graphite is a good conductor of electricity.

Thus, Statement I is correct and Statement II is incorrect.

35. (b) As we go down the group (top to bottom), ionisation energy decreases and reactivity increases. But lead appears to less reactive than expected. It is due to the surface coating of oxide and partly due to high overpotential for the reduction of H^+ to H_2.

With dil. H_2SO_4, lead forms insoluble coating of $PbSO_4$, hence it does not react further. Thus, Statement I and Statement II are correct but Statement II is not the correct explanation of Statement I.

Revisal Problems (JEE Advanced)

1. (c)

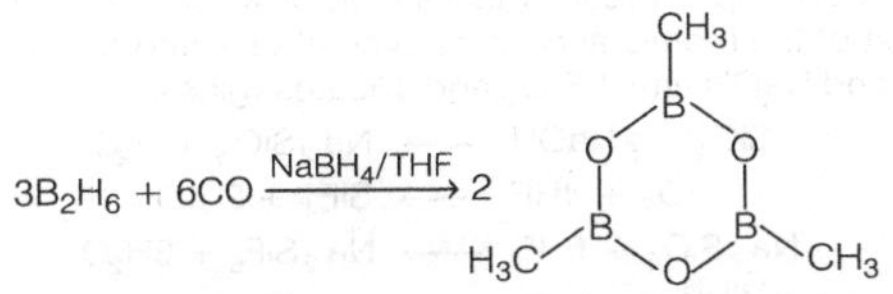

B-trimethyl boroxine

THF is tetrahydrofuran (solvent) and $NaBH_4$ is a reducing agent.

2. (b)

(a) Incorrect

(b) Silica is a network solid with tetrahedral coordination and is a giant molecule. Thus, correct.

(c) SnO_2, PbO_2 - amphoteric oxide
GeO_2 - basic oxide

3. (d)

(a) BH_3 and SiH_4 are volatile and catch fire on exposure to air. Thus, correct.

(b) B_2O_3 and SiO_2 are acidic oxides soluble in NaOH.

$$B_2O_3 + 6NaOH \longrightarrow 2Na_3BO_3 + 3H_2O$$
$$SiO_2 + 2NaOH \longrightarrow Na_2SiO_3 + H_2O$$

Thus, correct.

(c) Borates and silicates have tetrahedral structural units. Thus, correct.

4. (b) $CaC_2 + H_2O \longrightarrow Ca(OH)_2 + CH\equiv CH$

$$Ca_3P_2 + 6H_2O \longrightarrow 3Ca(OH)_2 + 2PH_3$$

CaC_2 and Ca_3P_2 in two different containers (with holes) are thrown into sea water. PH_3, some P_2H_4 and H_2 are formed which are spontaneously flammable gases. $CH\equiv CH$ formed catches fire immediately (even in water) and thus can guide the path for missing ships.

5. (a) $K_4[Fe(CN)_6] + 6H_2SO_4 + 6H_2O \xrightarrow{\Delta} 2K_2SO_4 + FeSO_4 + 3(NH_4)_2SO_4 + 6CO$

6. (b) Zeolites are naturally occurring aluminosilicates having an open three-dimensional structure with loosely held large metal cations and water molecules in spacious honey comb like cavities. The electrostatic binding of the cations increases with increasing charge (actually charge density) $M^{3+} > M^{2+} > M^{+}$. Thus, zeolite mineral like $Na_2[Al_2Si_3O_{10}]\cdot 2H_2O$ (natrolite) will exchange two Na^+ ions for one Ca^{2+}/Mg^{2+} ion present in hard water.

$$Na_2Ze + CaCl_2 \rightleftharpoons CaZe + 2NaCl$$

Note *The equilibrium may be reversed by using a high concentration of NaCl and the zeolite may be regenerated. This is the principle of water softening by zeolites.*

7. (b) Mixture of O_2 and CO_2 is called carbogen.

8. (d) Carbogen is used as an antidote to minimise poisonous effect of CO gas.

$$CO + O_2 \longrightarrow CO_2$$

Mixture of CO_2 and O_2 is called carbogen.

9. (a) Pb_3O_4 is used as a primer for iron to prevent it from rusting.

Pb_3O_4 is also called sindoor.

10. (b)

(a) Polyyne (discovered in 1995) - (sp) Thus, correct.

$$-\!\!\{C\equiv C-C\equiv C-C\equiv C-C\equiv C\}_n\!\!-$$

(b) Bucky ball - (sp^2) Thus, incorrect.

(c) Graphite - (sp^2) Thus, correct.

(d) Diamond - (sp^3) Thus, correct.

11. (d)

12. (c) If $(CH_3)_3SiCl$ is added then it blocks one terminal side completely for further reaction and thus, the chain length is controlled.

$$(CH_3)_3SiCl \xrightarrow{H_2O} (CH_3)_3Si(OH)$$

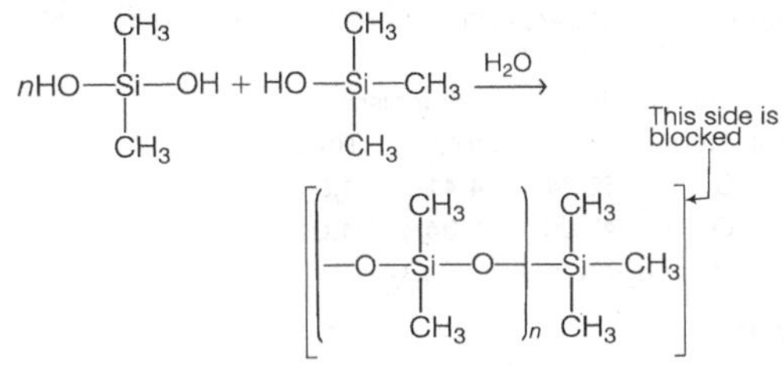

13. (b) $6CO_2 + 12H_2O \xrightarrow{\text{Sunlight}} 6O_2 + C_6H_{12}O_6 + 6H_2O$

All O_2 is from $12H_2O$ molecules.

If oxygen of H_2O is labelled by (^{18}O) then all the six O_2 molecules are ^{18}O.

14. (a,b,d)

(a) Graphite uses three electrons in bonding with sp^2 hybrid orbitals and fourth electron is delocalised in π-bonding. Thus, graphite is conducting.
Diamond uses four electrons in bonding with sp^3 hybrid orbitals and thus non-conducting. Thus, correct.

(b) Graphite consists of planar hexagonal layers of C-atoms held together by weak van der Waals' forces. Layers can slide over one another. This makes graphite a lubricant and slippery. Thus, correct.

(c) Based on (a). Thus, incorrect.

(d) Diamond $\xrightarrow{\Delta}$ Graphite
Thus, graphite is thermodynamically more stable than diamond.

15. (b,c) Silicon, germanium and boron are metalloids and used as semiconductors.

16. (a,c,d)

(a) PbS- insoluble (black precipitate)

(b) $Na_2[Pb(OH)_6]$ - soluble

(c) $PbSO_4$ - insoluble (white precipitate)

(d) $PbCl_2$ - insoluble (white precipitate)

Note *$PbCl_2$ is soluble in hot water.*

17. (b,d) When CO_2 goes into the atmosphere, it gets dissolved in H_2O also.

$$CO_2 + H_2O \rightleftharpoons H_2CO_3$$
$$H_2CO_3 \rightleftharpoons H^+ + HCO_3^-$$

Water becomes acidic. Hence, pH of water decreases.

Due to increase of acidic nature, fertility of soil decreases.

Note *Fertility of alkaline soil is better.*

18. (a,b)

(a) $HCOOH \xrightarrow[\text{Conc. } H_2SO_4]{373\text{ K}} H_2O + CO$

(b) $\begin{matrix} COOH \\ | \\ COOH \end{matrix} \xrightarrow{\Delta} CO_2 + CO + H_2O$

(c) $CH_4 + 2O_2 \xrightarrow{\Delta} CO_2 + 2H_2O$

(d) $C(\text{graphite}) + O_2 \longrightarrow CO_2$

19. (b,c,d) Carbon suboxide is C_3O_4.

(a) $\begin{matrix} CH_2COOH \\ | \\ CH_2COOH \end{matrix} \xrightarrow{P_2O_5} \begin{matrix} CH_2CO \\ | \\ CH_2CO \end{matrix} \!\!> O + H_2O$

Succinic acid

(b) $CH_2(COOH)_2 \xrightarrow{P_2O_5} C_3O_2 + H_2O$

Malonic acid

(c) $O{=}\underset{sp}{C}{=}\underset{sp}{C}{=}\underset{sp}{C}{=}O$

Each C-atom is *sp* -hybridised.

(d)

Atom	%	Moles	Ratio
C	52.94	4.41	1.5
O	47.06	2.94	1.0

Molecular formula = C_3O_2

20. (a,b)

(a) $CO + \underset{(Fused)}{NaOH} \xrightarrow{\Delta} \underset{\text{Sodium formate}}{HCOONa}$

(b) $CO + \underset{(Fused)}{CH_3ONa} \xrightarrow{\Delta} \underset{\text{Sodium acetate}}{CH_3COONa}$

(c) $2CO + B_2H_6 \longrightarrow \underset{\text{Borane carbonyl}}{2BH_3CO}$

(d) $2CO + 2Na \xrightarrow{NH_3} \underset{\text{Colourless crystalline salt}}{NaOC{\equiv}CONa}$

21. (b,c,d) Carbon lacks *d*-orbital hence, it cannot expand its valency beyond four. Silicon, tin and lead can use *d*-orbital and thus can expand their octet beyond eight.

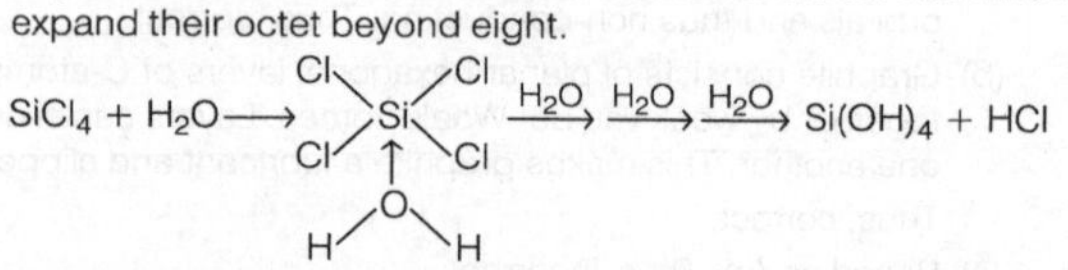

22. (a,b,c,d)

(a)

	$O{=}C{=}O$	$^-N{=}C{=}N^-$
Total electrons	22	22 Thus, isoelectronic.
Structure	Linear	Linear
θ	180°	180°
C-atom	*sp* -hybridised	*sp* -hybridised

(b) $BeC_2 + 2H_2O \longrightarrow Be(OH)_2 + \underset{\text{Ethyne}}{CH{\equiv}CH}$

$Mg_2C_3 + 4H_2O \longrightarrow 2Mg(OH)_2 + \underset{\text{Propyne}}{CH_3{-}C{\equiv}CH}$

Thus, correct.

(c) $Be_2C + 4H_2O \longrightarrow 2Be(OH)_2 + \underset{\text{Methane}}{CH_4}$

$Al_4C_3 + 12H_2O \longrightarrow 4Al(OH)_3 + \underset{\text{Methane}}{3CH_4}$

Both form CH_4 on reaction with H_2O, hence called methanides. Thus, correct.

(d) CaC_2 has NaCl type of lattice with Ca^{2+} replacing Na^+ and C_2^{2-} replacing Cl^-.

Thus, correct.

23. (a,b,c) (a) Correct (b) Correct (c) Correct (d) Incorrect

24. (b) Silica (SiO_2) is almost unreactive because of very high Si—O bond enthalpy. It resists the attack of $X_2(Cl_2, Br, I_2, F_2)$ and H_2, most of the metals and acids even at elevated temperature. But HF and NaOH attack SiO_2 and silicates (glass).

$SiO_2 + 2NaOH \longrightarrow Na_2SiO_3 + H_2O$

$SiO_2 + 4HF \longrightarrow SiF_4 + 2H_2O$

$\underset{\text{Glass}}{Na_2SiO_3} + 6HF \longrightarrow Na_2SiF_6 + 3H_2O$

Thus, Statement I and Statement II are correct but Statement II is not the correct explanation for Statement I.

Note *HF attacks glass,. This property of reaction of HF with glass is used to marking it.*
HF is stored in wax-vessel and not in glass vessel.

25. (b)

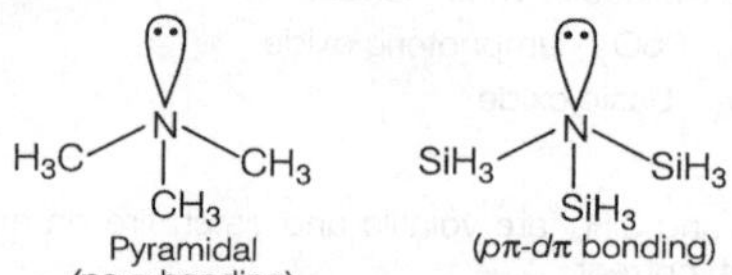

Pyramidal (no π-bonding) (pπ-dπ bonding)

In $N(CH_3)_3$, there is no *d*-orbital in C-atom thus, lone pair causes *lp-bp* repulsion and as NH_3, it is pyramidal in structure. In $N(SiH_3)_3$, lone pair of N-atom is transferred to the empty *d*-orbitals of silicon (*pπ- dπ* overlapping). Thus, repulsion is minimised. Structure is thus planar.

Thus, Statement I and Statement II are correct but Statement II is not the correct explanation of Statement I.

26. (a)

(i) → (u)
(ii) → (s)
(iii) → (p)
(iv) → (t)
(v) → (r)
(vi) → (q)

27. (a)

(i) CO_2 — dry ice
— fire extinguisher
— H_2CO_3 in H_2O is acidic
— C-atom is *sp* -hybridised
— Greenhouse effect (global warming)

Thus, (i) – (p,q,r,t,u)

(ii) $Al_2(SO_4)_3 + 6H_2O \longrightarrow 2Al(OH)_3 + \underset{(Strong)}{3H_2SO_4}$

Acidic solution due to hydrolysis, used in fire extinguisher device along with $NaHCO_3$.

Thus, (ii) → (q,r)

(iii) ($^-C{\equiv}N$)–ligand

(N, C) atoms are *sp*-hybridised.

Thus, (iii) → (s,t)

(iv) ($^-C{\equiv}C^-$) (*sp* -hybridised)

Thus, (iv) → (t)

(v) ($^{3-}C{-}C{\equiv}C^-$) (*sp*-hybridised)

Thus, (v) → (t)

Explanations for Q. Nos. (28-30)

Pb_3O_4 is a double oxide of PbO and PbO_2 in 2 : 1 ratio.

$$(2PbO)\cdot(PbO_2) \equiv Pb_3O_4$$

$$Pb_3O_4 + 3H_2SO_4 \longrightarrow 3PbSO_4 + 3H_2O + \frac{1}{2}O_2$$

$$Pb_3O_4 \longrightarrow 2PbO + PbO_2$$

$$2PbO + 2H_2SO_4 \longrightarrow 2PbSO_4 + 2H_2O$$

$$PbO_2 \longrightarrow PbO + \frac{1}{2}O_2$$

$$PbO + H_2SO_4 \longrightarrow PbSO_4 + H_2O$$

$$Pb_3O_4 + 4HNO_3 \longrightarrow 2Pb(NO_3)_2 + PbO_2 + 2H_2O$$

This reaction indicates that Pb_3O_4 is a double oxide. PbO_2 is not affected by HNO_3 while PbO dissolves in HNO_3 forming $Pb(NO_3)_2$.

28. (b)

29. (a)

30. (c)

Explanations for Q. Nos. (31 and 32)

Final product is NH_2CONH_2 (along with $CaCO_3$).

Thus, *C* is $CaCN_2$.

CO_2 turns *B* milky. Thus, *B* is $Ca(OH)_2$.

B is formed along with ($CH \equiv CH$). Thus, *A* is CaC_2.

$$CaC_2 + N_2 \longrightarrow CaCN_2 + C$$

Thus, *D* is carbon.

Mixture of $CaCN_2$ and C is called nitrolim.

Thus, *A* : CaC_2
B : $Ca(OH)_2$
C : $CaCN_2$ (cyanamide)
D : Carbon

31. (a) **32.** (c)

Note *CN_2^{2-} is isoelectronic of CO_2 and has linear structure.*

$$\bar{N}{=}C{=}\bar{N}$$

Explanations for Q. Nos. (33-35)

$$RMgCl + SiCl_4 \longrightarrow MgCl_2 + \underset{(A)}{RSiCl_3}$$

$$RSiCl_3 + RMgCl \longrightarrow \underset{(B)}{R_2SiCl_2} + MgCl_2$$

$$R_2SiCl_2 + RMgCl \longrightarrow \underset{(C)}{R_3SiCl} + MgCl_2$$

$$\underset{(A)}{RSiCl_3} + 3H_2O \longrightarrow \underset{\text{Trisilanol}}{RSi(OH)_3} + 3HCl$$

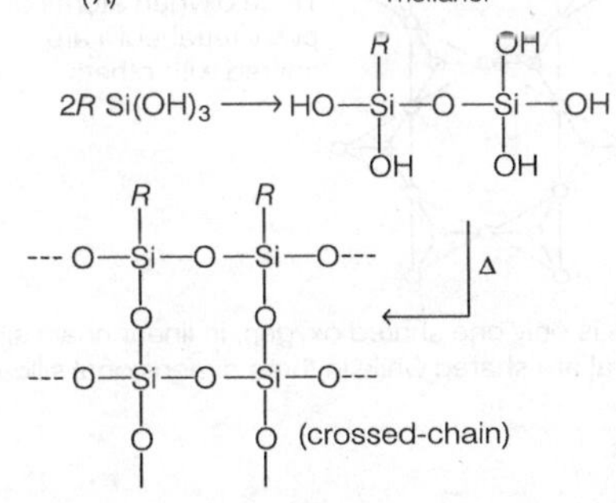

Thus, cross-linked polymer (silicone) is formed from *A*.

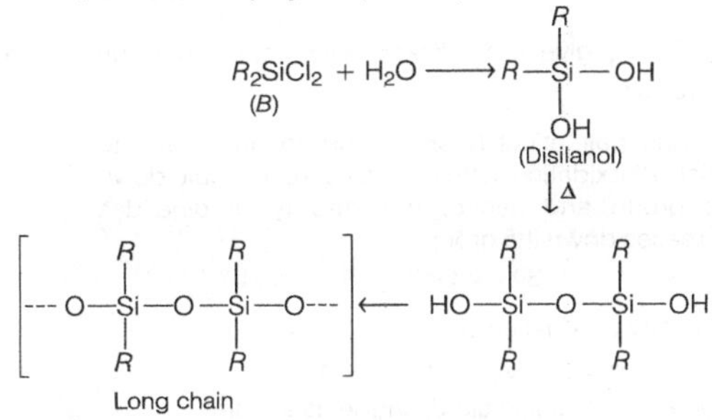

Thus, long-chain polymer is formed from *B*.

Cross-linked polymer is used in cement-work and making water proof fibre. Using $\left(C_6H_5{-}\overset{OH}{\underset{C_6H_5}{Si}}{-}OH\right)$, silicone resin is formed.

$$\underset{(C)}{R_3SiCl} + H_2O \longrightarrow \underset{\text{Silanol}}{R_3Si(OH)}$$

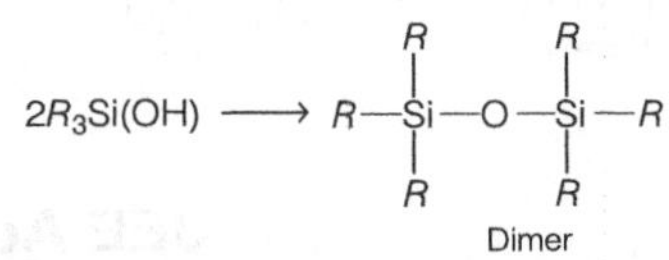

Thus, dimer is formed from *C*.

33. (c) **34.** (d) **35.** (c)

36. (8) C_{60} (fullerene – in a shape of soccer ball) has 20 hexagonal faces and 12 pentagonal faces.

Thus, $x = 20$
$y = 12$
$(x - y) = 8$

37. (5) CCl_4, $SiCl_4$, $GeCl_4$, $SnCl_4$, $PbCl_4$ are volatile liquids.

38. (8) Let CO = *x* L then $CO_2 = (10 - x)$ L

$$\underset{\substack{1\,L\\(10-x)}}{CO_2} + C \longrightarrow \underset{\substack{2L\\2(10-x)\,L}}{2CO}$$

CO_2 is reduced to CO. There is no effect on CO. Thus, total volume after reaction $= 2(10 - x) + x$

$$2(10 - x) + x = 12$$
$$x = 8\,L$$

Thus, $CO = 8\,L$

39. (4) Be_2C, Mg_2C_3, CaC_2 and Al_4C_3 are ionic carbides whereas SiC and B_4C_3 are covalent carbides.

WC and Cr_4C_3 are interstitial carbides.

40. (6) Photosynthesis reaction

$$6CO_2 + 12H_2O \xrightarrow{\text{Sunlight}} 6O_2 + \underset{\text{Glucose}}{C_6H_{12}O_6} + 6H_2O$$

Note *All the O_2 are from water, thus twelve H_2O molecules are consumed and six H_2O molecules are formed per glucose molecule synthesised.*

JEE Main & AIEEE Archive

1. (b) $RSiCl_3$ give cross-linked silicone polymer on hydrolysis.

2. (c) Inert pair effect becomes predominant due to which +2 oxidation state becomes more stable down the group and hence, the stability of dihalides increases down the group.

$$SiX_2 < GeX_2 < SnX_2 < PbX_2$$

3. (c) In MCl_2, oxidation state of $M = +2$

In MCl_4, oxidation state of $M = +4$

Higher the oxidation state, smaller is the size, greater the polarising power, greater is the covalent character.

Hence, MCl_4 is more covalent and MCl_2 is more ionic.

4. (a) A species is amphoteric if it is soluble in acid (behave as a base) as well as in base (behave as an acid).

$$\underset{\text{Base}}{SnO_2} + \underset{\text{Acid}}{4HCl} \longrightarrow SnCl_4 + 2H_2O$$

$$\underset{\text{Acid}}{SnO_2} + \underset{\text{Base}}{2NaOH} \longrightarrow Na_2SnO_3 + H_2O$$

5. (d) In SiO_2, each silicon atom is bonded to four oxygen atoms that are arranged tetrahedrally around it and each O-atom is attached to two silicon atoms by covalent bonds.

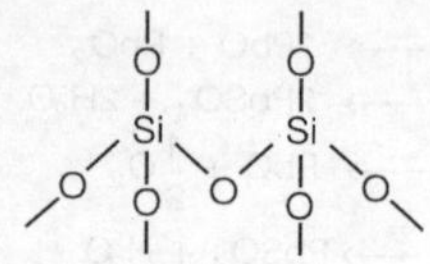

6. (a) As temperature decreases, white tin (β-form) changes to grey tin (α-form).

$$\alpha\text{-Sn} \xrightleftharpoons{13.2^\circ C} \beta\text{-Sn}$$

α-Sn has a much lower density.

7. (b) Glass is a supercooled liquid, which forms a non-crystalline solid. The common lime-soda glass used for bottles and jars is a supercooled mixture of sodium and calcium silicates.

8. (d) C-atoms of graphite form covalently bonded plates (layers). These layers are held together by weak forces of attraction, i.e. one layer can slide over other to cause lubricacy. It cannot be melted easily as large number of atoms being bonded strongly in the layer to form big entity.

JEE Advanced & IIT JEE Archive

1. (b,d) Diamond has a three dimensional network structure, a hard substance where graphite is soft due to layered structure.

In graphite, only three valence electrons are involved in bonding and one electron remain free giving electrical conductivity. In diamond, all the four valence electrons are covalently bonded hence, insulator.

Diamond is better thermal conductor than graphite. Electrical conductivity is due to availability of free electrons, thermal conduction is due to transfer of thermal vibrational energy from one atom to another atom. A compact and precisely aligned crystals like diamond thus facilitates better movement of heat.

In graphite, C—C bond acquires some double bond character, hence higher bond order than in diamond.

2. (3) $Be_nAl_2Si_6O_{18}$,

$$2n + 6 + 24 - 36 = 0$$

$$\Rightarrow \quad n = 3$$

3. (a) In group 13, 14, 15, as we descend down in group, the higher oxidation state becomes less stable due to inert pair effect. Therefore, lead show +2 as stable oxidation state. Hence, Pb^{4+} acts as a strong oxidising agent and itself reduced to Pb^{2+} very easily. Both Statement I and Statement II are correct and Statement II is the correct explanation of Statement I.

4. (c)

(i) Bi^{3+} hydrolysis to $(BiO)^+$. Thus, (i) $\rightarrow$ (q)

(ii) $[AlO_2]^-$ exists in basic medium. On acidification gives $Al(OH)_3$. Thus, (ii) $\rightarrow$ (r).

(iii) Orthosilicate (SiO_4^{4-}) on heating changes into pyrosilicate $Si_2O_7^{6-}$. Thus, (iii) $\rightarrow$ (p).

(iv) Tetraborate ion $[B_4O_7^{2-}]$ on treatment with dil. acid hydrolysis gradually to orthoboric acid. Thus, (iv) $\rightarrow$ (q,r)

5. (b) In sheet silicates, three out of four oxygen of SiO_4^{4-} unit are shared as shown below

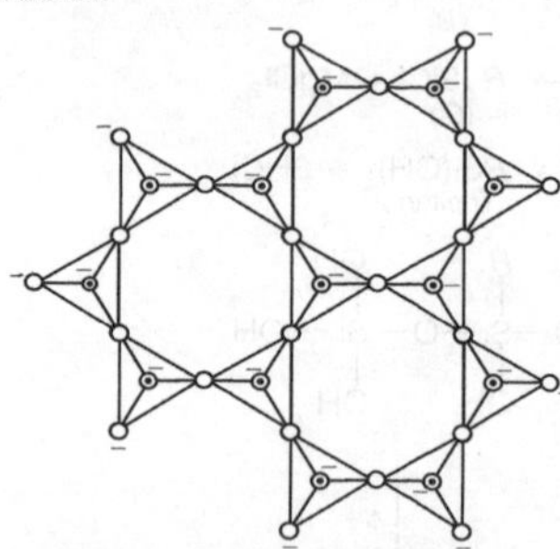

In pryosilicates, there is only one shared oxygen, in linear chain silicates, two oxygen per tetrahedral are shared while in three dimensional silicates, all four oxygens are shared.

Last 3 Years' Questions
JEE Main & Advanced
(2016-2018)

JEE Main

2016

1. Which one of the following statements about water is false? **[Hydrogen]**

(a) Water can act both as an acid and as a base
(b) There is extensive intramolecular hydrogen bonding in the condensed phase
(c) Ice formed by heavy water sinks in normal water
(d) Water is oxidised to oxygen during photosynthesis

2. The main oxides formed on combustion of Li, Na and K in excess of air respectively are **[Alkali Metals]**

(a) LiO_2, Na_2O_2 and K_2O
(b) Li_2O_2, Na_2O_2 and KO_2
(c) Li_2O, Na_2O_2 and KO_2
(d) Li_2O, Na_2O and KO_2

3. Which of the following atoms has the highest first ionisation energy? **[Periodic Table]**

(a) Na (b) K
(c) Sc (d) Rb

2017

4. Both lithium and magnesium display several similar properties due to the diagonal relationship; however, the one which is incorrect is **[Alkali Metals]**

(a) Both form basic carbonates
(b) Both form soluble bicarbonates
(c) Both form nitrides
(d) nitrates of both Li and Mg yield NO_2 and O_2 on heating

5. Sodium salt of an organic acid 'X' produces effervescence with conc. H_2SO_4. 'X' reacts with the acidified aqueous $CaCl_2$ solution to give a white precipitate which decolourises acidic solution of $KMnO_4$.'X' is **[Alkali Metals]**

(a) C_6H_5COONa (b) $HCOONa$
(c) CH_3COONa (d) $Na_2C_2O_4$

2018

6. Hydrogen peroxide oxidises $[Fe(CN)_6]^{4-}$ to $[Fe(CN)_6]^{3-}$ in acidic medium but reduces $[Fe(CN)_6]^{3-}$ to $[Fe(CN)_6]^{4-}$ in alkaline medium. The other products formed are, respectively. **[Hydrogen]**

(a) $(H_2O + O_2)$ and H_2O
(b) $(H_2O + O_2)$ and $(H_2O + OH^-)$
(c) H_2O and $(H_2O + O_2)$
(d) H_2O and $(H_2O + OH^-)$

7. When metal 'M' is treated with NaOH, a white gelatinous precipitate 'X' is obtained, which is soluble in excess of NaOH. Compound 'X' when heated strongly gives an oxide which is used in chromatography as an adsorbent. The metal 'M' is **[Group 13 Elements]**

(a) Zn (b) Ca
(c) Al (d) Fe

JEE Advanced

2016

1. The increasing order of atomic radii of the following Group 13 elements is **[Group 13 Elements, Type 1]**

(a) Al < Ga < In < Tl (b) Ga < Al < In < Tl
(c) Al < In < Ga < Tl (d) Al < Ga < Tl < In

2. The crystalline form of borax has **[Group 13 Elements, Type 1]**

(a) tetranuclear $[B_4O_5(OH)_4]^{2-}$ unit
(b) all boron atoms in the same plane
(c) equal number of sp^2 and sp^3 hybridised boron atoms
(d) one terminal hydroxide per boron atom

2017

3. Which of the following combination will produce H_2 gas? **[Hydrogen, Type 1]**

(a) Fe metal and conc. HNO_3
(b) Cu metal and conc. HNO_3
(c) Au metal and NaCN (*aq*) in the presence of air
(d) Zn metal and NaOH (*aq*)

4. The option(s) with only amphoteric oxides is(are) **[Periodic Table, Type 2]**

(a) NO, B_2O_3, PbO, SnO_2
(b) Cr_2O_3, CrO, SnO, PbO
(c) Cr_2O_3, BeO, SnO, SnO_2
(d) ZnO, Al_2O_3, PbO, PbO_2

Note ***Type-1*** *Only one correct option,* ***Type-2*** *More then one correct option,* ***Type-3*** *Single digit Answer,* ***Type-4*** *Numerical value based*

Answer *with* Explanations

JEE Main

1. *(b)* There is extensive intermolecular H-bonding in the condensed phase.

2. *(c)* $$2Li + \underset{\text{(Excess)}}{\frac{1}{2}O_2(g)} \longrightarrow Li_2O$$

$$2Na + \underset{\text{(Excess)}}{O_2(g)} \longrightarrow Na_2O_2;$$

$$K + \underset{\text{(Excess)}}{O_2(g)} \longrightarrow KO_2$$

3. *(c)* Order of first ionisation energy is $Sc > Na > K > Rb$. Due to poor shielding effect, removal of one electron from 4*s* orbital is difficult as compared to 3*s*-orbital.

4. *(a)* **Mg** can form basic carbonate while Li cannot.

$$5\,Mg^{2+} + 6\,CO_3^{2-} + 7H_2O \longrightarrow 4MgCO_3 \cdot Mg(OH)_2 \cdot 5H_2O + 2\,HCO_3^-$$

5. *(d)* The reaction takes place as follows

$$\underset{(X)}{Na_2C_2O_4} + \underset{\text{(conc.)}}{H_2SO_4} \longrightarrow Na_2SO_4 + H_2O \underset{\text{Effervescence}}{+ CO\uparrow + CO_2\uparrow}$$

$$\underset{(X)}{Na_2C_2O_4} + CaCl_2 \longrightarrow \underset{\text{White ppt.}}{CaC_2O_4} + 2NaCl$$

$$5CaC_2O_4 + \underset{\text{Purple}}{2KMnO_4} + 8H_2SO_4 \longrightarrow K_2SO_4 + 5CaSO_4 + \underset{\text{Colourless}}{2MnSO_4} + 10CO_2 + 8H_2O$$

Hence, X is $Na_2C_2O_4$.

6. *(c)* Both reactions in their complete format are written below

(i) In acidic medium,

$$[Fe^{2+}(CN)_6]^{4-} + H_2\overset{-1}{O_2} + 2H^+ \longrightarrow [Fe^{3+}(CN)_6]^{3-} + 2H_2\overset{-2}{O}$$

(ii) In alkaline medium,

$$[Fe^{3+}(CN)_6]^{3-} + H_2\overset{-1}{O_2} + 2OH^- \longrightarrow [Fe^{2+}(CN)_6]^{4-} + O_2 + 2H_2O$$

Hence, H_2O (for reaction (i)) and $O_2 + H_2O$ (for reaction (ii)) are produced as by product.

7. *(c)* Among the given metals Al forms white gelatinous ppt. with NaOH.

Hence, the probable metal can be Al. This ppt. is dissolved in excess of NaOH due to the formation of sodium metal Aluminate. Both the reactions are shown below.

$$Al^{3+} \xrightarrow{NaOH} \underset{\substack{\text{White gelatinous ppt.} \\ (X)\text{ of aluminium} \\ \text{hydroxide}}}{Al(OH)_3} \xrightarrow{\text{Excess of NaOH}} \underset{\substack{\text{Sodium} \\ \text{metaaluminate)} \\ \text{soluble}}}{NaAlO_2}$$

Aluminium hydroxide on strong heating gives alumina (Al_2O_3) which is used as an adsorbent in chromatography. This reaction can be seen as :

$$2Al(OH)_3 \xrightarrow{\Delta} Al_2O_3 + 3H_2O$$

Thus, metal M is Al.

Ca, being below sodium in electrochemical reactivity series, cannot displaces Na from its aqueous solution.

Zn reacts with NaOH to form sodium zincate which is a soluble compound.

Fe reacts with sodium hydroxide to form tetrahydroferrate (II) sodium which is again a soluble complex.

JEE Advanced

1. *(b)* Due to poor shielding of *d*-orbital in Ga, atomic radius of Ga is smaller than that of Al. Thus, Ga < Al < In < Tl.

2. *(a, c, d)* $Na_2B_4O_7 \cdot 10H_2O$ (borax) is actually made of two tetrahedral and two triangular units, and is actually written as $Na_2[B_4O_5(OH)_4] \cdot 5H_2O$.

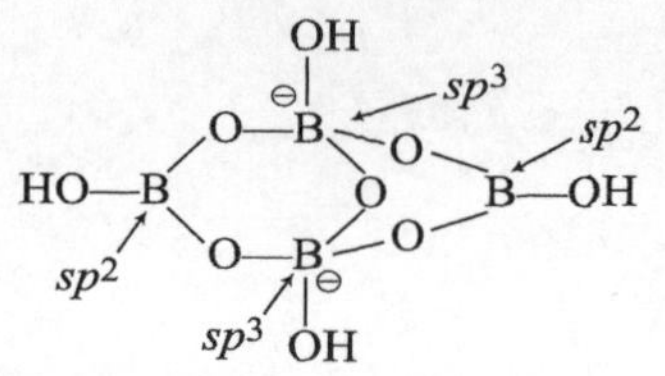

(a) Thus, correct.
(b) Boron atoms are in different planes thus, incorrect.
(c) Two sp^2 and two sp^3-hybridised B atoms thus, correct.
(d) Each boron has one —OH group thus, correct.

3. *(d)* $\underset{\text{Amphoteric}}{Zn} + 2NaOH \longrightarrow Na_2ZnO_2 + H_2$

4. *(a, b)* (c) is incorrect because NO is neutral oxide.
(d) is incorrect because CrO is basic oxide.